Soil Biology

Volume 37

Series Editor
Ajit Varma, Amity Institute of Microbial Technology,
Amity University Uttar Pradesh, Noida, UP, India

For further volumes:
http://www.springer.com/series/5138

Ricardo Aroca
Editor

Symbiotic Endophytes

 Springer

Editor
Ricardo Aroca
Departamento de Microbiología
  del Suelo y Sistemas Simbióticos
Estación Experimental del Zaidín (CSIC)
Granada
Spain

ISSN 1613-3382   ISSN 2196-4831 (electronic)
ISBN 978-3-642-39316-7 ISBN 978-3-642-39317-4 (eBook)
DOI 10.1007/978-3-642-39317-4
Springer Heidelberg New York Dordrecht London

Library of Congress Control Number: 2013948624

© Springer-Verlag Berlin Heidelberg 2013
This work is subject to copyright. All rights are reserved by the Publisher, whether the whole or part of the material is concerned, specifically the rights of translation, reprinting, reuse of illustrations, recitation, broadcasting, reproduction on microfilms or in any other physical way, and transmission or information storage and retrieval, electronic adaptation, computer software, or by similar or dissimilar methodology now known or hereafter developed. Exempted from this legal reservation are brief excerpts in connection with reviews or scholarly analysis or material supplied specifically for the purpose of being entered and executed on a computer system, for exclusive use by the purchaser of the work. Duplication of this publication or parts thereof is permitted only under the provisions of the Copyright Law of the Publisher's location, in its current version, and permission for use must always be obtained from Springer. Permissions for use may be obtained through RightsLink at the Copyright Clearance Center. Violations are liable to prosecution under the respective Copyright Law.
The use of general descriptive names, registered names, trademarks, service marks, etc. in this publication does not imply, even in the absence of a specific statement, that such names are exempt from the relevant protective laws and regulations and therefore free for general use.
While the advice and information in this book are believed to be true and accurate at the date of publication, neither the authors nor the editors nor the publisher can accept any legal responsibility for any errors or omissions that may be made. The publisher makes no warranty, express or implied, with respect to the material contained herein.

Printed on acid-free paper

Springer is part of Springer Science+Business Media (www.springer.com)

Preface

Plants have evolved in contact with a huge number of microorganisms, especially soil microorganisms. This contact during millions of years has forced a coevolution between plants and microorganisms. Some microorganisms are considered pathogens to plants, and both pathogens and plants have codeveloped mechanisms of infection and defense, respectively. However, not all microorganisms are pathogens; there are also beneficial microorganisms which improve the performance of the plants under both optimal and stressful conditions. Some of these beneficial microorganisms live inside the plant tissues and they are called endophytes. At the same time, most of these endophytes can be considered symbionts in the wider meaning of the term: mutualist association between two living organisms.

The most recent knowledge about all plant endophytes known so far is summarized in this Soil Biology volume: rhizobial and actinorrhizal bacteria, endophytic plant growth-promoting rhizobacteria, arbuscular mycorrhizal fungi, and other endophytic fungi including yeasts.

Some chapters deal with the presymbiotic communication and establishment of the symbiosis between rhizobial, actinorhizal, and arbuscular mycorrhizal symbionts. Other chapters examine the abiotic and biotic factors that may alter the symbiotic relationships, especially the nitrogen fixation by rhizobial and actinorhizal bacteria and the nutrient uptake by arbuscular mycorrhizal fungi. Finally, some chapters are devoted to the functional diversity of these symbionts.

Overall, the book updates all the information available about these symbiotic endophytes, and it will serve also for specialized researchers as well as for graduate students, since each chapter is a compendium of basic concepts and the most advance knowledge. It will satisfy everyone who reads it.

Granada, Spain   Ricardo Aroca

Contents

Part I Rhizobial Symbiosis

1 **Journey to Nodule Formation: From Molecular Dialogue to Nitrogen Fixation** 3
Tessema Kassaw and Julia Frugoli

2 **A Roadmap Towards a Systems Biology Description of Bacterial Nitrogen Fixation** 27
Marie Lisandra Zepeda-Mendoza and Osbaldo Resendis-Antonio

3 **Carbon Metabolism During Symbiotic Nitrogen Fixation** 53
Emmanouil Flemetakis and Trevor L. Wang

4 **Genomic and Functional Diversity of the Sinorhizobial Model Group** 69
Alessio Mengoni, Marco Bazzicalupo, Elisa Giuntini, Francesco Pini, and Emanuele G. Biondi

Part II Actinorhizal Symbiosis

5 **Establishment of Actinorhizal Symbioses** 89
Alexandre Tromas, Nathalie Diagne, Issa Diedhiou, Hermann Prodjinoto, Maïmouna Cissoko, Amandine Crabos, Diaga Diouf, Mame Ourèye Sy, Antony Champion, and Laurent Laplaze

6 **Abiotic Factors Influencing Nitrogen-Fixing Actinorhizal Symbioses** 103
Hiroyuki Tobita, Ken-ichi Kucho, and Takashi Yamanaka

7 **Diversity of *Frankia* Strains, Actinobacterial Symbionts of Actinorhizal Plants** 123
Maher Gtari, Louis S. Tisa, and Philippe Normand

Part III Endophytic Plant Growth-Promoting Rhizobacteria (PGPR)

8 **Abiotic Stress Tolerance Induced by Endophytic PGPR** 151
   Patricia Piccoli and Rubén Bottini

9 **Fighting Plant Diseases Through the Application of *Bacillus* and *Pseudomonas* Strains** 165
   Sonia Fischer, Analía Príncipe, Florencia Alvarez, Paula Cordero, Marina Castro, Agustina Godino, Edgardo Jofré, and Gladys Mori

10 **Functional Diversity of Endophytic Bacteria** 195
   Lucía Ferrando and Ana Fernández-Scavino

Part IV Arbuscular Mycorrhizal Symbiosis

11 **Chemical Signalling in the Arbuscular Mycorrhizal Symbiosis: Biotechnological Applications** 215
   Juan A. López-Ráez and María J. Pozo

12 **Carbon Metabolism and Costs of Arbuscular Mycorrhizal Associations to Host Roots** 233
   Alex J. Valentine, Peter E. Mortimer, Aleysia Kleinert, Yun Kang, and Vagner A. Benedito

13 **Arbuscular Mycorrhizal Fungi and Uptake of Nutrients** 253
   M. Miransari

14 **Arbuscular Mycorrhizal Fungi and the Tolerance of Plants to Drought and Salinity** 271
   Mónica Calvo-Polanco, Beatriz Sánchez-Romera, and Ricardo Aroca

15 **Root Allies: Arbuscular Mycorrhizal Fungi Help Plants to Cope with Biotic Stresses** 289
   María J. Pozo, Sabine C. Jung, Ainhoa Martínez-Medina, Juan A. López-Ráez, Concepción Azcón-Aguilar, and José-Miguel Barea

Part V Other Endophytic Fungi

16 **Fungal Endophytes in Plant Roots: Taxonomy, Colonization Patterns, and Functions** 311
   Diana Rocío Andrade-Linares and Philipp Franken

17 **Endophytic Yeasts: Biology and Applications** 335
   Sharon Lafferty Doty

Index 345

Part I
Rhizobial Symbiosis

Chapter 1
Journey to Nodule Formation: From Molecular Dialogue to Nitrogen Fixation

Tessema Kassaw and Julia Frugoli

1.1 Introduction

1.1.1 Nitrogen Fixation and Legumes

The use of legumes in agricultural rotations was documented by Pliny the Elder in 147 BC (Crawford et al. 2000). The Leguminosae family is taxonomically categorized into three subfamilies which encompass nodulating plants, the Caesalpinioideae, Mimosoideae, and Papilionoideae (Doyle and Luckow 2003), although most nodulation occurs in the Papilionoideae. They are the third largest family of angiosperms consisting of more than 650 genera and over 18,000 species (Lewis et al. 2005) and are second only to Gramineae in their importance as human food accounting for 27 % of the world's crop production and contributing 33 % of the dietary protein nitrogen needs of humans (Graham and Vance 2003). Legumes include a large number of domesticated species harvested as crops for human and animal consumption as well as for oils, fiber, fuel, fertilizers, timber, medicinals, chemicals, and horticultural varieties (Lewis et al. 2005). One of the reasons legumes are so popular is their ability to satisfy their nutritional needs by communicating and establishing symbiotic relationships with microbes in the rhizosphere. Like 80 % of the terrestrial plant species, legumes form arbuscular mycorrhizal (AM) associations, where the fungus colonizes the cortical cells to access carbon supplied by the plant and the fungus helps the plant in the transfer of mineral nutrients, particularly phosphorus, from the soil (Smith and Read 2008). AM is a very ancient symbiosis more than 400 million years old. In contrast, the endosymbiosis of plants with nitrogen-fixing bacteria is limited to only a few plant families and is more recent evolutionarily, approximately 60 million years old (Godfroy et al. 2006). Around 88 % of legumes examined to date form nodules in association

T. Kassaw • J. Frugoli (✉)
Department of Genetics & Biochemistry, Clemson University, Clemson, SC, USA
e-mail: tessemk@clemson.edu; jfrugol@clemson.edu

with rhizobia, and important agricultural legumes alone contribute about 40–60 million metric tons of fixed N_2 annually while another 3–5 million metric tons are fixed by legumes in natural ecosystems (de Faria et al. 1989; Smil 1999).

The legume–rhizobia symbiosis has been investigated using *Pisum sativum* (pea), *Vicia sativa* (vetch), *Medicago sativa* (alfalfa), *Medicago truncatula* (barrel medic), *Glycine max* (soybean), *Phaseolus vulgaris* (bean), *Sesbania rostrata* (sesbania), and *Lotus japonicus* (lotus). The large genome size and low efficiency of transformation of many crop legumes combined with the advent of genomic research resulted in a concentration on two symbiotic models, *M. truncatula* and *L. japonicus* (Oldroyd and Geurts 2001; Udvardi and Scheible 2005). These systems provide an opportunity for researchers to study both bacterial and fungal symbioses not supported by the well-studied model plant, *Arabidopsis thaliana*, at the molecular level. Both species have all the tools for a model system such as a small diploid genome, self-fertility, ease of transformation, short life cycle, high level of natural diversity, and a wealth of genomic resources (Handberg and Stougaard 1992; Cook 1999). Both belong to the Papilionoideae subfamily, and based on the nature of the nodule they develop, they are classified as determinate (*L. japonicus*) and indeterminate (*M. truncatula*) nodulators. Determinate nodules are characterized by a nonpersistent meristem which ceases development at early stage. This results in round nodules with a homogeneous central fixation zone composed of infected rhizobia-filled cells interspersed with some uninfected cells. In contrast, indeterminate nodules are cylindrical and consist of a gradient of developmental zones with a persistent apical meristem that supports indeterminate growth, an infection zone, a fixation zone, and zone of senescence (Fig. 1.1). The two model plants represent the two nodule development strategies, and most findings discussed in this chapter come from these systems.

1.1.2 Early Steps in Legume–Rhizobia Symbiosis

1.1.2.1 Overview

Soil bacteria belonging to the genera Rhizobium, Bradyrhizobium, Sinorhizobium, Allorhizobium, and Mesorhizobium establish a unique beneficial interaction with most legumes and a few nonleguminous plants in the family of Ulmaceae (*Parasponia* sp.). The interaction between rhizobia and the host plant results in the formation of N_2-fixing nodules. Within these nodules, bacteria are provided with a carefully regulated oxygen and carbon supply which makes it possible for the bacteria to reduce nitrogen efficiently for the plant. The early steps of the symbiosis begin with the exchange of discrete signals, a molecular dialogue, between the bacteria and the plants (Shaw and Long 2003). Plants produce and release chemicals, mainly flavonoids and isoflavonoids, into the rhizosphere. These molecular signals initiate root nodulation by the induction of *nod* genes in rhizobia, promoting bacterial movement towards the plant and enhancing the growth of the

1 Journey to Nodule Formation: From Molecular Dialogue to Nitrogen Fixation

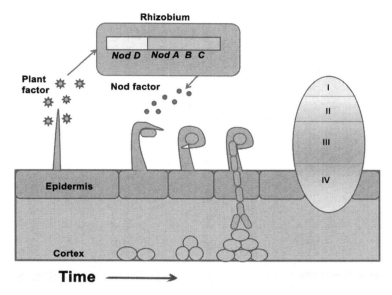

Fig. 1.1 Initial phases in the legume–rhizobium symbiosis. The interaction between rhizobia and legume microsymbionts is determined by two specific steps in the mutual signal exchange. First, bacterial nodulation (*nod*) genes are activated in response to plant-secreted signal molecules (plant factors), especially flavonoids, resulting in biosynthesis and secretion of Nod factors by rhizobia bacteria. In the second step, Nod factors elicit two simultaneous processes in the host plant roots, triggering the infection process and nodule formation (cortical cell division). The infection process includes curling of the root hair around the attached bacteria, infection thread formation, and release of rhizobia from the infection thread into the dividing cortical cells while nodule formation includes mitotic activation of the inner cortical cells, division, and establishment of a meristem zone (I) and infection zone (II), a nitrogen fixation zone (III), and in a senescence zone (IV)

bacterial cells (Phillips and Tsai 1992). The plant factors are recognized by rhizobial NodD proteins, transcriptional regulators that bind directly to a signaling molecule, and are able to activate downstream *nod* genes (Mulligan and Long 1985). Rhizobial *nod* genes are responsible for the production and secretion of species-specific Nod factors, lipochitooligosaccharidic signaling molecules (Zhu et al. 2006). Upon exposure to Nod factors, the plant root hair cells growth is altered, a periodic calcium spiking is induced, a preinfection thread structure is formed, gene expression is altered, and inner cortical cells in the root are mitotically activated, which together leads to the formation of nodule primordia (Ane et al. 2004; Kuppusamy et al. 2004; Mitra et al. 2004; Geurts et al. 2005; Middleton et al. 2007). The infection thread housing the bacteria advances through this actively dividing zone of cells to the nodule primordia. The subsequent release of the bacteria into individual cortical cells by endocytosis results in the enclosure of the bacteria within a plant membrane called the peribacteroid or symbiosome membrane. The peribacteroid membrane effectively isolates the bacteria from the host cell cytoplasm while controlling transport of selected metabolites in both directions (Puppo et al. 2005). The bacteria inside the symbiosome membrane differentiate into bacteroids that produce nitrogenase for nitrogen fixation (Lodwig et al. 2003). However, for effective nitrogen fixation, nitrogenase needs a low oxygen environment, while at the same time rapid transport of oxygen to the sites of respiration must be ensured. These conflicting demands are met by the presence of millimolar concentrations of the oxygen-binding protein leghemoglobin within the cytoplasm of nodule cells (Ott et al. 2005). Recent work using leghemoglobin RNA interference lines in *L. japonicus* showed altered bacterial and plant cell differentiations, decreased amino acid levels in nodules, and a defect in nitrogen fixation (Ott et al. 2009). The resulting physiological and morphological changes in the host plant lead to the formation of nodules, a suitable environment for bacterial nitrogen fixation (Fig. 1.1). The fixed nitrogen obtained by the plant is not without cost, as the plants in return must contribute a significant amount of energy in the form of carbon skeletons to the bacteria.

1.1.2.2 Plant-Derived Signals

The plant starts the molecular dialogue by releasing flavonoid and isoflavonoid compounds to the rhizosphere (Redmond et al. 1986; Kosslak et al. 1987). Flavonoids have multiple roles in rhizobia–legume symbiosis. These compounds serve as chemoattractants for the rhizobial symbiont and trigger the biosynthesis and release of Nod factors from the bacteria. They do so by acting as a signaling molecule, binding to the bacterial transcription factor *NodD*. *NodD* in turn activates the expression of rhizobial *nod* genes, which are responsible for the production of Nod factors (lipochitin oligosaccharides). The perception of Nod factors by a receptor in the legume host triggers a sequence of events, including curling of root hairs around the invading rhizobia, the entry of the rhizobia into the plant through infection threads, and the induction of cell division in the root cortex that marks formation of the nodule primordium. The recognition of specific flavonoids secreted by the root by compatible rhizobia is the earliest step in determining host specificity.

Flavonoids are also involved in the initiation of the nodule through their action on the plant hormone auxin and could thus play a developmental role in addition to their action as *Nod* gene inducers (Hirsch 1992). RNA interference (RNAi) in *M. truncatula* used to silence the enzyme that catalyzes the first committed step of the flavonoid pathway, chalcone synthase (CHS), reduced the level of flavonoids, and the silenced roots were unable to initiate nodules, even though normal root hair curling was observed (Wasson et al. 2006). In addition, Wasson et al. (2006) rescued nodule formation and flavonoid accumulation by supplementing plants with the precursor flavonoids naringenin and liquiritigenin. Subramanian et al. (2006) used a similar RNAi-mediated approach to silence isoflavone synthase (the entry point enzyme for isoflavone biosynthesis in soybean). Isoflavonoid levels in these plants were below detection, and a major decrease in nodulation was observed suggesting that endogenous isoflavones are essential for the establishment of symbiosis between soybean and *Bradyrhizobium japonicum*. In *M. truncatula*, RNA interference-mediated suppression of two flavone synthase II (MtFNSII) genes, the key enzymes responsible for flavone biosynthesis, resulted in flavone depleted roots and significantly reduced nodulation providing genetic evidence that flavones are important for nodulation in *M. truncatula* as well (Zhang et al. 2007). Combined, this genetic evidence reinforces the importance of flavonoids in nodule initiation and establishment.

Even though flavonoids are the most potent *Nod* gene inducers, other non-flavonoid compounds such as jasmonates, aldonic acid, betaines, and xanthones can also induce the expression of *nod* genes in rhizobia but only at high concentrations (Phillips et al. 1992; Gagnon and Ibrahim 1998; Mabood and Smith 2005). In addition to their role in defense response against pathogens, both jasmonic acid and methyl jasmonate strongly induced the expression of *nod* genes in *B. japonicum* (Mabood and Smith 2005). *B. japonicum* inoculants preincubated with jasmonic acid and methyl jasmonate can accelerate nodulation, nitrogen fixation, and plant growth of soybean under controlled environment conditions (Mabood and Smith 2005). While *Lupinus albus* secretes diverse compounds into the rhizosphere, the majority are other non-flavonoid compounds, aldonic acids. The family of aldonic, erythronic, and tetronic acids (4-C sugar acids) induced the expression of *nod* genes in several bacteria (such as *Rhizobium lupini, Mesorhizobium loti,* and *Sinorhizobium meliloti*) and led to low but significant increases in β-galactosidase activities (Gagnon and Ibrahim 1998). In addition to the flavonoids, alfalfa (*M. sativa*) releases two betaines, trigonelline and stachydrine, that induce *nod* genes in *Rhizobium meliloti* (Phillips et al. 1994). These compounds are secreted in large quantities by germinating alfalfa seeds. Another plant-derived signal important for bacterial attachment to the plant root is plant lectin. Lectins are glycoproteins secreted from the tip of root hairs which mediate specific recognition of the bacterial surface carbohydrate molecules. Several experiments have shown the host specificity of plant lectin-mediated bacterial attachment by expressing plant lectin genes from one legume species in another and cross-inoculated with noncompatible rhizobia (Diaz et al. 1989; van Rhijn et al. 1998). Diaz et al. (1995) also reported the sugar binding activity of pea lectin in white clover and the localization on the external surface of elongating epidermal cells and tips of emerging root hairs, similar to the result observed in pea.

1.1.2.3 Bacterial-Derived Signals

Rhizobia establish the nodulation symbiosis in different legume plants by exchanging chemical signals with their legume partners. The molecular communication begins on the bacterial side with the recognition of the flavonoids by rhizobial NodD proteins (NodD1, D2, and D3). These proteins are transcriptional regulators which bind directly to a signaling molecule and activate downstream *nod* genes (Oldroyd and Downie 2004; Mandal et al. 2010). Upon activation of the *nod* genes by the plant signal, the bacteria release species-specific Nod factors to the

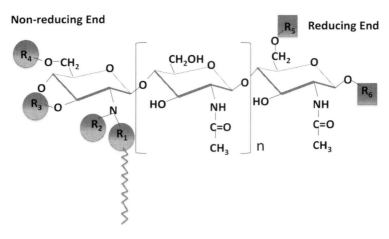

Fig. 1.2 Generalized structure of Nod factors. Species-specific Nod factors have a variety of substitutions at the positions noted by shape. At the nonreducing end (*circles*), R1 is an acyl group with 16–20 carbons in a chain with various levels of unsaturation. R2 is a hydrogen or a methyl group, R3 a hydrogen or carbonyl, and R4 a hydrogen, carbonyl, or acetyl group. At the reducing end (*squares*), R5 can be a hydrogen, sulfate, acetate, fucosyl, sulfo-methylfucosyl, acetyl-methylfucosyl, or arabinose, while R6 is a hydrogen or glycerol. The degree of oligomerization (n) can vary from 1 to 4

rhizosphere. Nod factor molecules are lipochitooligosaccharides consisting of three to five β-1, 4 linked *N*-acetyl-glucosamine residues that are acylated with a fatty acid of 16–20 C-atoms in length on the amino group of the nonreducing glucosamine (Fig. 1.2 and Price and Carlson 1995). The common *nod* genes *nodABC* are structural *nod* genes important for the biosynthesis of the core backbone of all the Nod factors and have a pivotal role in infection and nodulation process (Spaink et al. 1993). The *nod*ABC operons are structurally conserved and functionally interchangeable among different rhizobia without altering the host range (Martinez et al. 1990). This common core structure may, however, be modified by a number of species-specific substituents on the distal or reducing terminal residues which make each bacterial factor unique for each host plant. The substituents include acetate, sulfate, carbamoyl, glycosyl, methyl, arabinose, fucose, and mannose groups. Therefore, the host-specific *nod* genes (*nodHPQGEFL*) are important to specify the different substitution present on the backbone of Nod factors, allowing nodulation of a specific host plant (Brelles-Marino and Ane 2008). Mutation of these particular genes leads to an extended or altered host range (Djordjevic et al. 1985). In general, a correct chemical structure is required for induction of a particular plant response and Nod factor-induced signal transduction cascade.

Nod factors act as signal molecules to simultaneously initiate the nodule formation process programmed in the host plant as well as to trigger the infection process (Kouchi et al. 2010). But several other bacterial molecules are important in the legume–rhizobial interaction. For example, rhizobial extracellular polysaccharides (EPS) are host plant-specific molecules involved in signaling or in root hair attachment. Extracellular polysaccharides are species- or strain-specific polysaccharide molecules with a large diversity in structure and are secreted into the environment or retained at the bacterial surface as a capsular polysaccharide (Laus et al. 2005). EPS-deficient mutants are impaired in efficient induction of root hair curling and especially in infection thread formation which finally leads to the formation of ineffective nodules (Pellock et al. 2000; van Workum et al. 1998). K-antigen polysaccharides are among the most studied acidic polysaccharides involved in nodulation (Becker et al. 2005). The mutation on both *rkpJ* and *rkpU* genes of *S. fredii* HH103 which are vital for production of K-antigen polysaccharides led to reduced nodulation and starvation for nitrogen; their expression was unaffected by flavonoids (Hidalgo et al. 2010). Hence, other bacterial molecules besides Nod factors play a critical role in the progression of the rhizobia–legume interaction.

1.1.3 Nod Factor Signal Transduction Pathway

Genetic studies in the model legumes *M. truncatula* and *L. japonicus* led to the identification of plant genes involved in the early steps in nodulation (Limpens and Bisseling 2003; Levy et al. 2004). A series of mutant screens identified a number of key regulators essential for Nod factor (NF) signaling. Similar set of genes have been found for the two model systems and are described below in spatial/temporal order from the surface of the root hair, based on their mutant phenotypes and diagrammed in Fig. 1.3.

At the cell surface are the LysM-RK receptor kinases (LYK3 and NFP in *M. truncatula* and their counterparts NFR1 and NFR5 in *L. japonicus*) which perceive Nod factors and trigger the signal transduction cascade essential for all early symbiotic events (Limpens et al. 2005; Smit et al. 2007; Broghammer et al. 2012). *MtNFP* is orthologous to *LjNFR5*, and a knockout mutation in this gene causes complete loss of Nod factor-inducible responses (Amor et al. 2003). *M. truncatula* LYK3, a putative high-stringency receptor that mediates bacterial infection, has been localized in a punctate distribution at the cell periphery, consistent with plasma membrane localization and upon inoculation co-localizes with FLOTILLIN4 (FLOT4) tagged with mCherry, another punctate plasma membrane-associated protein required for infection (Haney et al. 2011). Catoira et al. (2001) reported that the *hair curling (hcl)* mutants in *M. truncatula* altered the formation of signaling centers that normally provide positional information for the reorganization of the microtubular cytoskeleton in epidermal and cortical cells. Genetic analysis of calcium spiking in *hcl* mutants showed wild type calcium spiking in response to NF suggesting *HCL* acting downstream of earlier NF signaling events (Wais et al. 2000). Using a weak *hcl* allele, *hcl-4*, Smit et al. (2007) found that *hcl* mutants were defective in *LYK3* (LysM receptor kinase) and act as Nod factor entry receptor important for both root hair curling and infection thread formation. Since both *MtNFP* and *MtLYK3* encode transmembrane

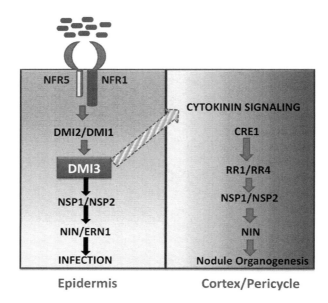

Fig. 1.3 Nod factor signaling cascade in legume–rhizobia symbiosis. The Nod factor receptors NFR5 and NFR1, consisting of extracellular LysM domains and intracellular kinase domain, are positioned at the surface of the root cells to perceive the Nod factor signal from the bacteria and trigger the downstream events. Downstream genes common to both mycorrhizal and bacterial symbiosis (DMI1, DMI2, DMI3) and the mostly rhizobial symbiosis-specific downstream transcription factors (GRAS protein, NSP1 and NSP2, and NIN) are activated upon Nod factor perception. The two processes simultaneously triggered by Nod factor are the infection process in epidermal cells and nodule organogenesis in cortical cells opposite to the xylem poles

receptors containing LysM domains, they were proposed to be good candidates for binding the chitin backbone of NF (Limpens et al. 2003; Arrighi et al. 2006). Recently, two groups reported that the Lotus orthologues of NFP (NFR5) and LYK3 (NFR1) make a heterodimer and perceive the rhizobial lipochitin oligosaccharide signal molecules by direct binding (Madsen et al. 2011; Broghammer et al. 2012). A Rho-like small GTPase (*Lt*ROP6) from *L. japonicus* was also identified as an NFR5-interacting protein both in vitro and in planta but did not interact with NFR1 (Ke et al. 2012).

Subsequently, *does not make infections genes* (*DMI1*, *DMI2*, and *DMI3*) and the *GRAS*-type transcription regulators *nodulation-signaling pathway genes* (*NSP1* and *NSP2*) in *M. truncatula* (Geurts et al. 2005) as well as the *SYMRK, CASTOR, POLLUX, Nup85, Nup133*, and *CCaMK* genes of *L. japonicus* are involved (Ane et al. 2004; Paszkowski 2006; Zhu et al. 2006; Murakami et al. 2007). Plants carrying a single mutation in one of these genes are defective for most of the early responses of Nod factor signaling as well as mycorrhization (Paszkowski 2006; Zhu et al. 2006) with the exception of NSP1 which is nodulation specific (Maillet et al. 2011), indicating that both mycorrhizal and rhizobial symbiosis rely on partially overlapping genetic programs that regulate both signaling pathways.

Since the arbuscular mycorrhizal symbiosis is a very ancient association while the legume/rhizobia symbiosis is relatively more recent (Godfroy et al. 2006), the existence of common genes led to a hypothesis that ancestral legumes may have co-opted part of the signaling machinery of this ancient symbiosis to facilitate the more recent symbiosis with nitrogen-fixing rhizobia (Udvardi and Scheible 2005; Zhu et al. 2006).

As reported by Geurts et al. (2005) and Udvardi and Scheible (2005), among the seven key genes identified in *M. truncatula*, the *DMI1*, *DMI2*, and *DMI3* downstream genes are shared between both rhizobial and fungal symbiosis in the *DMI* dependent signaling pathways. The *dmi1*, *dmi2*, and *dmi3* mutants do not show root hair deformation, gene expression, or mitotic induction of cortical cells but do show swelling at the tip of the root hairs in response to Nod factor and are blocked early in the establishment of mycorrhizal association (Wais et al. 2000; Oldroyd and Long 2003; Mitra et al. 2004). The DMI1 and DMI3 proteins are highly conserved in most land plants, in contrast to the less conserved DMI2 protein. The *DMI2* gene encodes a receptor-like kinase with extracellular leucine-rich repeats, a transmembrane domain, and intracellular kinase domain (Endre et al. 2002; Levy et al. 2004). DMI2 is called NORK for *n*odule *r*eceptor *k*inase in *M. sativa* and the corresponding orthologue in *L. japonicus* is called *SYMRK* (*sym*biosis *r*eceptor *k*inase) (Endre et al. 2002; Stracke et al. 2002). The only interacting partner for DMI2 reported thus far is 3-hydroxy-3-methyl-glutaryl-CoA reductase (*Mt*HMGR1) (Kevei et al. 2007). Mutagenesis and deletion analysis showed that the interaction requires the cytosolic active kinase domain of DMI2 and the cytosolic catalytic domain of *Mt*HMGR1 (Kevei et al. 2007). Several interacting partners for SYMRK have been reported. These include *S*ymRK-*i*nteracting *p*roteins SIP1 and SIP2, which have essential roles in the early symbiosis signaling and nodule organogenesis (Zhu et al. 2008; Chen et al. 2012), and a *S*ymRK-*i*nteracting *E*3 ubiquitin ligase (SIE3) shown to bind and ubiquitinate SymRK in vitro and in planta (Yuan et al. 2012). The *DMI1* gene encodes an ion channel-like protein which mediates the early ion fluxes observed in root hairs responding to Nod factors (Ane et al. 2004; Zhu et al. 2006). DMI1 and its orthologues are important either to trigger the opening of calcium release channels or compensate for the charge release during the calcium efflux as counter ion channels. DMI1 is required for the generation of Nod factor-induced, nucleus-associated Ca^{2+} spikes that are critical for nodule initiation, and protein localization to the nuclear envelope of *M. truncatula* root hair cells correlates with the nuclear association of Ca^{2+} spiking (Peiter et al. 2007). Both DMI1 and DMI2 are upstream of calcium spiking, and plants with mutations in these genes are blocked for calcium spiking and downstream nodulation events (Shaw and Long 2003). The DMI1 orthologues CASTOR and POLLUX were initially reported to localize in the plastids of pea root cells and onion epidermal cells (Imaizumi-Anraku et al. 2005). However, a functional DMI1::GFP fusion protein localized to the nuclear envelope in *M. truncatula* roots when expressed both under a constitutive 35S promoter and a native *DMI1* promoter (Riely et al. 2007). Recently, immunogold labeling localized the endogenous CASTOR protein to the nuclear envelope of *L. japonicus* root cells, consistent with a role of CASTOR and POLLUX in modulating the nuclear envelope membrane potential (Charpentier et al. 2008).

Calcium spiking is a central component of the common symbiotic pathway. Recent work using calcium chameleon reporters in *M. truncatula* roots suggests that tightly regulated Ca^{2+}-mediated signal transduction is key to reprogramming root cell development at the critical stage of commitment to endosymbiotic infection (Sieberer et al. 2012). Two nucleoporin genes (*NUP133* and *NUP85*) have been identified at an equivalent position in the Nod factor signaling pathway with the DMI1 protein and are required for the common symbiotic pathways and for calcium spiking responses (Kanamori et al. 2007; Saito et al. 2007). *NUP133* encodes a protein that has sequence similarity to human nucleoporin Nup133 and localizes in the nuclear envelope, indicating that both NUP133 and NUP85 are members of the nuclear pore complex in legumes (Kanamori et al. 2007). Both genes are required for the calcium spiking that is induced in response to Nod factors, but further research is required to clarify the roles of *NUP133* and *NUP85* in leguminous plants.

DMI3 acts immediately downstream of calcium spiking in the nodulation-signaling pathway and is required for both nodulation and mycorrhizal infection (Levy et al. 2004). In contrast to the other mutants mentioned, calcium spiking and root hair swelling in response to Nod factor are wild type in a *dmi3* mutant background whereas symbiotic gene expression or cell divisions for nodule formation are defective (Mitra et al. 2004). The *DMI3* gene encodes a Ca^{2+} and calmodulin-dependent protein kinase (CCaMK) that responds to the Ca^{2+} signal (Mitra et al. 2004; Geurts et al. 2005). These protein families are multifunctional, with a kinase domain, a calmodulin (CaM)-binding domain and a Ca^{2+}-binding domain with three EF hands (Oldroyd and Downie 2004). The CCaMKs have the capacity to bind calcium in two ways, either by direct binding to the three EF hands or by forming a complex with calmodulin to regulate the kinase activity. The interaction of the Ca^{2+} with the C-terminal EF hands results in autophosphorylation of the CCaMK and allows CaM binding, which leads to substrate phosphorylation (Cook 2004; Levy et al. 2004). In general, CCaMK perceives the calcium spiking signature and transduces this to induction of the downstream genes involved in mycorrhizal or rhizobial symbiosis. Split yellow fluorescent protein complementation and yeast-2-hybrid systems demonstrated that the highly conserved nuclear protein IPD3 is an interacting partner of DMI3 and that the interaction is through a C-terminal coiled-coil domain (Messinese et al. 2007). In separate report, characterization of three independent retrotransposon Tos17 insertion lines of rice *OsIPD3* upon AM fungus *Glomus intraradices* inoculation revealed that the *Osipd3* mutants were unable to establish a symbiotic association with *G. intraradices* confirming the role of this CCaMK in root symbiosis with AM fungi (Chen et al. 2008).

Beyond this point, in the common symbiotic pathway transcriptional regulators of the *NIN, GRAS (NSP1, NSP2)*, and *ERF* families are required for upregulation of nodulation-expressed genes and initiation of nodulation (Madsen et al. 2010). Both *NSP1* and *NSP2* encode putative transcriptional regulators of the GRAS protein family (Smit et al. 2005; Kalo et al. 2005). The NSP1 protein has been localized in the nucleus similar to the upstream gene DMI3 (Smit et al. 2005). NSP2, however, migrates from its original location in the nuclear envelope and endoplasmic reticulum into the nucleus where it regulates the transcription of early nodulin genes after Nod factor elicitation (Kalo et al. 2005). Since both NSP1 and NSP2 form a complex (Hirsch et al. 2009) and are genetically downstream of DMI3, at least one of these genes is the target of DMI3 action. Cross species complementation studies also showed *NSP1* and *NSP2* functions are conserved in nonlegumes. *OsNSP1* and *OsNSP2* from rice were able to fully rescue the root nodule symbiosis-defective phenotypes of the mutants of corresponding genes in the model legume, *L. japonicus* (Yokota et al. 2010). Recently, Liu et al. (2011) reported that NSP1 and NSP2 are also a vital component of strigolactone biosynthesis in *M. truncatula* and rice. Mutations in both genes reduced expression of DWARF27, a gene essential for strigolactone biosynthesis (Liu et al. 2011). Downstream of NSP1 and NSP2, another putative transcription factor, NIN, first identified in *L. japonicus*, is essential for coordinating nodule organogenesis and bacterial entry (Marsh et al. 2007). *NIN* encodes a transmembrane transcriptional regulator with homology to Notch of Drosophila (Schauser et al. 1999). Early NF-induced gene expression using an *ENOD11:GUS* reporter fusion in the *Mtnin-1* mutant showed that *MtNIN* is not essential for early Nod factor signaling but may function downstream of the early NF signaling pathway to coordinate and regulate temporal and spatial formation of root nodules (Marsh et al. 2007). The perception of Nod factors also leads to the activation of another transcription factor with DNA-binding capability, ERN1, an AP2-like transcription factor in the ERF subfamily, which is necessary for nodulation and functions in early Nod factor signaling (Middleton et al. 2007). Mutations in *ERN* block the initiation and development of infection threads and thus block nodule invasion by the bacteria. *ERN1* is induced rapidly after *S. meliloti* inoculation and is necessary for Nod factor-induced gene expression. Unlike wild type plants, *ern1* mutants do not form spontaneous nodules when transformed with activated calcium- and calmodulin-dependent protein kinase, and Nod factor application does not induce *ENOD11:GUS* expression (Middleton et al. 2007). A second ERF transcription factor, EFD (for *e*thylene response *f*actor required for nodule *d*ifferentiation), is required for the differentiation of functional Fix+nodules and may participate in an ethylene-independent feedback inhibition of the nodulation process as well as regulating the expression of the primary cytokinin response regulator MtRR4 (Vernie et al. 2008).

The formation of nodule primordia involves dedifferentiation and reactivation of cortical root cells to establish the nodule primordium, a mass of rapidly proliferating undifferentiated cells, opposite to the protoxylem poles (Timmers et al. 1999; Penmetsa and Cook 1997). Both gain-of-function and loss-of-function mutants have shown that cytokinin signaling through the cytokinin receptor kinase (LHK1) is important for reactivation of cortical cells (Tirichine et al. 2007; Murray et al. 2007). Plet et al. (2011) also reported cytokinin signaling in *M. truncatula* integrates bacterial and plant cues to coordinate symbiotic nodule organogenesis in an *MtCRE1* dependent manner. Simultaneous with the formation of the primordia is the endocytic-like entry of the bacteria to the plant root cells, associated with the plant driven infection thread formation, bacterial cell division within the infection thread, progression of the infection threads towards the dividing nodule primordia, and finally the invasion of the developing nodule. Several genes important to Nod factor recognition have been reported to affect infection thread initiation and growth. The *M. truncatula* Hair Curling (*HCL*) gene encodes the LYK3 receptor-like kinase a specific function of which is to initiate the infection thread on Nod factor recognition (Limpens et al. 2003; Smit et al. 2007). Besides its role in nodule initiation, the receptor kinase DMI2 plays a key role in symbiosome formation and is expressed both on the host cell plasma membrane and the membrane surrounding the infection thread (Limpens et al. 2005). Also required for both infection thread growth in root hair cells and the further development of nodule primordia are the orthologous *LIN/CERBERUS* genes in *M. truncatula* and *L. japonicus*, which encode predicted E3 ubiquitin ligases containing a highly conserved U-box and WD40 repeat domains functions and function at an early stage of the rhizobial symbiotic process (Kiss et al. 2009 Yano et al. 2009). The Vapyrin (*VPY*) gene is essential for the establishment of the arbuscular mycorrhizal symbiosis and is also important for rhizobial colonization, and nodulation. *VPY* acts downstream of the common signaling pathway (Murray et al. 2011). In addition, flotillin (FLOT2 and FLOT4) and remorin (*Mt*SYMREM1) proteins, which promote trafficking and aggregation of membrane proteins, are required for infection by rhizobia, possibly by acting as scaffolds for recruitment of membrane proteins involved in nodulation signaling [for review, see Oldroyd et al. (2011)].

1.1.4 Nodule Autoregulation

The symbiosis between leguminous plants and rhizobia under conditions of nitrogen limitation leads to the development of new plant organs, the N_2-fixing nodules that are usually formed on roots but also on stems in a few plants. The bacteria require energy and a suitable environment for nitrogenase, the enzyme important for nitrogen fixation (Crawford et al. 2000). Hence, the nodulation process is energy intensive, and the plants need to maintain a balance between cost and benefit by limiting the number of nodules that form. Plants use local and long-distance or systemic signaling to coordinate and adjust the number of nodulation events (Kosslak and Bohlool 1984; Caetano-Anolles and Gresshoff 1991). This negative feedback inhibition, in which the earlier nodulation events suppress the subsequent development of nodules in young tissues, is called autoregulation of nodulation (AON) (Pierce and Bauer 1983; Searle et al. 2003; Oka-kira and Kawaguchi 2006). AON employs root-derived and shoot-derived long-distance signals. The root-derived signal is generated in roots in response to rhizobial infection and then translocated to the shoot, while the shoot-derived signal is generated in shoot and then translocated back to the root to restrict further nodulation. AON is activated upon the perception of Nod factor in the elongation zone of the root with emerging root hairs where rhizobial infection occurs being most affected (Bhuvaneswari et al. 1981).

AON is under both environmental and developmental controls and appears to be universally used by legumes to control the extent of nodulation. Mutations affecting AON lead to supernodulation or hypernodulation, associated with root developmental defects. Genetic analysis of AON began with the isolation of supernodulating mutants, which have lost their ability to autoregulate nodule numbers. AON mutants are characterized by forming an excessive number of nodules and a short root compared to their wild type counterparts. For instance, *Glycine max nts* (Carroll et al. 1985); *Lotus japonicus har1* (Krusell et al. 2002), *tml* (Magori and Kawaguchi 2009), and *klavier* (Oka-Kira et al. 2005); *Pisum sativum sym29* (Krusell et al. 2002) and *nod3* (Sidorova and Shumnyi 2003); *Medicago truncatula sunn* (Penmetsa et al. 2003; Schnabel et al. 2005), *lss* (Schnabel et al. 2010), and *rdn1* (Schnabel et al. 2011) mutants are defective in autoregulation and thus form an excessive number of nodules. The *NTS/NARK, HAR1, SYM29,* and *SUNN* genes encode a leucine-rich repeat receptor kinase with homology to Arabidopsis *CLAVATA1* (Searle et al. 2003; Krusell et al. 2002; Schnabel et al. 2005). *KLAVIER* also encodes a different LRR receptor kinase (Miyazawa et al. 2010) while *NOD3* and *RDN1* encode proteins of unknown function (Schnabel et al. 2011). Since autoregulation is mediated through a long-distance signaling involving shoot and root, shoot to root reciprocal grafting studies using wild type and autoregulation defective mutants revealed that there are both shoot controlled as well as root controlled supernodulators. It is believed that the shoot-controlled supernodulators are impaired in either in the perception of the root-derived infection signal or in the transmission of the shoot-derived autoregulation signal. On the other hand, the root controlled mutants are thought to be impaired in either the transmission of the root-derived infection signal or in the perception of the shoot-derived autoregulatory signal.

Nodule initiation and development are also determined by physiological conditions and phytohormones. Successful nodule formation and subsequent nitrogen fixation occur normally only under nitrogen-limiting conditions (Schultze and Kondorosi 1998). However, the mutants defective in AON are partially nitrate tolerant (Caba et al. 1998; Carroll et al. 1985). This suggests that at some stage, the autoregulation signal and the nitrate signal talk to each other to inhibit nodule formation. In addition, the gaseous phytohormone, ethylene, is also a negative regulator of nodule organogenesis. The *M. truncatula* mutant, *skl*, encoding an EIN2 orthologue (Penmetsa et al. 2008), is insensitive to ethylene and shows a tenfold increases in nodule number relative to the wild type (Penmetsa and Cook 1997). Rhizobial inoculation and exogenous ACC induce ethylene synthesis and thereby lead to suppressed nodule and root development in *sunn* mutants (Penmetsa et al. 2003). Similarly, the addition of the ethylene inhibitors like L-α-aminoethoxyvinylglycine enhanced nodule development in common bean and pea (Guinel and Sloetjes 2000; Tamimi and Timko 2003). Another phytohormone, auxin, is mostly produced in younger plant shoots and moves long distance to the root tip following an auxin concentration gradient to trigger root, nodules, and other plant organ development (Pacios-Bras et al. 2003; Schnabel and Frugoli 2004). In fact, IAA produced by the rhizobia is reported to increase nodule formation (Pii et al. 2007). In uninoculated roots of *sunn* mutant plants, auxin transport from shoot to root is approximately three times higher than the wild type (van Noorden et al. 2006), and auxin transport inhibitors such as NPA significantly reduce nodulation in wild type plants and *sunn* mutants but not *skl* mutants (van Noorden et al. 2006; Prayitno et al. 2006), suggesting a role for auxin in regulating nodule number as well.

The plant hormone, cytokinin, is also implicated in nodulation. Exogenous application of cytokinins to legume roots induced responses similar to rhizobial Nod factors, including cortical cell division, amyloplast deposition, and induction of early nodulin gene expression (Bauer et al. 1996). Gonzalez-Rizzo et al. (2006) identified a *M. truncatula* homologue of Arabidopsis Cytokinin Response1 (CRE1), a cytokinin receptor histidine kinase. Using RNA interference to downregulate *MtCRE1*, they demonstrated that MtCRE1 acts as a negative regulator of lateral root formation and as a positive regulator of nodulation. Expression analysis of genes downstream in cytokinin signaling, *MtRR1* and *MtRR4*, and the early nodulin *MtNIN1* in *M. truncatula* suggests that these three genes are involved in crosstalk between Nod factor and cytokinin signaling pathways depending on MtCRE1 (Gonzalez-Rizzo et al. 2006). Cytokinin activation of MtCLE13, a short peptide involved in nodulation (Mortier et al. 2010), depends on CRE1 and NIN but not on NSP2 and ERN1, suggesting two parallel pathways triggered by cytokinin in the root cortex, activation of cortical cell division and activation of MtCLE13 to inhibit further nodulation (Mortier et al. 2012). In addition, CLE genes (12–13 amino acid long secreted peptides) which are involved in both shoot and root meristem homeostasis, vascular differentiation, and nodulation comprise a gene family of up to 40 members and play a role in either activation of the root-derived AON signal or have the potential to interact with leucine-rich repeat receptor kinases such as SUNN (Mortier et al. 2010). Identification of these AON genes in combination with phytohormones and other growth regulators and the intensive study of the nodulation-signaling cascade will aid understanding of the fascinating and complex events leading to legume nodule formation and regulation and plant development in general.

1.1.5 Nodule Senescence

Functional nodules are not maintained throughout the life cycle of the host plant. The peak nitrogen fixation period in both determinate and indeterminate nodules is restricted to between 3 and 5 weeks after infection (Lawn and Burn 1974; Puppo et al. 2005). During the vegetative growth stage and flowering, nodules are the major carbon sink in legumes. In the course of pod filling, however, the seed is the strongest sink for photosynthate, and nodules start to gradually senesce. The first symptoms of senescence are the deterioration of leghemoglobin, resulting in a pink to green color change in the nodule, and loss of turgidity in old nodules (Perez Guerra et al. 2010). In mature indeterminate nodules, the senescence zone starts in zone IV (Fig. 1.1). Upon aging, this senescence zone gradually moves in a proximal–distal direction until it reaches the apical part and leads to nodule degeneration (Puppo et al. 2005; Van de Velde et al. 2006). On the other hand, in determinate nodules it expands radially from the center to the periphery. The primary targets for nodule senescence are symbiosomes, in the same manner as chloroplasts in leaf senescence, with several common senescence associated genes both up- and downregulated in leaf and nodule senescence suggesting a shared pathway (Van de Velde et al. 2006). Since symbiosomes are derived from the uptake of prokaryotic cells that fix nitrogen, and chloroplasts are postulated to have originated from the uptake of cells that fix carbon, this common pathway is not surprising. Using transmission electron microscopy, two consecutive stages were distinguished during nodule senescence: a first stage, characterized by bacteroid degradation with a few dying plant cells and a more advanced stage of nodule senescence, during which cells had completely resorbed their symbiosomes and started to decay and collapse (Van de Velde et al. 2006). Hence, the final fate of the bacteria and the plant cells that form the nitrogen-fixing organelle is death. Plant cysteine proteinases are important in controlling nodule senescence. An Asnodf32 protein which encodes a nodule-specific cysteine proteinase in *Astragalus sinicus* was reported to play an important role in the regulation of root nodule senescence. In Asnodf32-silenced hairy roots, the period of bacteroid active nitrogen fixation was significantly extended and enlarged nodules were also observed (Li et al. 2008). Recently, an *M. truncatula* transcription factor, MtNAC969, was also reported to participate in nodule senescence. MtNAC969 is induced by nitrate treatment and, similar to senescence markers, was antagonistically affected by salt in roots and nodules; MtNAC969 RNAi silenced nodules accumulated amyloplasts in the nitrogen-fixing zone and were prematurely senescent (de Zelicourt et al. 2012). Nodule senescence is an active process programmed in development; thus, reactive oxygen species, antioxidants, hormones, and proteinases also play a key role (Puppo et al. 2005).

1.1.6 Concluding Remarks

Both the plant and the bacteria have coevolved a complex series of signals and responses to establish the symbiosis. Understanding of the plant side of the symbiosis has increased rapidly in recent years and continues to accelerate as genomic and molecular tools are brought to bear on the problem. Interestingly, the majority of signal transduction molecules on the plant side of the symbiosis are not exclusive to legumes, suggesting that the bacteria have co-opted existing plant pathways to establish the symbiosis and regulate nodulation. This suggests that in the future, it may be possible to establish nitrogen-fixing symbiosis in plants that currently do not nodulate.

References

Amor BB, Shaw SL, Oldroyd GE, Maillet F, Penmetsa RV, Cook D, Long SR, Denarie J, Gough C (2003) The *NFP* locus of *Medicago truncatula* controls an early step of Nod factor signal transduction upstream of a rapid calcium flux and root hair deformation. Plant J 34:495–506

Ane JM, Kiss GB, Riely BK, Penmetsa RV, Oldroyd GE, Ayax C, Levy J, Debelle F, Baek JM, Kalo P, Rosenberg C, Roe BA, Long SR, Denarie J, Cook DR (2004) *Medicago truncatula DMI1* required for bacterial and fungal symbioses in legumes. Science 303(5662):1364–1367

Arrighi JF, Barre A, Ben Amor B, Bersoult A, Soriano LC, Mirabella R, de Carvalho-Niebel F, Journet EP, Gherardi M, Huguet T et al (2006) The *Medicago truncatula* lysine motif-receptor-like kinase gene family includes *NFP* and new nodule-expressed genes. Plant Physiol 142:265–279

Bauer P, Ratet P, Crespi M, Schultze M, Kondorosi A (1996) Nod factors and cytokinins induce similar cortical cell division, amyloplast deposition and Msenod12A expression patterns in alfalfa roots. Plant J 10:91–105

Becker A, Fraysse N, Sharypova L (2005) Recent advances in studies on structure and symbiosis-related function of rhizobial K antigens and lipopolysaccharides. Mol Plant Microbe Interact 18:899–905

Bhuvaneswari TV, Bhagwat AA, Bauer WD (1981) Transient susceptibility of root cells in four common legumes to nodulation by Rhizobia. Plant Physiol 68:1144–1149

Brelles-Marino GQ, Ane JM (2008) Nod factors and the molecular dialogue in the rhizobia–legume interaction. In: Couto GN (ed) Nitrogen fixation research progress. Nova Science, New York, NY, pp 173–227

Broghammer A, Krusell L, Blaise M, Sauer J, Sullivan JT, Maolanon N, Vinther M, Lorentzen A, Madsen EB, Jensen KJ, Roepstorff P, Thirup S, Ronson CW, Thygesen MB, Stougaard J (2012) Legume receptors perceive the rhizobial lipochitin oligosaccharide signal molecules by direct binding. Proc Natl Acad Sci USA 109(34):13859–13864

Caba JM, Recalde L, Ligero F (1998) Nitrate-induced ethylene biosynthesis and the control of nodulation in alfalfa. Plant Cell Environ 21:87–93

Caetano-Anolles G, Gresshoff PM (1991) Efficiency of nodule initiation and autoregulatory responses in a supernodulating soybean mutant. Appl Environ Microbiol 57:2205–2210

Carroll BJ, McNeil DL, Gresshoff PM (1985) Isolation and properties of soybean (*Glycine max* L. Merr) mutants that nodulate in the presence of high nitrate concentrations. Proc Natl Acad Sci USA 82:4162–4166

Catoira R, Timmers AC, Maillet F, Galera C, Penmetsa RV, Cook D, Denarie J, Gough C (2001) The HCL gene of *Medicago truncatula* controls Rhizobium-induced root hair curling. Development 128:1507–1518

Charpentier M, Bredemeier R, Wanner G, Takeda N, Schleiff E, Parniske M (2008) *Lotus japonicus* CASTOR and POLLUX are ion channels essential for perinuclear calcium spiking in legume root endosymbiosis. Plant Cell 20:3467–3479

Chen C, Ane JM, Zhu H (2008) OsIPD3, an ortholog of the *Medicago truncatula* DMI3 interacting protein IPD3, is required for mycorrhizal symbiosis in rice. New Phytol 180:311–315

Chen T, Zhu H, Ke D, Cai K, Wang T, Gou H, Hong Z, Zhang Z (2012) A MAP kinase kinase interacts with SymRK and regulates nodule organogenesis in *Lotus japonicus*. Plant Cell 24:823–838

Cook DR (1999) *Medicago truncatula* – a model in the making. Curr Opin Plant Biol 2:301–304

Cook DR (2004) Unraveling the mystery of Nod factor signaling by a genomic approach in *Medicago truncatula*. Proc Natl Acad Sci USA 101(13):4339–4340

Crawford N, Kahn M, Leustek T, Long S (2000) Nitrogen and sulfur. American Association of Plant Biologists, Rockville, MD

de Faria SM, Lewis GP, Sprent JI, Sutherland JM (1989) Occurrence of nodulation in the leguminosae. New Phytol 111:607–619 de Zelicourt A, Diet A, Marion J, Laffont C, Ariel F, Moison M, Zahaf O, Crespi M, Gruber V, Frugier F (2012) Dual involvement of a *Medicago truncatula* NAC transcription factor in root abiotic stress response and symbiotic nodule senescence. Plant J 70:220–230

Diaz CL, Melchers LS, Hooykaas PJJ, Lugtenberg BJJ, Kijne JW (1989) Root lectin as a determinant of host-plant specificity in the Rhizobium-legume symbiosis. Nature 338:579–581

Diaz CL, Logman TJJ, Stam HC, Kijne JW (1995) Sugar-binding activity of pea lectin expressed in white clover root hairs. Plant Physiol 109:1167–1177

Djordjevic MA, Schofield PR, Rolfe BG (1985) Tn-5 mutagenesis of *Rhizobium trifolii* host specific nodulation genes results in mutants with altered host range ability. Mol Gen Genet 200:463–471

Doyle JJ, Luckow MA (2003) The rest of the iceberg: legume diversity and evolution in a phylogenetic context. Plant Physiol 131:900–910

Endre G, Kereszt A, Kevei Z, Mihacea S, Kalo P, Kiss GB (2002) A receptor kinase gene regulating symbiotic nodule development. Nature 417:962–996

Gagnon H, Ibrahim RK (1998) Aldonic acids: a novel family of nod gene inducers of *Mesorhizobium loti*, *Rhizobium lupini*, and *Sinorhizobium meliloti*. Mol Plant Microbe Interact 11(10):988–998

Geurts R, Fedorova E, Bisseling T (2005) Nod factor signaling genes and their function in the early stages of Rhizobium infection. Curr Opin Plant Biol 8(4):346–352

Godfroy O, Debelle F, Timmers T, Rosenberg C (2006) A rice calcium- and calmodulin-dependent protein kinase restores nodulation to a legume mutant. Mol Plant Microbe Interact 19(5):495–501

Gonzalez-Rizzo S, Crespi M, Frugier F (2006) The *Medicago truncatula* CRE1 cytokinin receptor regulates lateral root development and early symbiotic interaction with *Sinorhizobium meliloti*. Plant Cell 18:2680–2693

Graham PH, Vance CP (2003) Legumes: importance and constraints to greater use. Plant Physiol 131:872–877

Guinel FC, Sloetjes LL (2000) Ethylene is involved in the nodulation phenotype of *Pisum sativum R50* (sym 16), a pleiotropic mutant that nodulates poorly and has pale green leaves. J Exp Bot 51:885–894

Handberg K, Stougaard J (1992) *Lotus japonicus*, an autogamous, diploid legume species for classical and molecular genetics. Plant J 2:487–496

Haney CH, Riely BK, Tricoli DM, Cook DR, Ehrhardt DW, Long SR (2011) Symbiotic rhizobia bacteria trigger a change in localization and dynamics of the *Medicago truncatula* receptor kinase LYK3. Plant Cell 23:2774–2787

Hidalgo A, Margaret I, Crespo-Rivas JC, Parada M, Murdoch PS, Lopez A, Buendıa-Claverıa AM, Moreno J, Albareda M, Gil-Serrano AM, Rodrıguez-Carvajal MA, Palacios JM, Ruiz-Sainz JE, Vinardell JM (2010) The *rkpU* gene of *Sinorhizobium fredii* HH103 is required for bacterial K-antigen polysaccharide production and for efficient nodulation with soybean but not with cowpea. Microbiology 156:3398–3411

Hirsch AM (1992) Developmental biology of legume nodulation. New Phytol 122:211–237

Hirsch S, Kim J, Munoz A, Heckmann AB, Downie JA, Oldroyd GED (2009) GRAS proteins form a DNA binding complex to induce gene expression during nodulation signaling in *Medicago truncatula*. Plant Cell 21:545–557

Imaizumi-Anraku H, Takeda N, Charpentier M, Perry J, Miwa H, Umehara Y, Kouchi H, Murakami Y, Mulder L, Vickers K, Pike J, Downie JA, Wang T, Sat S, Asamizu E, Tabata S, Yoshikawa M, Murooka Y, Wu GJ, Kawaguchi M, Kawasaki S, Parniske M, Hayashi M (2005) Plastid proteins crucial for symbiotic fungal and bacterial entry into plant roots. Nature 433:527–531

Kalo P, Gleason C, Edwards A, Marsh J, Mitra RM, Hirsch S, Jakab J, Sims S, Long SR, Rogers J, Kiss GB, Downie JA, Oldroyd GE (2005) Nodulation signaling in legumes requires NSP2, a member of the GRAS family of transcriptional regulators. Science 308(5729):1786–1789

Kanamori N, Madsen LH, Radutoiu S, Frantescu M, Quistgaard EM, Miwa H, Downie JA, James EK, Felle HH, Haaning LL, Jensen TH, Sato S, Nakamura Y, Saito K, Yoshikawa M, Yano K, Miwa H, Uchida H, Asamizu E, Sato S, Tabata S, Imaizumi-Anraku H, Umehara Y, Kouchi H, Murooka Y, Szczyglowski K, Downie JA, Parniske M, Hayashi M, Kawaguchi M (2007) NUCLEOPORIN85 is required for calcium spiking, fungal and bacterial symbioses, and seed production in *Lotus japonicus*. Plant Cell 19:610–624

Ke D, Fang Q, Chen C, Zhu H, Chen T, Chang X, Yuan S, Kang H, Ma L, Hong Z et al (2012) The small GTPase ROP6 interacts with NFR5 and is involved in nodule formation in *Lotus japonicus*. Plant Physiol 159:131–143

Kevei Z, Lougnon G, Mergaert P, Horvath GV, Kereszt A, Jayaraman D, Zaman N, Marcel F, Regulski K, Kiss GB, Kondorosi A, Endre G, Kondorosi E, Ane JM (2007) 3-hydroxy-3-methylglutaryl coenzyme a reductase 1 interacts with NORK and is crucial for nodulation in *Medicago truncatula*. Plant Cell 19:3974–3989

Kiss E, Olah B, Kalo P, Morales M, Heckmann AB, Borbola A, Lozsa A, Kontar K, Middleton P, Downie JA, Oldroyed GE, Endre G (2009) LIN, a novel type of U-box/WD40 protein, controls early infection by rhizobia in legumes. Plant Physiol 151:1239–1249

Kosslak RM, Bohlool BB (1984) Suppression of nodule development of one side of a split-root system of soybeans caused by prior inoculation of the other side. Plant Physiol 75:125–130

Kosslak RM, Bookland R, Barkei J, Paaren HE, Applebaum ER (1987) Induction of *Bradyrhizobium japonicum* common nod genes by isoflavones isolated from *Glycine max*. Proc Natl Acad Sci USA 84:7428–7432

Kouchi H, Imaizumi-Anraku H, Hayashi M, Hakoyama T, Nakagawa T, Umehara Y, Suganuma N, Kawaguchi M (2010) How many peas in a pod? Legume genes responsible for mutualistic symbioses underground. Plant Cell Physiol 51:1381–1397

Krusell L, Madsen LH, Sato S, Aubert G, Genua A, Szczyglowski K, Duc G, Kaneko T, Tabata S, De Bruijn FJ, Pajuelo E, Sandal N, Stougaard J (2002) Shoot control of root development and nodulation is mediated by a receptor-like kinase. Nature 420:422–426

Kuppusamy KT, Endre G, Prabhu R, Penmetsa RV, Veereshlingam H, Cook DR, Dickstein R, VandenBosch KA (2004) LIN, a *Medicago truncatula* gene required for nodule differentiation and persistence of rhizobial infections. Plant Physiol 136(3):3682–3691

Laus MC, van Brussel AA, Kijne JW (2005) Role of cellulose fibrils and exopolysaccharides of *Rhizobium leguminosarum* in attachment to and infection of *Vicia sativa* root hairs. Mol Plant Microbe Interact 18:533–538

Lawn RJ, Burn WA (1974) Symbiotic nitrogen fixation in soybeans. I. Effect of photosynthetic source-sink manipulations. Crop Sci 14:11–16

Levy J, Bres C, Geurts R, Chalhoub B, Kulikova O, Duc G, Journet EP, Ane JM, Lauber E, Bisseling T, Denarie J, Rosenberg C, Debelle F (2004) A Putative Ca2+ and calmodulin-dependent protein kinase required for bacterial and fungal symbioses. Science 303(5662):1361–1364

Lewis G, Schrire B, Mackind B, Lock M (2005) Legumes of the world. Royal Botanic Gardens, Kew

Li Y, Zhou L, Li Y, Chen D, Tan X, Lei L, Zhou J (2008) A nodule-specific plant cysteine proteinase, *AsNODF32*, is involved in nodule senescence and nitrogen fixation activity of the green manure legume *Astragalus sinicus*. New Phytol 180:185–192

Limpens E, Bisseling T (2003) Signaling in symbiosis. Curr Opin Plant Biol 6(4):343–350

Limpens E, Franken C, Smit P, Willemse J, Bisseling T, Geurts R (2003) LysM domain receptor kinases regulating rhizobial Nod factor-induced infection. Science 302:630–633

Limpens E, Mirabella R, Fedorova E, Franken C, Franssen H, Bisseling T, Geurts R (2005) Formation of organelle-like N_2-fixing symbiosomes in legume root nodules is controlled by *DMI2*. Proc Natl Acad Sci USA 102(29):10375–10380

Liu W, Kohlen W, Lillo A, Op den Camp R, Ivanov S, Hartog M, Limpens E, Jamil M, Smaczniak C, Kaufmann K, Yangb WC, Hooivelde GJ, Charnikhovac T, Bouwmeesterc HJ, Bisseling T, Geurts R (2011) Strigolactone biosynthesis in *Medicago truncatula* and rice requires the symbiotic GRAS-type transcription factors NSP1 and NSP2. Plant Cell 23:3853–3865

Lodwig EM, Hosie AHF, Bourdes A, Findlay K, Allaway D, Karunakaran R, Downie JA, Poole PS (2003) Amino-acid cycling drives nitrogen fixation in the legume–Rhizobium symbiosis. Nature 422:722–726

Mabood F, Smith DL (2005) Pre-incubation of *Bradyrhizobium japonicum* with jasmonates accelerates nodulation and nitrogen fixation in soybean (*Glycine max*) at optimal and suboptimal root zone temperatures. Physiol Plant 125:311

Madsen LH, Tirichine L, Jurkiewicz A, Sullivan JT, Heckmann AB, Bek AS, Ronson CW, James EK, Stougaard J (2010) The molecular network governing nodule organogenesis and infection in the model legume *Lotus japonicus*. Nat Commun 1:10

Madsen EB, Antolín-Llovera M, Grossmann C, Ye J, Vieweg S, Broghammer A, Krusell L, Radutoiu S, Jensen ON, Stougaard J, Parniske M (2011) Autophosphorylation is essential for the in vivo function of the *Lotus japonicus* Nod factor receptor 1 and receptor-mediated signalling in cooperation with Nod factor receptor 5. Plant J 65:404–417

Magori S, Kawaguchi M (2009) Long-distance control of nodulation: molecules and models. Mol Cells 27:129–134

Maillet F, Poinsot V, André O, Puech-Pages V, Haouy A, Gueunier M, Cromer L, Giraudet D, Formey D, Niebel A, Martinez EA, Driguez H, Becard G, Denarie J (2011) Fungal lipochitooligosaccharide symbiotic signals in arbuscular mycorrhiza. Nature 469:58–63

Mandal SM, Chakraborty D, Dey S (2010) Phenolic acids act as signaling molecules in plant-microbe symbioses. Plant Signal Behav 5(4):359–368

Marsh JF, Rakocevic A, Mitra RM, Brocard L, Sun J, Eschstruth A, Long SR, Schultze M, Ratet P, Oldroyd GE (2007) *Medicago truncatula NIN* is essential for rhizobial-independent nodule organogenesis induced by autoactive calcium/calmodulin-dependent protein kinase. Plant Physiol 144:324–335

Martinez E, Romero D, Palacios R (1990) The Rhizobium genome. Crit Rev Plant Sci 9:59–93

Messinese E, Mun J, Yeun L, Jayaraman D, Rouge P, Barre A, Lougnon G, Schornack S, Bono J, Cook D, Ane J (2007) A novel nuclear protein interacts with the symbiotic *DMI3* calcium and calmodulin dependent protein kinase of *Medicago truncatula*. Mol Plant Microbe Interact 20:912–921

Middleton PH, Jakab J, Penmetsa RV, Starker CG, Doll J, Kalo P, Prabhu R, Marsh JF, Mitra RM, Kereszt A, Dudas B, VandenBosch K (2007) An *ERF* transcription factor in *Medicago truncatula* that is essential for Nod factor signal transduction. Plant Cell 19:1221–1234

Mitra RM, Gleason CA, Edwards A, Hadfield J, Downie JA, Oldroyd GE, Long SR (2004) A Ca2+/calmodulin-dependent protein kinase required for symbiotic nodule development: gene identification by transcript-based cloning. Proc Natl Acad Sci USA 101(13):4701–4705

Miyazawa H, Oka-Kira E, Sato N, Takahashi H, Wu GJ, Sato S, Hayashi M, Betsuyaku S, Nakazono M, Tabata S, Harada K, Sawa S, Fukuda H, Kawaguchi M (2010) The receptor-like kinase *KLAVIER* mediates systemic regulation of nodulation and non-symbiotic shoot development in *Lotus japonicus*. Development 137:4317–4325

Mortier V, Den Herder G, Whitford R, Van de Velde W, Rombauts S, D'Haeseleer K, Holsters M, Goormachtig S (2010) CLE peptides control *Medicago truncatula* nodulation locally and systemically. Plant Physiol 153:222–237

Mortier V, De Waver E, Vuylsteke M, Holsters M, Goormachtig S (2012) Nodule numbers are governed by interaction between CLE peptides and cytokinin signaling. Plant J 70:367–376

Mulligan JT, Long SR (1985) Induction of *Rhizobium meliloti nodC* expression by plant exudate requires *nodD*. Proc Natl Acad Sci USA 82:6609–6613

Murakami Y, Miwa H, Imaizumi-Anraku H, Kouchi H, Downie JA, Kawaguchi M, Kawasaki S (2007) Positional cloning identifies *Lotus japonicus NSP2*, a putative transcription factor of the GRAS family, required for *NIN* and *ENOD40* gene expression in nodule initiation. DNA Res 13(6):255–265

Murray JD, Karas BJ, Sato S, Tabata S, Amyot L, Szczyglowski K (2007) A cytokinin perception mutant colonized by Rhizobium in the absence of nodule organogenesis. Science 315:101–104

Murray JD, Muni RR, Torres-Jerez I, Tang Y, Allen S, Andriankaja M, Li G, Laxmi A, Cheng X, Wen J, Vaughan D, Schultze M, Sun J, Charpentier M, Oldroyd G, Tadege M, Ratet P, Mysore KS, Chen R, Udvardi MK (2011) *Vapyrin*, a gene essential for intracellular progression of arbuscular mycorrhizal symbiosis, is also essential for infection by rhizobia in the nodule symbiosis of *Medicago truncatula*. Plant J 65:244–252

Oka-Kira E, Kawaguchi M (2006) Long-distance signaling to control root nodule number. Curr Opin Plant Biol 9:496–502

Oka-Kira E, Tateno K, Miura K, Haga T, Hayashi M, Harada K, Sato S, Tabata S, Shikazono N, Tanaka A, Watanabe Y, Fukuhara I, Nagata T, Kawaguchi M (2005) klavier (klv), a novel hypernodulation mutant of *Lotus japonicus* affected in vascular tissue organization and floral induction. Plant J 44:505–515

Oldroyd GE, Downie JA (2004) Calcium, kinases and nodulation signaling in legumes. Nat Rev Mol Cell Biol 5(7):566–576

Oldroyd G, Geurts R (2001) *Medicago truncatula*, going where no plant has gone before. (Meeting report for the 4th Workshop on *Medicago truncatula*, July 7th–10th 2001, Madison, WI). Trends Plant Sci 6:552–554

Oldroyd GE, Long SR (2003) Identification and characterization of nodulation-signaling pathway 2, a gene of *Medicago truncatula* involved in Nod factor signaling. Plant Physiol 131:1027–1032

Oldroyd GED, Murray JD, Poole PS, Downie JA (2011) The rules of engagement in the legume-rhizobial symbiosis. Annu Rev Genet 45:119–144

Ott T, van Dongen JT, Gunther C, Krusell L, Desbrosses G, Vigeolas H, Bock V, Czechowski T, Geigenberger P, Udvardi MK (2005) Symbiotic leghemoglobins are crucial for nitrogen fixation in legume root nodules but not for general plant growth and development. Curr Biol 15(6):531–535

Ott T, Sullivan J, James EK, Flemetakis E, Gunther C, Gibon Y, Ronson C, Udvardi MK (2009) Absence of symbiotic leghemoglobins alters bacteroid and plant cell differentiation during development of *Lotus japonicus* root nodules. Mol Plant Microbe Interact 22:800–808

Pacios-Bras C, Schlaman HRM, Boot K, Admiraal P, Langerak JM, Stougaard J, Spaink HP (2003) Auxin distribution in *Lotus japonicus* during root nodule development. Plant Mol Biol 52:1169–1180

Paszkowski U (2006) A journey through signaling in arbuscular mycorrhizal symbioses. New Phytol 172(1):35–46

Peiter E, Sun J, Heckmann AB, Venkateshwaran M, Riely BK, Otegui MS, Edwards A, Freshour G, Hahn MG, Cook DR, Sanders D, Oldroyd GE, Downie JA, Ane JM (2007) The *Medicago truncatula* DMI1 protein modulates cytosolic calcium signaling. Plant Physiol 145:192–203

Pellock BJ, Cheng HP, Walker GC (2000) Alfalfa root nodule invasion efficiency is dependent on *Sinorhizobium meliloti* polysaccharides. J Bacteriol 182:4310–4318

Penmetsa RV, Cook DR (1997) A legume ethylene-insensitive mutant hyperinfected by its rhizobial symbiont. Science 275:527–530

Penmetsa RV, Frugoli J, Smith L, Long SR, Cook D (2003) Genetic evidence for dual pathway control of nodule number in *Medicago truncatula*. Plant Physiol 131:998–1008

Penmetsa RV, Uribe P, Anderson JP, Lichtenzveig J, Gish JC, Nam YW, Engstrom E, Xu K, Siskel G, Pereira M, Baek JM, Lopez-Meyer M, Long SR, Harrison MJ, Singh KB, Kiss GB, Cook DR (2008) The *Medicago truncatula* ortholog of Arabidopsis EIN2, sickle, is a negative regulator of symbiotic and pathogenic microbial associations. Plant J 55:580–595

Perez Guerra JC, Coussens G, De Keyser A, De Rycke R, De Bodt S, Van de Velde W, Goormachtig S, Holsters M (2010) Comparison of developmental and stress-induced nodule senescence in *Medicago truncatula*. Plant Physiol 152:1574–1584

Phillips DA, Tsai SM (1992) Flavonoids as plant signals to rhizosphere microbes. Mycorrhiza 1(2):55–58

Phillips DA, Joseph CM, Maxwell CA (1992) Trigonelline and stachydrine released from alfalfa seeds activate Nodd2 protein in *Rhizobium meliloti*. Plant Physiol 99:1526–1531

Phillips DA, Dakora FD, Sande E, Joseph CM, Zon J (1994) Synthesis, release, and transmission of alfalfa signals to rhizobial symbionts. Plant Soil 161:69–80

Pierce M, Bauer WD (1983) A rapid regulatory response governing nodulation in soybean. Plant Physiol 73:286–290

Pii Y, Crimi M, Cremonese G, Spena A, Pandolfini T (2007) Auxin and nitric oxide control indeterminate nodule formation. BMC Plant Biol 7:21

Plet J, Wasson A, Ariel F, Le Signor C, Baker D, Mathesius U, Crespi M, Frugier F (2011) MtCRE1-dependent cytokinin signaling integrates bacterial and plant cues to coordinate symbiotic nodule organogenesis in *Medicago truncatula*. Plant J 65:622–633

Prayitno J, Rolfe BG, Mathesius U (2006) The ethylene-insensitive sickle mutant of *Medicago truncatula* shows altered auxin transport regulation during nodulation. Plant Physiol 142:168–180

Price NP, Carlson RW (1995) Rhizobial lipo-oligosaccharide nodulation factors: multidimensional chromatographic analysis of symbiotic signals involved in the development of legume root nodules. Glycobiology 5(2):233–242

Puppo A, Groten K, Bastian F, Carzaniga R, Soussi M, Lucas MM, de Felipe MR, Harrison J, Vanacker H, Foyer CH (2005) Legume nodule senescence: roles for redox and hormone signalling in the orchestration of the natural aging process. New Phytol 165:683–701

Redmond JW, Batley M, Djordjevic MA, Innes RW, Kuempel PL, Rolfe BG (1986) Flavones induce expression of nodulation genes in Rhizobium. Nature 323:632–634

Riely BK, Lougnon G, Ane JM, Cook DR (2007) The symbiotic ion channel homolog DMI1 is localized in the nuclear membrane of *Medicago truncatula* roots. Plant J 49:208–216

Saito K, Yoshikawa M, Yano K, Miwa H, Uchida H, Asamizu E, Sato S, Tabata S, Imaizumi-Anraku H, Umehara Y, Kouchi H, Murooka Y, Szczyglowski K, Downie JA, Parniske M, Hayashi M, Kawaguchi M (2007) NUCLEOPORIN85 is required for calcium spiking, fungal and bacterial symbioses, and seed production in *Lotus japonicus*. Plant Cell 19:610–624

Schauser L, Roussis A, Stiller J, Stougaard J (1999) A plant regulator controlling development of symbiotic root nodules. Nature 402:191–195

Schnabel EL, Frugoli JF (2004) The *PIN* and *LAX* families of auxin transport genes in *Medicago truncatula*. Mol Genet Genomics 272:420–432

Schnabel E, Journet EP, de Carvalho-Niebel F, Duc G, Frugoli J (2005) The *Medicago truncatula SUNN* gene encodes a *CLV1*-like leucine-rich repeat receptor kinase that regulates nodule number and root length. Plant Mol Biol 58:809–822

Schnabel E, Mukherjee A, Smith L, Kassaw T, Long S, Frugoli J (2010) The *lss* supernodulation mutant of *Medicago truncatula* reduces expression of the *SUNN* gene. Plant Physiol 154:1390–1402

Schnabel EL, Kassaw TK, Smith LS, Marsh JF, Oldroyd GE, Long SR, Frugoli JA (2011) The ROOT DETERMINED NODULATION1 gene regulates nodule number in roots of *Medicago truncatula* and defines a highly conserved, uncharacterized plant gene family. Plant Physiol 157:328–340

Schultze M, Kondorosi A (1998) Regulation of symbiotic root nodule development. Annu Rev Genet 32:33–57

Searle IR, Men AE, Laniya TS, Buzas DM, Iturbe-Ormaetxe I, Carroll BJ, Gresshoff PM (2003) Long-distance signaling in nodulation directed by a CLAVATA1-like receptor kinase. Science 299:109–112

Shaw SL, Long SR (2003) Nod factor inhibition of reactive oxygen efflux in a host legume. Plant Physiol 132:2196–2204

Sidorova KK, Shumnyi VK (2003) A collection of symbiotic mutants in pea (*Pisum sativum* L.): creation and genetic study. Russ J Genet 39:406–413

Sieberer BJ, Chabaud C, Fournier J, Timmers ACJ, Barker DG (2012) A switch in Ca^{2+} spiking signature is concomitant with endosymbiotic microbe entry into cortical root cells of *Medicago truncatula*. Plant J 69:822–830

Smil V (1999) Nitrogen in crop production: an account of global flows. Global Biogeochem Cycles 13:647–662

Smit P, Raedts J, Portyanko V, Debelle F, Gough C, Bisseling T, Geurts R (2005) NSP1 of the GRAS protein family is essential for rhizobial nod factor-induced transcription. Science 308 (5729):1789–1791

Smit P, Limpens E, Geurts R, Fedorova E, Dolgikh E, Gough C, Bisseling T (2007) Medicago LYK3, an entry receptor in rhizobial nodulation factor signaling. Plant Physiol 145:183–191

Smith SE, Read DJ (2008) Mycorrhizal symbiosis. Academic, San Diego, CA

Spaink HP, Wijfjes AHM, van Vilet TB, Kijne JW, Lugtenberg JJ (1993) Rhizobial lipo-oligosaccharide signals and their role in plant morphogenesis; are analogous lipophilic chitin derivatives produced by the plant? Aust J Plant Physiol 20:381–392

Stracke S, Kistner C, Yoshida S, Mulder L, Sato S, Kaneko T, Tabata S, Sandal N, Stougaard J, Szczyglowski K, Parniske M (2002) A plant receptor-like kinase required for both bacterial and fungal symbiosis. Nature 27:959–962

Subramanian S, Stacey G, Yu O (2006) Endogenous isoflavones are essential for the establishment of symbiosis between soybean and *Bradyrhizobium japonicum*. Plant J 48:261–273

Tamimi SM, Timko MP (2003) Effects of ethylene and inhibitors of ethylene synthesis and action on nodulation in common bean (*Phaseolus vulgaris* L.). Plant Soil 257:125–131

Timmers AC, Auriac MC, Truchet G (1999) Refined analysis of early symbiotic steps of the Rhizobium-Medicago interaction in relationship with microtubular cytoskeleton rearrangements. Development 126:3617–3628

Tirichine L, Sandal N, Madsen LH, Radutoiu S, Albrektsen AS, Sato S, Asamizu E, Tabata S, Stougaard J (2007) A gain-of-function mutation in a cytokinin receptor triggers spontaneous root nodule organogenesis. Science 315:104–107

Udvardi MK, Scheible WR (2005) GRAS genes and the symbiotic green revolution. Science 308(5729):1749–1750

Van de Velde W, Guerra JC, De Keyser A, De Rycke R, Rombauts S, Maunoury N, Mergaert P, Kondorosi E, Holsters M, Goormachtig S (2006) Aging in legume symbiosis. A molecular view on nodule senescence in *Medicago truncatula*. Plant Physiol 141:711–720

van Noorden GE, Ross JJ, Reid JB, Rolfe BG, Mathesius U (2006) Defective long-distance auxin transport regulation in the *Medicago truncatula* super numeric nodules mutant. Plant Physiol 140:1494–1506

van Rhijn P, Goldberg RB, Hirsch AM (1998) *Lotus corniculatus* nodulation specificity is changed by the presence of a soybean lectin gene. Plant Cell 10:1233–1250

van Workum WAT, van Slageren S, van Brussel AAN, Kijne JW (1998) Role of exopolysaccharides of *Rhizobium leguminosarum* bv. viciae as host plant-specific molecules required for infection thread formation during nodulation of *Vicia sativa*. Mol Plant Microbe Interact 11:1233–1241

Vernie T, Moreau S, de Billy F, Plet J, Combier JP, Rogers C, Oldroyd G, Frugier F, Niebel A, Gamas P (2008) EFD is an ERF transcription factor involved in the control of nodule number and differentiation in *Medicago truncatula*. Plant Cell 20:2696–2713

Wais RJ, Galera C, Oldroyd G, Catoira R, Penmetsa RV, Cook D, Gough C, Denarie J, Long SR (2000) Genetic analysis of calcium spiking responses in nodulation mutants of *Medicago truncatula*. Proc Natl Acad Sci USA 97:13407–13412

Wasson AP, Pellerone FI, Mathesius U (2006) Silencing the flavonoid pathway in *Medicago truncatula* inhibits root nodule formation and prevents auxin transport regulation by Rhizobia. Plant Cell 18:1617–1629

Yano K, Shibata S, Chen WL, Sato S, Kaneko T, Jurkiewicz A, Sandal N, Banba M, Imaizumi-Anraku H, Kojima T, Ohtomo R, Szczyglowski K, Stougaard J, Tabata S, Hayashi M, Kouchi H, Umehara Y (2009) CERBERUS, a novel U-box protein containing WD-40 repeats, is required for formation of the infection thread and nodule development in the legume-Rhizobium symbiosis. Plant J 60:168–180

Yokota K, Soyano T, Kouchi H, Hayashi M (2010) Function of GRAS proteins in root nodule symbiosis is retained in homologs of a non-legume, rice. Plant Cell Physiol 51(9):1436–1442

Yuan S, Zhu H, Gou H, Fu W, Liu L, Chen T, Ke D, Kang H, Xie Q, Hong Z, Zhang Z (2012) A ubiquitin ligase of symbiosis receptor kinase involved in nodule organogenesis. Plant Physiol 160(1):106–117

Zhang J, Subramanian S, Zhang Y, Yu O (2007) Flavone synthases from *Medicago truncatula* are flavanone-2-hydroxylases and are important for nodulation. Plant Physiol 144:741–751

Zhu HY, Riely BK, Burns NJ, Ane JM (2006) Tracing nonlegume orthologs of legume genes required for nodulation and arbuscular mycorrhizal symbioses. Genetics 172:2491–2499

Zhu H, Chen T, Zhu M, Fang Q, Kang H, Hong Z, Zhang Z (2008) A novel ARID DNA-binding protein interacts with SymRK and is expressed during early nodule development in *Lotus japonicus*. Plant Physiol 148:337–347

Chapter 2
A Roadmap Towards a Systems Biology Description of Bacterial Nitrogen Fixation

Marie Lisandra Zepeda-Mendoza and Osbaldo Resendis-Antonio

2.1 Introduction

Nitrogen fixation is a fundamental natural process in which atmospheric nitrogen is reduced to ammonia. Several types of microorganisms that live in a variety of physiological conditions can perform nitrogen fixation. These microorganisms include bacteria such as *Rhizobium etli* (Masson-Boivin et al. 2009), *Klebsiella oxytoca* (Luftu-Cakmakci et al. 1981), *Frankia alni* (Schwmtzer and Tjepkema 1990), and cyanobacteria (Berman-Frank et al. 2003), as well as archaea such as *Methanococcus thermolithotrophicus* (Belay et al. 1984) and *Methanosarcina barkeri* (Bomar et al. 1985). Genome information from these microorganisms, in combination with data from high-throughput technologies, provides valuable material for elucidating the biological principles that characterize nitrogen fixation. Notably, the advent of high-throughput technologies has advanced our global understanding of the way that transcriptional regulatory and metabolic networks work together to support this biological process. While this task may sound easy, success is far from being a direct enterprise. The development of new paradigms is central to understand their basic mechanics and to make practical advancements in crop improvements. Thus, systems biology is a new field that can make notable contributions to these goals.

The central aim of this chapter is to present a conceptual view of how a systems biology description can be useful to construct hypotheses to improve our understanding of nitrogen fixation and to use the in silico modeling to perform a systematic and quantitative analysis of this biological phenomenon. We hope that

---

M.L. Zepeda-Mendoza
Licenciatura en Ciencias Genómicas-UNAM, Av. Universidad s/n, Col. Chamilpa, Cuernavaca, Morelos C.P. 62210, Mexico

O. Resendis-Antonio (✉)
Instituto Nacional de Medicina Genomica (INMEGEN), Periferico Sur 4809, Arenal Tepepan, Tlalpan 14610 Mexico City, Mexico
e-mail: oresendis@inmegen.gob.mx

the present study stimulates interest in this new scientific frontier and show that these computational methodologies can be useful to integrate information and generate knowledge for systematically uncovering the underlying metabolic activity during bacterial nitrogen fixation in *R. etli*.

2.1.1 Nitrogen Fixation

Nitrogen fixation can be performed by *rhizobia* soil bacteria in symbiosis with legume plants. This process has been extensively studied, and the entire genome sequences of select *Rhizobiaceae* bacteria (such as *Azorhizobium, Allorhizobium, Bradyrhizobium, Mesorhizobium, Rhizobium*, and *Sinorhizobium*) have been reported. During nitrogen fixation, the bacteria use the nitrogenase enzyme to transform atmospheric N_2 into ammonia. In addition to its crucial role in the nitrogen cycle, there are also important agricultural and environmental reasons for studying this process. For example, modern agriculture relies on inefficient industrial fertilizers to maximize crop production. The use of chemical fertilizers severely damages the environment. Large quantities of fossil fuels are needed for nitrogenous fertilizer production and fertilizer decomposition releases highly active greenhouse gases (Crutzen et al. 2007). Furthermore, fertilizer loss due to leaching causes waterway eutrophication (Graham and Vance 2003). However, nitrogen fixation can provide a clean and natural strategy for improving field crops, thereby avoiding or reducing environmental pollution and making strides towards sustainable agriculture. Taken together, these aspects highlight the importance of optimizing nitrogen input through its natural mechanisms. However, nitrogen fixation is a highly complex biochemical process that requires active signaling and metabolic interchanges between the plant and its symbiotic bacteria. This biological process can be divided into three main phases, which can be briefly explained as follows:

- *Bacterial attraction*: Symbiosis starts when the roots of the plant excrete phenolic flavonoid compounds. Bacteria expressing NodD proteins recognize these compounds and are attracted to the roots (Redmond et al. 1986).
- *Nodule formation*: Once bacteria are localized into the root, they produce strain-specific chito-oligosaccharides, known as nod factors, which induce nodule formation. Nodules are special plant structures that house the bacteria while they are in symbiosis with the plant (Caetano-Anolles and Gresshoff 1991; Ferguson et al. 2010). Then, the bacteria enter the plant through its root hairs. This process typically occurs at the root tips, but the bacteria can also enter through cracks in the epidermal tissue of the root. Bacterial entry causes cellular-level ionic changes in the plant (Felle et al. 1999) and the root is deformed to promote cortical cell divisions. Legumes have two types of nodules: determinate and indeterminate. The host plant determines the nodule type; some physical and biological properties are listed for each case in Table 2.1.

Table 2.1 Characteristics of the two types of nodules

Characteristic	Indeterminate nodule	Determinate nodule
Initial cell divisions	First anticlinically in the inner cortex and then periclinically in the endodermis and pericycle	First subepidermically in the outer cortex
Nodule shape	Cylindrical with a persistent meristem	Spherical, lacking a persistent meristem
Nodule vascular system	Less branched compared to determinate nodules	More branched compared to indeterminate nodules
Bacteroid population	Heterogeneous due to continued cell division activity One bacteroid per symbiosome	Homogeneous, because bacterial differentiation into bacteroids occurs synchronously and then the bacteroids undergo senescence Multiple bacteroids per symbiosome
Bacteroid	Enlarged shape because of having multiple genome amplification; branched by having membrane modifications; with low viability (Vasse et al. 1990; Mergaert et al. 2006)	Normal rod size with normal genome content; with high viability (Mergaert et al. 2006)
Plant examples	Alfalfa (*Medicago sativa*), clover (Trifolium), pea (*Pisum sativum*), barrel medic (*Medicago truncatula*) (Bond 1948; Libbenga and Harkes 1973; Newcomb 1976; Newcomb et al. 1979)	Soybean (*Glycine max*), bean (*Phaseolus vulgaris*), *Pongamia*, *Lotus japonicus* (Bond 1948; Libbenga and Harkes 1973; Newcomb 1976; Newcomb et al. 1979; Turgean and Bauer 1982; Calvert et al. 1984; Gresshoff and Delves 1986; Rolfe and Gresshoff 1988; Mathews et al. 1989)

– *Nitrogen-fixing bacteroids*: The bacteria continue migrating into the plant via infection threads until they reach the inner cortex and the nodule primordium. The bacteria are released into an infection droplet that is excreted near the growing tip of the infection thread. Then, in a process resembling endocytosis, the bacteria are surrounded by a plant-derived membrane, called the peribacteroid membrane, to form a symbiosome (Udvardi and Day 1997). Inside the nodule, the bacteria differentiate into bacteroids. At this stage, the bacteroids are distinctly different from the free-living form of bacteria (see Table 2.1). Once inside the mature nodule, a bacteroid is capable of fixing atmospheric nitrogen by maintaining an ammonia–carbon source exchange with the plant. This nitrogen-fixing capacity is the result of global gene expression changes that give the bacteria highly specialized metabolic activities.

The nitrogenase enzyme is responsible for atmospheric nitrogen reduction. This enzyme is highly oxygen sensitive, but the nodule protects it by providing a microaerobic environment. While nitrogenase performs the nitrogen reduction, the biochemical reaction also depends on the coordinated participation of other metabolic pathways to produce the required substrates. In other words, the nitrogenase enzyme participates in an essential reaction, but many other metabolic reactions are necessary to maintain the symbiotic nitrogen fixation. Hence, optimal nitrogen fixation also requires active metabolite exchange between the plant and the bacteroid. Amino acid transport between plant and bacteroid is essential to support the nitrogen fixation process (Lodwig et al. 2003). Thus, there is evidence that glutamate, or one of its derivatives, is provided by the plant whereas aspartate and alanine are secreted by the bacteroid (Day et al. 2001).

2.1.2 Symbiotic Relationship Between R. etli and Phaseolus vulgaris

It is important to clarify that symbiotic nitrogen fixation is a highly selective interaction; i.e., all legumes do not attract all *rhizobia*. Therefore, the bacterium and plant that are used to study and model nitrogen fixation must be carefully selected. For example, several studies have focused on the interaction between alfalfa and *Sinorhizobium meliloti*, which is one of the best-described interactions in symbiotic nitrogen fixation (Jones et al. 2007). Given this plant–bacteria specificity requirement, this chapter will focus on the metabolic analysis of symbiosis between the common bean (*P. vulgaris*) and *R. etli*.

We have chosen to focus on *R. etli* and *P. vulgaris* because of the large amount of physiological, molecular, and genetic information available on this symbiotic relationship. Furthermore, *P. vulgaris* is highly important to the agricultural market. The common bean represents 50 % of worldwide legume consumption (McClean et al. 2004), which is approximately 23 million tons, according to the Food and Agriculture Organization of the United Nations (http://www.fao.org/corp/statistics/en). In addition, *P. vulgaris* is a promiscuous legume because it can form nodules with various *rhizobia* (*R. etli*, *R. leguminosarum*, *R. propici*, *R. gallicum*, and *R. giardini*). However, *R. etli* is the most abundant symbiont for the wild-type bean. Taken together, we believe that this specific interaction is an appropriate model for studying legume–*Rhizobium* symbiosis with clear implications in agriculture technology.

2.2 Systems Biology of Bacterial Nitrogen Fixation

Nitrogen fixation involves a variety of signaling and metabolic pathways that coexist in a complex fashion. Consequently, understanding this process requires computational algorithms that can survey its complexity in a systematic and coherent fashion and can also drive additional and improved experimental study designs. While the metabolic and genetic aspects of this symbiosis have been studied extensively, many intermediate and late-stage events remain poorly understood. Notably, the recent advent of high-throughput technologies has produced extensive data on gene expression patterns and metabolic activities that occur during nitrogen fixation. However, interpreting data from multiple high-throughput technologies (e.g., fluxomics, sequencing, microarrays, protein interaction data, and metabolomics) is not a trivial task due to the varying levels of biological description and heterogeneity of these datasets. In this context, genome-scale computational methods must be developed in a systems biology framework to systematically integrate this myriad of data and construct biological hypotheses. Thus, a systems biology framework can contribute to the achievement of three important goals (1) data integration and interpretation, (2) computational modeling of the metabolic phenotypes associated with nitrogen fixation, and (3) experimental assessment of the in silico predictions. A metabolic phenotype refers to the set of metabolic reactions that are required to support nitrogen fixation.

Systems biology has two main schemes: top-down and bottom-up. The top-down scheme provides a descriptive data integration approach, enables genome-scale monitoring of cellular activity, and allows the user to focus on specific areas of interest in terms of cellular activity. The top-down approach also allows researchers to discover patterns, referred to as emergent properties that can only be observed when looking at the system as a whole (Bhalla and Iyengar 1999; Rodriguez-Plaza et al. 2012). In comparison, a bottom-up scheme provides a more systemic and quantitative analysis of the ways that specific external perturbations can affect biological networks such as transcriptional, signaling, and metabolic networks (Fig. 2.1).

Although these two approaches can be viewed as separate strategies, a combination of the bottom-up and top-down schemes is required to completely study a biological system. A computational algorithm known as constraint-based modeling has been developed for this purpose and has been successfully applied to a variety of biological systems (Palsson 2006; Larhlimi and Bockmayr 2007, 2009; Resendis-Antonio et al. 2007). This framework enables the observation of genotype–phenotype relationships for a selected organism. The steps to form a constraint-based model are as follows (see Fig. 2.2). First, an organism's genomic data are obtained. Next, a metabolic network is reconstructed with the use of bioinformatic tools and literature reviews. Then, computational simulations are conducted to predict the organism's potential response to external perturbations. A more detailed description of this process is provided in the following sections.

2.2.1 Metabolic Network Reconstruction

The first step in reconstructing a metabolic network is data collection. All of the relevant information regarding the process of interest must be collected. There are many different sources of biological information, ranging from laboratory experiments and literature research to new high-throughput technologies. The

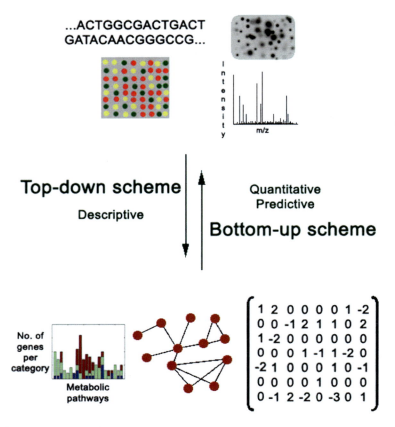

Fig. 2.1 The two schemes in systems biology. The top-down scheme starts with data coming from different sources (microarrays, DNA or RNA sequencing, mass spectrometry, etc.) to provide a description of the cellular activity in a genomic scale. The bottom-up scheme works with integrated data to build models and carry out in silico simulations to do predictions and guide hypothesis testing

amount of physiological data has grown exponentially in recent years in a variety of Rhizobiaceae; in particular high-throughput technologies have provided information across a wide range of biological levels such as transcriptome, proteome, and metabolome (see, for instance, Sarma and Emerich 2005, 2006; Resendis-Antonio et al. 2011, 2012). All of this physiological information needs to be collected to warranty a high-quality microorganism's metabolic network reconstruction.

Once the necessary information has been collected, we can systematically integrate the data by reconstructing a metabolic network. At this stage, we have a set of metabolic interactions that are supported by experimental evidence, bioinformatic predictions, or both. In turn, each reaction is associated with basic genome information such as which enzyme carries out the reaction, which genes encode the enzyme, the processed mRNA sequence, and the maximum reaction flux. However,

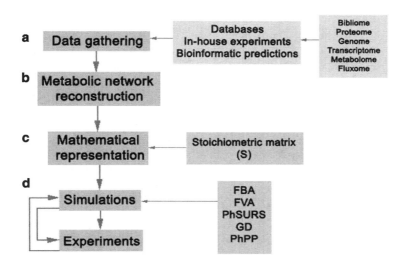

Fig. 2.2 Steps of the constraint-based modeling. (**a**) *Data gathering*. Data can come from many different sources, such as metabolomic, transcriptomic, proteomic, fuxomic, and genomic experiments carried out in-house or stored in public databases, as well as from bioinformatic predictions and from the literature. (**b**) *Metabolic network reconstruction*. The gathered data has to be integrated in a coherent fashion by taking into account all the available information related to every reaction. This integrated data constitutes the metabolic reconstruction. (**c**) *Mathematical representation of the network*. Once it has been reconstructed, a metabolic network is mathematically represented through an $m \times n$ matrix, called stoichiometric matrix (S), in which the m rows are the metabolites and the n columns are the reactions. The entry value in (m,n) is the stoichiometric coefficient of the metabolite m in the reaction n, and the value is positive if the metabolite is produced in the reaction and negative if it is consumed. (**d**) *Simulations and experimental cross talk*. In silico simulations are made based on the model by changing some parameters of the components to make predictions and lead future hypothesis testing experiments. Thus a retroactive process between laboratory experiments and computational analysis is established. Some of the most used and important simulations are flux balance analysis (FBA), flux variability analysis (FVA), phenotypic space uniform random sampling (PhSURS), gene deletion (GD), and phenotypic phase-plane (PhPP)

given that our knowledge is incomplete, this metabolic reconstruction frequently contains gaps that can limit the in silico phenotype analysis. In order to fill these missing metabolic reactions it is also crucial to include bioinformatic predictions or alternative computational methods (Orth and Palsson 2010). Two types of missing metabolic reactions can be distinguished: gap and orphan reactions. The first ones are those without experimental results confirming its presence but with the metabolic context suggesting it. The second ones include all the reactions that we expect to exist, preferably with experimental evidence suggesting it, but there is no clear evidence to which genes they are associated (Orth and Palsson 2010). For example, if experimental data were not available for all the enzymes in the glycolysis pathway, the missing enzymes could be predicted based on a variety of techniques that include the completely annotated pathway in a closely related species with some bioinformatics and experimental methods. Thus, bioinformatic predictions can fill any gaps in a pathway. This data collection and gap-filling process are fundamental to ensure a proper metabolic reconstruction that will increase the predictive capabilities of the model.

2.2.2 Mathematical Representation of the Metabolic Reconstruction

Once the data have been integrated into a metabolic reaction set, the reconstruction can be mathematically represented as a $m \times n$ matrix, called the stoichiometric matrix (S), in which each of the rows represents a metabolite and each of the columns represents a reaction. If the metabolite m is produced in the reaction n, the entry value of (m,n) in S is the metabolite's stoichiometric coefficient in that reaction. If a metabolite is consumed, its entry (m,n) is the negative of its stoichiometric coefficient. As an example, we present the metabolic network reconstruction of the glycolysis pathway. The glycolytic reactions and their corresponding enzymes are shown in Table 2.2. The corresponding S is shown in Fig. 2.3.

The organization of the reconstructed network can also be represented graphically. Cellular metabolism involves two sets of objects: reactions and metabolites. Given this bipartite nature, metabolic networks can be separated into their two components and represented as a reaction or as a metabolite graph (Montañez et al. 2010). In a reaction graph, each node is an enzyme that carries out a given reaction where links connect any two nodes that share a common metabolite. In a metabolite graph, the compounds are the nodes and any two nodes are linked if they participate in the same reaction. An example of these complementary representations is shown in Fig. 2.4.

The type of question asked in a study determines which graphical representation is the most appropriate. For example, if the question is how the network metabolites work together to optimize a biological process, then the metabolite graph should be used (Resendis-Antonio et al. 2012). However, if the question is how the metabolic reactions work together to optimize a biological process, then the reaction graph should be used. Although the questions may sound similar, they have important intrinsic differences. In the first question, the focus is solely on the metabolite and does not involve which reaction it belongs to, i.e., the metabolite can be linked to many different reactions. In the second scenario, the focus is on the enzyme and not its associated metabolites.

2 A Roadmap Towards a Systems Biology Description of Bacterial Nitrogen Fixation

Table 2.2 Glycolysis metabolic pathway reconstruction

Reaction	Enzyme
Glucose + $ATP^{4-} \rightarrow G6P^{2-} + ADP^{3-} + H^+$	Hexokinase (HK)
$G6P^{2-} \leftrightarrow F6P^{2-}$	Phosphoglucose isomerase (PGI)
$F6P^{2-} + ATP^{4-} \rightarrow F1,6BP^{4-} + ADP^{3-} + H^+$	Phosphofructokinase (PFK-1)
$F1,6BP^{4-} \leftrightarrow DHAP^{2-} + G3P^{2-}$	Fructose-bisphosphate aldolase (ALDO)
$DHAP^{2-} \leftrightarrow G3P^{2-}$	Triosephosphate isomerase (TPI)
$G3P^{2-} + P_i^{2-} + NAD^+ \leftrightarrow 1,3BPG^{4-} + NADH + H^+$	Glyceraldehyde phosphate dehydrogenase (GAPDH)
$1,3BPG^{4-} + ADP^{3-} \leftrightarrow 3PG^{3-} + ATP^{4-}$	Phosphoglycerate kinase (PGK)
$3PG^{3-} \leftrightarrow 2PG^{3-}$	Phosphoglycerate mutase (PGM)
$2PG^{3-} \leftrightarrow PEP^{3-} + H_2O$	Enolase (ENO)
$PEP^{3-} + ADP^{3-} + H^+ \rightarrow Pyr^- + ATP^{4-}$	Pyruvate kinase (PK)

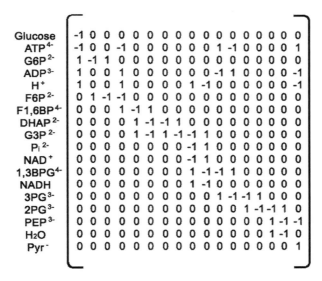

Fig. 2.3 Representation of the glycolysis pathway through a stoichiometric matrix. In this figure we have used the following notations: glucose-6-phosphate (G6P), fructose-6-phosphate (F6P), fructose-1,6-bisphosphate (F1,6BP), dihydroxyacetone phosphate (DHAP), glyceraldehyde-3-phosphate (G3P), 1,3-bisphosphoglycerate (1,3BPG), 3-phosphoglycerate (3PG), 2-phosphoglycerate (2PG), phosphoenolpyruvate (PEP), and pyruvate (Pyr). From left to right the reactions are in the same order as shown in Table 2.2

2.2.2.1 Topological Analysis of the Genome-Scale Metabolic Reconstruction

Graph theory can be used to perform topological studies on metabolic networks. Thus, topological analysis is important for uncovering cellular organizational principles and understanding its mechanisms for coordinating biological functions. Several classic topological properties are explained below.

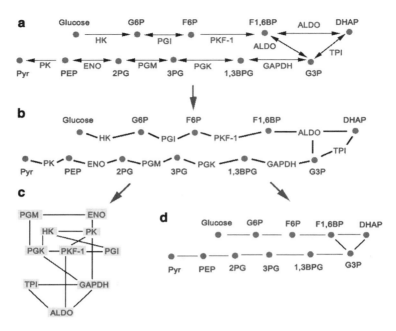

Fig. 2.4 Graphical representations of a metabolic network. (**a**) Conventional biochemical representation of the glycolysis pathway. This pathway can also be represented as a bipartite graph. (**b**) The bipartite graph of the glycolysis pathway consists of two different sets, enzymes and metabolites, represented by two different types of nodes. Nodes of one type can only be linked to nodes of the other type, so here a metabolite node is linked to an enzyme node if it participates in the reaction carried out by that enzyme. This multipartite graph can be separated into two graphs, each one with one set of data. (**c**) The reaction graph is built with the enzymes set. In this graph each node is an enzyme and a link connects two nodes if they share any metabolite. (**d**) The metabolite graph is built with the metabolites set. Here each node represents a metabolite and the links connecting them are the metabolic reactions that transform one metabolite to another. Here graphs (**b**), (**c**), and (**d**) are undirected because they are only for the purpose of representing general relationships, not taking into account the causality as directed networks do when the question of the study has been defined

- *Degree distribution*: The degree or connectivity of a node describes its number of nearest neighbors, i.e., its number of links. The degree values of all of the nodes in the network form a distribution; this distribution describes the number of nodes per degree (Newman 2003). The probability distribution of metabolic networks can usually fit a power law, reflecting a scale-free topology (Barabasi 2009). Scale-free networks are characterized by a few components with high connectivity (called hubs, e.g., ATP) and numerous components with low connectivity. There is evidence that this characteristic yields network robustness against the random loss of components (Han et al. 2004; Barabasi and Oltvai 2004; Jeong et al. 2001). However, other studies have suggested that this finding should be carefully reviewed in biological systems (Stumpf and Porter 2012).

- *Clustering coefficient* (*C*): The clustering coefficient describes the degree to which the neighbors of a given node are connected to one another. In a social context, this is similar to asking "how many of my friends are also friends?" The clustering coefficient has a value between 0 and 1, with 1 indicating that all of the neighbors are connected to one another (Watts and Strogatz 1998).
- *Shortest path length*: The shortest path length is defined as the average of the minimum number of links (the shortest path) between every two nodes in the network. This measure is typically very small in metabolic networks (Jeong et al. 2000; Wagner and Fell 2001), although not all studies are in agreement (Arita 2004).
- *Module*: A module is a subnetwork of nodes with high clustering coefficients between one another. In other words, it is a set of nodes that are more closely related to one another than to the rest of the network. Nodes that share a common biological function are called functional modules. If there is no implicated biological function among the nodes, they are referred to as structural modules (Barabasi et al. 2011). However, there is evidence that functional modules can be associated to topological modules (Resendis-Antonio et al. 2012).

2.2.3 Computational Simulations

After the metabolic network reconstruction has been mathematically represented as S, the metabolic capabilities of the organism can be evaluated. However, before continuing with this aim, several issues must be solved. Dynamic modeling of a genome-scale metabolic reconstruction using ordinary differential equations (ODE) or partial differential equations (PDE) requires specific knowledge of the associated enzyme kinetics. The lack of specific information on metabolic reactions is a central problem in systems biology. Thus, there is a need to develop alternative methods that can overcome these limitations (Orth et al. 2011; Palsson 2011). In order to overcome such limitations, some frameworks have been developed that contribute to the exploration of the metabolic responses of a microorganism at steady-state behavior without an exhaustive number of kinetic parameters. Constraint-based modeling is a systemic and comprehensive systems biology approach that addresses these issues by imposing physical, biological, and thermodynamic constraints on the entire set of metabolic reactions under steady-state conditions (Fig. 2.5a) (Palsson 2006; Price et al. 2004). In this approach, an organism's metabolic capacities are entirely defined by the properties of the stoichiometric matrix S and the enzymatic constraints of each metabolic reaction. Thus, it is possible to analyze the phenotypic space of an organism from its genome-scale metabolic reconstruction.

While the constraints define and limit the metabolic responses of a microorganism, a large number of possible metabolic states still exist. Therefore, developing computational methods to survey the properties of this space is a priority. To this end, a variety of methods that link the metabolic activity of a network to the

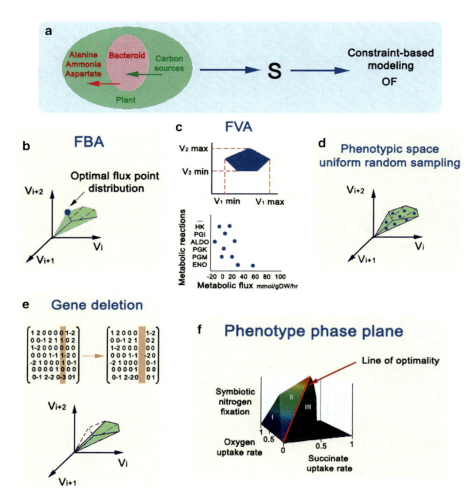

Fig. 2.5 Constraint-based modeling and in silico simulations. (**a**) *Constraint-based modeling*. Once the *S* is constructed and the objective function (OF) selected, an approach to analyze the physiological capabilities of the bacteroid during nitrogen fixation is constraint-based modeling. It works by studying the results of in silico simulations that impose a variety of biological and thermodynamic constraints subject to a steady-state condition. (**b**) *Flux balance analysis (FBA)*. From all the space of possible flux values (the solution space), FBA finds the steady-state flux distribution subject to mass balance and thermodynamics constraints that maximizes an OF, in our case the bacterial nitrogen fixation. To meet the steady-state assumption, the equality $Sv = 0$ is set, with v containing the flux values of all the reactions in the network. (**c**) *Flux variability analysis (FVA)*. FVA is used when there is more than one optimal flux distribution. It computes the minimal and maximal flux of each reaction such that a fixed OF is produced. (**d**) *Phenotypic space uniform random sampling*. This method is used to characterize the solution space. It samples some points of the space in a uniform random way, thus allowing an unbiased characterization of the solution space. Some aspects that can be studied with this unbiased method are the flux correlation between reactions and the characterization of the size and shape of the solution space. (**e**) *Gene deletions*. Gene deletions can be simulated by deleting from the *S* the reactions associated to the deleted gene product. Theoretically speaking, it results in a reduction of the solution space. It can be used to investigate gene dispensability and mutant phenotypes. (**f**) *Phenotypic phase plane (PhPP)*.

2 A Roadmap Towards a Systems Biology Description of Bacterial Nitrogen Fixation

phenotype of a microorganism have been developed. These methods address questions aimed at understanding genotype–phenotype relationships for a biological process of interest. Before describing examples of these methods, two key concepts are explained: solution space and objective function.

- *Solution space*: The metabolic capacities of the model organism during the process of interest together with the enzymatic constraints of each metabolic reaction constitute a phenotypic space called the solution space. The solution space contains all of the possible steady-state metabolite fluxes in the reconstructed metabolic network. Consequently, the metabolic activities identified in different regions of the solution space can be associated with a wide range of phenotypes proper to diverse physiological conditions, including growth rates, metabolite production rates, and the rate of bacterial nitrogen fixation. Mathematically speaking, to represent a cellular metabolism at a steady state, the matrix S must be multiplied by the vector x such that the product equals 0, in other words

$$S \cdot x = 0$$

Because S is a matrix with m rows and n columns, the set of all vectors x that satisfy this equation forms an n-dimensional solution space. It is important to mention that the size of the solution space is *constrained* by the minimum and maximum flux values assigned to each reaction, see Fig. 2.5b.

- *Objective function (OF)*: The objective function represents a linear combination of the metabolite fluxes that are needed for a metabolic process to occur. In other words, the OF is the sum of the fluxes of metabolites, each of which is multiplied by a coefficient that indicates their weighted contribution to the metabolic process. Mathematically speaking, this function is represented as

$$OF = c^T \cdot v$$

where c and v are a weight and metabolic flux vector respectively. The flux vector contains the flux information of all the metabolites included in the reconstruction, it indicates the flux of the metabolites participating in the OF. It is important to select an OF that represents the specific physiological conditions in the organism of interest. In the particular case of nitrogen fixation for *R. etli*, the OF can be defined by taking into account various information sources, such as literature

Fig. 2.5 (continued) This method characterizes the steady-state solution space projected in two or three dimensions to divide the steady-state flux distributions into regions with similar metabolic flux patterns and a similar OF response if certain metabolites were additionally supplied to the network. The line of optimality corresponds to the conditions where metabolic activity is organized in the most efficient way to maximize the OF

reviews and metabolome data generated during the nitrogen fixation stage of *R. etli* (Resendis-Antonio et al. 2012).

Now that we have described these key concepts, we explain below some of the methods used in constraint-based modeling for exploring the phenotype capacities of genome-scale metabolic reconstructions.

2.2.3.1 Flux Balance Analysis

Flux balance analysis (FBA) uses optimization principles to identify the metabolic state of a microorganism and works as explained next. First, an objective function (OF) is defined to simulate the microorganism's physiological state in the process of interest. In order to reduce the number of kinetic parameters required to simulate the metabolic activity, the biological process is assumed to occur at a steady state while maximizing the OF. Finally, the metabolic flux distribution that maximizes the OF in the solution space is computed by linear programming (Fig. 2.5b). Thus, by maximizing the OF, one can identify the metabolic activity along the metabolic pathways required to produce an efficient biological process which, in our case, is nitrogen fixation.

FBA has implicit advantages and disadvantages. On one hand, it requires a limited number of parameters to predict the metabolic state associated with a phenotype. Furthermore, it represents an elegant formalism for analyzing genome-scale metabolic reconstructions while encompassing cross talk from high-throughput technologies (Orth et al. 2010; Resendis-Antonio et al. 2012). However, these advantages can also be seen as a drawback. Due to a lack of kinetic parameters, FBA cannot predict metabolic concentrations and its outputs are only valid under a steady-state assumption (Orth et al. 2010). The temporal behavior of metabolic concentrations and fluxes can be analyzed by dynamic flux balance analysis or genome-scale linear perturbation models (Varma and Palsson 1994; Resendis-Antonio 2009; Jamshidi and Palsson 2008). However, these issues remain open to further investigation in systems biology.

2.2.3.2 Flux Variability Analysis

Given the high dimensionality of the optimization problem described in Sect. 2.2.3.1, the metabolic phenotype that maximizes the OF is usually not unique. This result is due to redundant mechanisms underlying the metabolic network. Flux variability analysis (FVA) is a computational tool that is useful for quantifying this redundancy by identifying the set of metabolic phenotypes that result in an equivalent maximal (or minimal) OF. This analysis is extremely important for identifying biochemical pathways that can potentially generate the same phenotype. The results are fundamental for a proper interpretation of the in silico outputs. From a computational point of view, FVA computes the minimal and maximal fluxes of each reaction for a fixed OF (Fig. 2.5c). This analysis identifies the feasible range of flux values for each reaction, thus representing the different metabolic phenotypes that a microorganism can use to adapt to its environment. In other words, this method does not identify all the optimal solutions but rather identifies the range of flux variability that is possible within any given solution (Mahadevan and Schilling 2003).

2.2.3.3 Phenotypic Space Uniform Random Sampling

A method called phenotypic space uniform random sampling can be used to survey the range of possible biochemical states without the bias of an OF. This analysis enables characterization of the size and shape of the solution space and provides information about the variety of a microorganism's metabolic phenotype states. The method consists of sampling the solution space in a random and uniform manner to extract interesting information about its characteristics. For example, pairwise correlation coefficients can be calculated between all of the metabolic fluxes to identify the degree of correlation between each pair of reactions (Fig. 2.5d) (Price et al. 2004).

2.2.3.4 Gene Deletions

Gene deletions can be simulated in silico by deleting the columns from S associated with the gene or genes product of interest or, alternatively, by constraining its corresponding upper and lower bound in the flux vector. Hence, these in silico gene deletions result in a reduction of the solution space (Fig. 2.5e). This type of study can be used to investigate gene dispensability (Duarte et al. 2004), to examine the evolution of wild-type strains towards new optimal states under selection pressures (Fong et al. 2003), and to predict the metabolic phenotypes of mutant strains (Resendis-Antonio et al. 2007).

2.2.3.5 Phenotypic Phase Plane

The phenotypic phase plane is a method for characterizing a projection of the steady-state solution space in two or three dimensions. In this analysis, steady-state flux distributions can be divided into a finite number of regions based on similar metabolic flux patterns and shadow prices. A shadow price describes how the OF would change if additional metabolites were supplied to the network (Edwards et al. 2001). Thus, the regions of the plane are classified based on the extent to which metabolite availability limits the process of interest and which biochemical activity is the optimal to induce a specific phenotype, in this case nitrogen fixation (Fig. 2.5f).

2.2.4 Experimental Cross Talk

The described methods above are useful for examining a microorganism's metabolic capabilities in a given physiological state. These methods allow us to link a cell's metabolic activity to a specific phenotype. However, in silico predictions can differ from experimental results due to network incompleteness introduced during the metabolic reconstruction, an improperly defined OF, or an inaccurate description of the cell's environmental conditions. Thus, an iterative, retroactive process is needed to reconcile laboratory data and computational predictions.

Simulation results can lead to hypothesis that in order to be verified require the design of new experiments. These experiments can provide additional data, resulting in information that contributes to construct a more realistic computational representation of nitrogen fixation. For instance, this approach could be used to refine the OF definition. The corrected OF could then be used as input for computational methods, the results of which will either further confirm the metabolic activity of previous simulations or lead to the identification of other pathways for future research.

2.3 Systems Biology Description of *R. etli* and *P. vulgaris* Symbiosis

A systems biology analysis of the symbiotic interactions between *R. etli* and *P. vulgaris* was conducted for nitrogen fixation. Top-down and bottom-up strategies were used to achieve the following goals (1) genome-scale metabolic reconstruction and topological analysis of the *R. etli*, (2) constraint-based metabolic modeling of the nitrogen-fixing bacteroid, and (3) construction of a computational platform capable to predict the metabolic phenotype in *R. etli* and compare between our computational predictions and the experimental results.

2.3.1 Metabolic Reconstruction of R. etli

The metabolic reconstruction of *R. etli* for simulating the symbiotic interaction with *P. vulgaris* was based on the integration of several levels of biological data. This information came from various papers reported in the literature and data obtained from high-throughput technologies. This latter source including proteomic, transcriptomic, and metabolomic experiments (Resendis-Antonio et al. 2007, 2011, 2012). In addition other information sources were used, including the *R. etli* genome annotation from KEGG (Kyoto Encyclopedia of Genes and Genomes) and available scientific literature (e.g., biochemical textbooks and previous reports). In total, the current version of the *R. etli* metabolic reconstruction, identified as iOR450, consists of 450 genes, 405 reactions, and 377 metabolites. Although *R. etli* contains approximately 7,000 genes and thus 450 genes may sound too small to be representative of its genome, this reconstruction is based on well-curated and reliable experimental data.

Functional classification of these 450 genes revealed metabolic pathways with an important role in nitrogen fixation, including glycolysis, the TCA cycle, the pentose phosphate pathway, oxidative phosphorylation, amino acid production, glycogen and poly-β-hydroxybutyrate (PHB) biosynthesis, nitrogen reduction, secretion systems, and fatty acid metabolism. This version of the reconstruction is the basic platform for expansion and improvement in future versions. For instance, a significant number of the identified proteins are related to transport (e.g., the transport of small molecules), thus reflecting extensive metabolic cross talk between the plant and the bacteroid (Resendis-Antonio et al. 2007, 2011, 2012). Our metabolic reconstruction will include this physiological information and these processes will be experimentally verified. Currently, this reconstruction is a cornerstone in exploring the metabolic capacities of *R. etli* when facing different environmental conditions.

The graphical representation of the network provides information to elucidate how the metabolite organization within the network supports bacterial nitrogen fixation. For instance, structural modules were identified based on a purely topological criterion to elucidate their relationship with *R. etli*'s biological functions (Ravasz et al. 2002; Resendis-Antonio et al. 2005). Hence, based on a topological criterion (defined as the inverse square of the minimal path length between every pair of nodes), there were reported nine topological modules whose metabolic composition fell into three main groups: nucleic acids, peptides, and lipids. Notably, the functional composition of these modules and the quantitative metabolome data have contributed to establish hypotheses of how the production of metabolites is required for sustaining an optimal nitrogen fixation (Resendis-Antonio et al. 2012). This kind of analysis is an adequate scheme to survey the basic organizational principles by which a biological process can happen in nature. Thus, by correlating the metabolites of modules with the high-throughput data, it was observed that the concentrations of most of the metabolites in each nodule upregulate their concentration during nitrogen fixation compared to the free-living bacteria. This finding supports the idea that a coherent functional biological activity is required at a functional level (Hartwell et al. 1999; Resendis-Antonio et al. 2005, 2012).

2.3.2 *Constraint-Based Modeling and Simulation Results*

Once the metabolic reconstruction was complete, a constraint-based model allowed us to explore the metabolic capacities of the network for *R. etli*. To this end, first we have constructed an appropriate OF for mimicking the metabolic activity during nitrogen fixation. This function was defined based on data from literature review and high-throughput experiments, specially taking into account the following information:

- Plant–bacteroid exchange of certain amino acids may be a general mechanism in *rhizobia* (Prell and Poole 2006).
- Sixteen ATP molecules are required to reduce one N_2 molecule into two ammonium molecules, which are subsequently exported to the plant (Patriarca et al. 2002; Lodwig and Poole 2003).
- Glycogen and PHB accumulate during nitrogen fixation and serve as carbon storage (Bergersen and Turner 1990; Lodwig et al. 2005; Sarma and Emerich 2006; Trainer and Charles 2006).
- Mutations in the biosynthesis of branched chain amino acids, such as L-valine, produce defective nodule formations (De las Nieves Peltzer et al. 2008), which indicates that this metabolite is an essential component. At a similar level, we considered that L-histidine is a central compound in nitrogen fixation (Dixon and Kahn 2004). In agreement with this experimental findings, both amino acids where included in the OF.
- Metabolome experiments have identified new metabolites with statistically significant changes during bacteroid activity compared to free-living bacteria. In order to give a better constraint-based analysis, there has been suggested that these metabolites can be used to define a more complete OF. These metabolites include CMP, 3-phospho-D-glycerate, and 2-oxoglutarate (Resendis-Antonio et al. 2012).

By integrating these findings, an OF representing the metabolic flux of key metabolites in nitrogen fixation was defined as follows:

$$Z^{Fix} = \text{glycogen} + \text{hist[c]} + \text{lys[c]} + \text{phb[c]} + \text{val[c]} + \text{ala[e]} + \text{asp[e]} + \text{nh4[e]} \\ + \text{mal[c]} + \text{trp[c]} + \text{arg[c]} + \text{cit[c]} + \text{cmp[c]} + \text{fum[c]} + \text{3pg[c]} + \text{akg[c]}$$

where glycogen, histidine, lysine, polyhydroxybutyrate, valine, alanine, aspartate, and ammonium are denoted as glycogen, hist[c], lys, phb[c], val[c], ala[e], asp[e], and nh4[e], respectively. Similarly, mal, trp, arg, cit, cmp, fum, 3 pg, and akg denote malate, tryptophan, arginine, citrate, CMP, fumarate, 3-phospho-D-glycerate, and 2-oxoglutarate, respectively. All these metabolites are required to support an effective symbiotic nitrogen fixation, and the location of the metabolites is represented by [c] or [e] for cytoplasmic or external compounds, respectively. Once the OF was defined, a series of studies and simulations were performed to explore, characterize, and predict bacteroid metabolic phenotypes.

Thus, FBA was performed to examine the metabolic fluxes associated with nitrogen fixation. With the FBA results and based on the gene products associated with these reactions, one can predict the set of essential genes for carrying out nitrogen fixation. Furthermore, FVA was used to characterize the core metabolic activity within the set of alternative solutions. The reactions with zero range in flux variability were selected as part of central metabolism in bacterial nitrogen fixation.

The experimental data had to be related to the computational results to biologically support the model; consequently the agreement between the computational interpretations and the experimental data must be quantified to make this comparison. To this end, a consistency coefficient has been defined as the fraction of genes and enzymes that FBA predicted as active compared to those that were detected by transcriptome and proteome technologies. Based on this parameter, the consistency coefficients for the simulation of nitrogen fixation were reported to be 0.61 for genes and 0.71 for enzymes. To ensure the quality of the reconstruction and evaluate the coherence of the computational simulations, we proceed to assess the capacity of the model to predict physiological knowledge when gene deletion occurs (see Fig. 2.5e). To evaluate capacity of the model to predict physiological behaviors after gene silencing, in silico gene deletion analysis was performed on the genes encoding PHB synthase, glycogen synthase, arginine deiminase, myo-inositol dehydrogenase, pyruvate carboxylase, citric acid cycle enzymes, PEP carboxykinase, bisphosphate aldolase, and nitrogenase. These simulations were in qualitative agreement with the experimental counterpart in a variety of Rhizobiaceae (Resendis-Antonio et al. 2007, 2010, 2012).

On the other hand, the robustness of the topological structure of metabolic network (e.g., its functional modules) was explored by changing its external environment, specifically the uptake rates of succinate and inositol. The effects of these changes were evaluated by constructing a phenotypic phase plane. Then, a subset of 20 points, corresponding to select uptake rate conditions within the metabolic phase plane, was subjected to FBA. The metabolic subnetworks were graphically represented based on the FBA results. The topological variations between these subnetworks were defined by the fraction of overlapping metabolites. This study concluded that the metabolic profile required for optimizing nitrogen fixation does not significantly change for a wide range of succinate or inositol uptake rates. This finding suggests that the metabolic network supporting nitrogen fixation is robust to environmental changes (Resendis-Antonio et al. 2012).

2.3.3 Comparative Analysis of Simulations and Experimental Data

A detailed comparison between the in silico simulation results and the high-throughput data highlighted several important aspects of bacterial nitrogen fixation; such processes are described in Table 2.3. Overall, in this chapter we support that a systems biology approach enable the reconstruction and comprehensive analysis of *R. etli* metabolism during symbiosis with *P. vulgaris*, providing a systematic framework to reach a broader view of this important interaction. The current metabolic description can be improved by continually assessing the discrepancies between the computational results and the experimental data, whereas its

Table 2.3 Comparison of the computational and experimental results

Biological process	Computational results	Experimental results
Entner–Doudoroff and pentose phosphate pathways	Besides gluconeogenesis, the existence of a fueling pathway based on pentoses is predicted in the bacteroid	Metabolome experiments detected metabolites of the pentose phosphate and the Entner–Doudoroff pathways (Resendis-Antonio et al. 2012)
Nitrogen fixation enzymes	As expected, if viewing this finding as a positive control for the OF, these enzymes were predicted by the model to be central in the nitrogen fixation	As expected, the *nif* and *nifx* genes and coded proteins that are involved in nitrogen fixation were identified by the microarrays and the proteome experiments
Nucleotides metabolism	Some enzymes of the purine and the pyrimidine pathways actively participate for an optimal nitrogen fixation	Several of the enzymes that participate in the purine and pyrimidine pathways were identified in the proteome data (Resendis-Antonio et al. 2011)
Oxidative phosphorylation	Nonzero fluxes through oxidative phosphorylation and removal of all cytochrome oxidase reactions result in total loss of nitrogen fixation	The essentiality of respiration for nitrogen fixation has been previously reported (Lodwig and Poole 2003; Batut and Boistard 1994)
TCA cycle	The model predicts incomplete use of the TCA cycle	Experiments are not always able to detect all the TCA cycle enzymes. Besides, mutants in the TCA cycle in *B. japonicus* are still able to fix nitrogen, suggesting that a complete set of TCA cycle enzymes is not required for fixation (Lodwig and Poole 2003; Green and Emerich 1997)
Gluconeogenesis pathway	The model concluded that the gluconeogenesis pathway is active in nitrogen fixation	It is known for *R. etli* that the gluconeogenesis pathway is active in nitrogen fixation (Lodwig and Poole 2003). This finding was also identified in the proteome experiment (Resendis-Antonio et al. 2011)
Ammonium assimilation	FBA predicted that there is no activity in the ammonium assimilation pathway during symbiosis	Ammonium assimilation is not observed during nitrogen fixation and an increase in ammonium assimilation negatively affects nodulation (Mendoza et al. 1995)
PHB and glycogen accumulation	Simulations of deletion of PHB synthase predict that symbiotic nitrogen fixation increases	In *R. etli* PHB synthase deletion causes increase in nitrogen fixation (Cevallos et al. 1996). A similar result is observed upon glycogen synthase deletion in *R. tropici* (Marroqui

(continued)

Table 2.3 (continued)

Biological process	Computational results	Experimental results
		et al. 2001). Furthermore, it has been suggested that inhibition of one of the polymers results in accumulation of the other (Cevallos et al. 1996)
Arginine deiminase pathway	Arginine deiminase deletion predicts a decrease in symbiotic nitrogen fixation	Nitrogen fixation is reduced upon arginine deiminase deletion (D'Hooghe et al. 1997)
Myo-inositol catabolic pathway	The model predicts that the activity of the enzyme increases nitrogen fixation and that its deletion decreases the fixation activity	Mutation of myo-inositol dehydrogenase in *Sinorhizobium fredii* increases nitrogen fixation (Jiang et al. 2001). Besides, myo-inositol, 2-dehydrogenase proteins were detected in *R. etli* (Resendis-Antonio et al. 2011)
Fatty acids metabolism	Analysis of the functional modules suggests that the fatty acids are important for nitrogen fixation (Resendis-Antonio et al. 2012)	Fatty acid metabolism can play a significant role in nitrogen fixation in *R. etli* (Resendis-Antonio et al. 2011). An explanation could be that it can supply a variety of precursors to the bacteroid, such as components of the rhizobial membrane, lipopolysaccharides, and coenzymes required in signal transduction

agreements can lead to the design of new experiments (Resendis-Antonio et al. 2012).

2.4 Conclusion

High-throughput technologies provide valuable data for describing the global landscape of cellular activity. However, these techniques do not have the full capacity for describing and predicting the integrated functions of biological processes. Using a systems biology approach, we were able to construct a proper computational framework that serves as a guide for integrating various types of "-omics" data (Palsson 2006, 2011). With this computational framework, we could describe and predict metabolic activities, as well as design experiments that explore genotype–phenotype relationships involved in nitrogen fixation. The combination of experimental data and computational modeling advances our understanding of the main metabolic mechanisms that support bacterial nitrogen fixation. This achievement will undoubtedly have important effects in developing sustainable agricultural programs by optimizing cost-effective crop improvements and ultimately diminishing the pollution effects of chemical fertilizers.

Acknowledgments Osbaldo Resendis-Antonio thanks the financial support of the Research Chair on Systems Biology-INMEGEN, FUNTEL-Mexico. The authors are grateful for comments and suggestions from a referee during the review process of this chapter.

References

Arita M (2004) The metabolic world of *Escherichia coli* is not small. Proc Natl Acad Sci USA 101 (6):1543–1547
Barabasi AL (2009) Scale-free networks: a decade and beyond. Science 325:412–413
Barabasi AL, Oltvai ZN (2004) Network biology: understanding the cell's functional organization. Nat Rev Genet 5:101–113
Barabasi AL, Gulbahce N, Loscalzo J (2011) Network medicine: a network-based approach to human disease. Nat Rev Genet 12(1):56–68
Batut J, Boistard P (1994) Oxygen control in Rhizobium. Antonie Van Leeuwenhoek 66:129–150
Belay N, Sparling R, Daniels L (1984) Dinitrogen fixation by a thermophilic methanogenic bacterium. Nature 312:286–288
Bergersen FJ, Turner GL (1990) Bacteroids from soybean root nodules: accumulation of poly-b-hydroxybutyrate during supply of malate and succinate in relation to N_2 fixation in flow-chamber reactions. Proc R Soc Lond B Biol Sci 240:39–59
Berman-Frank I, Lundgren P, Falkowski P (2003) Nitrogen fixation and photosynthetic oxygen evolution in cyanobacteria. Res Microbiol 154:157–164
Bhalla US, Iyengar R (1999) Emergent properties of biological signaling pathways. Science 283 (5400):381–387
Bomar M, Knoll K, Widdel F (1985) Fixation of molecular nitrogen by *Methanosarcina barkeri*. FEMS Microbiol Lett 31:47–55
Bond L (1948) Origin and developmental morphology of root nodules of *Pisum sativum*. Bot Gaz 109:411–434
Caetano-Anolles G, Gresshoff PM (1991) Plant genetic control of nodulation. Annu Rev Microbiol 45:345–382
Calvert HE, Pence MK, Pierce M, Malik NSA, Bauer WD (1984) Anatomical analysis of the development and distribution of Rhizobium infection in soybean roots. Can J Bot 62:2375–2384
Cevallos MA, Encarnacion S, Leija A, Mora Y, Mora J (1996) Genetic and physiological characterization of a *Rhizobium etli* mutant strain unable to synthesize poly-beta-hydroxybutyrate. J Bacteriol 178:1646–1654
Crutzen P, Mosier AR, Smith KA, Winiwarter W (2007) NO release from agro-fuel production negates global warming reduction by replacing fossil fuels. Atmos Chem Phys Discuss 7:11191–11205
D'Hooghe I, Vander Wauven C, Michiels J, Tricot C, de Wilde P et al (1997) The arginine deiminase pathway in *Rhizobium etli*: DNA sequence analysis and functional study of the arcABC genes. J Bacteriol 179:7403–7409
Day DA, Poole PS, Tyerman SD, Rosendahl L (2001) Ammonia and amino acid transport across symbiotic membranes in nitrogen-fixing legume nodules. Cell Mol Life Sci 58:61–71
De las Nieves Peltzer M, Roques N, Poinsot V, Aguilar OM, Batut J, Capela D (2008) Auxotrophy accounts for nodulation defect of most *Sinorhizobium meliloti* mutants in the branched-chain amino acid biosynthesis. Mol Plant Microbe Interact 21(9):1232–1241

Dixon R, Kahn D (2004) Genetic regulation of biological nitrogen fixation. Nat Rev Microbiol 2:621–631

Duarte NC, Herrgard MJ, Palsson BØ (2004) Reconstruction and validation of *Saccharomyces cerevisiae* iND750, a fully compartmentalized genome-scale metabolic model. Genome Res 14:1298–1309

Edwards JS, Ramakrishna R, Palsson BØ (2001) Characterizing the metabolic phenotype: a phenotype phase plane analysis. Biotechnol Bioeng 77:27–36

Felle HH, Kondorosi E, Kondorosi A, Schultze M (1999) Elevation of the cytosolic free [Ca2+] is indispensable for the transduction of the nod factor signal in alfalfa. Plant Physiol 121:273–279

Ferguson BJ, Indrasumunar A, Hayashi S, Lin MH, Lin YH, Reid DE, Gresshoff PM (2010) Molecular analysis of legume nodule development and autoregulation. J Integr Plant Biol 52 (1):61–76

Fong SS, Marciniak JY, Palsson BØ (2003) Description and interpretation of adaptive evolution of *Escherichia coli* K-12 MG1655 by using a genome-scale in silico metabolic model. J Bacteriol 185:6400–6408

Graham PH, Vance CP (2003) Legumes: importance and constraints to greater use. Plant Physiol 131:872–877

Green LS, Emerich DW (1997) The formation of nitrogen-fixing bacteroids is delayed but not abolished in soybean infected by an alpha-ketoglutarate dehydrogenase-deficient mutant of *Bradyrhizobium japonicum*. Plant Physiol 114:1359–1368

Gresshoff PM, Delves AC (1986) Plant genetic approaches to symbiotic nodulation and nitrogen fixation in legumes. In: Blonstein AD, King PJ (eds) Plant gene research III. A genetical approach to plant biochemistry. Springer, Wien, pp 159–206

Han JD, Bertin N, Hao T, Goldberg DS, Berriz GF, Zhang LV, Dupuy D, Walhout AJ, Cusick ME, Roth FP, Vidal M (2004) Evidence for dynamically organized modularity in the yeast protein-protein interaction network. Nature 430(6995):88–93

Hartwell H, Hopfield JJ, Leibler S, Murray AW (1999) From molecular to modular cell biology. Nature 402:C47–C52

Jamshidi N, Palsson BØ (2008) Formulating genome-scale kinetic models in the post-genome era. Mol Syst Biol 4:171

Jeong H, Tombor B, Albert R, Oltvai ZN, Barabási AL (2000) The large-scale organization of metabolic networks. Nature 407:651–654

Jeong H, Mason SP, Barabasi AL, Oltvai ZN (2001) Lethality and centrality in protein networks. Nature 411(6833):41–42

Jiang G, Krishnan AH, Kim YW, Wacek TJ, Krishnan HB (2001) A functional myo-inositol dehydrogenase gene is required for efficient nitrogen fixation and competitiveness of *Sinorhizobium fredii* USDA191 to nodulate soybean (*Glycine max* [L.] Merr.). J Bacteriol 183:2595–2604

Jones KM, Kobayashi H, Davies BW, Taga ME, Walker GC (2007) How rhizobial symbionts invade plants: the Sinorhizobium-Medicago model. Nat Rev Microbiol 5(8):619–633

Larhlimi A, Bockmayr A (2007) Constraint-based analysis of gene deletion in a metabolic network. Workshop on constraint based methods for bioinformatics, WCB'07, Porto, pp 48–55

Larhlimi A, Bockmayr A (2009) A new constraint-based description of the steady-state flux cone of metabolic networks. Discrete Appl Math 157:2257–2266

Libbenga KR, Harkes PAA (1973) Initial proliferation of cortical cells in the formation of root nodules in *Pisum sativum* L. Planta 114:17–28

Lodwig E, Poole P (2003) Metabolism of Rhizobium bacteroid. Crit Rev Plant Sci 22:37–78

Lodwig EM, Hosie AHF, Bourdès A, Findlay K, Allaway D, Karunakaran R, Downie JA, Poole PS (2003) Amino-acid cycling drives nitrogen fixation in the legume-Rhizobium symbiosis. Nature 422:722–726

Lodwig EM, Leonard M, Marroqui S, Wheeler TR, Findlay K et al (2005) Role of polyhydroxybutyrate and glycogen as carbon storage compounds in pea and bean bacteroids. Mol Plant Microbe Interact 18:67–74

Luftu-Cakmakci M, Evans HJ, Seidler RJ (1981) Characteristics of nitrogen-fixing *Klebsiella oxytoca* isolated from wheat roots. Plant Soil 61:53–63

Mahadevan R, Schilling CH (2003) The effects of alternate optimal solutions in constraint-based genome-scale metabolic models. Metab Eng 5:264–276

Marroqui S, Zorreguieta A, Santamaria C, Temprano F, Soberon M, Megias M, Downie JA (2001) Enhanced symbiotic performance by *Rhizobium tropici* glycogen synthase mutants. J Bacteriol 183:854–864

Masson-Boivin C, Giraud E, Perret X, Batut J (2009) Establishing nitrogen-fixing symbiosis with legumes: how many rhizobium recipes? Trends Microbiol 17(10):458–466

Mathews A, Carroll BJ, Gresshoff PM (1989) Development of Bradyrhizobium infection in supernodulating and non-nodulating mutants of soybean (*Glycine max* [L.] Merrill). Protoplasma 150:40–47

McClean P, Kami J, Gepts P (2004) Genomic and genetic diversity in common bean. In: Wilson RF, Stalker HT, Brummer EC (eds) Legume crop genomics. AOCS, Champaign, IL, pp 60–82

Mendoza A, Leija A, Martinez-Romero E, Hernandez G, Mora J (1995) The enhancement of ammonium assimilation in *Rhizobium etli* prevents nodulation of *Phaseolus vulgaris*. Mol Plant Microbe Interact 8:584–592

Mergaert P, Uchiumi T, Alunni B, Evanno G et al (2006) Eukaryotic control on bacterial cell cycle and differentiation in the Rhizobium-legume symbiosis. Proc Natl Acad Sci USA 103 (13):5230–5235

Montañez R, Medina MA, Sole RV, Rodriguez-Caso C (2010) When metabolism meets topology: reconciling metabolite and reaction networks. Bioessays 32:246–256

Newcomb W (1976) A correlated light and electron microscopic study of symbiotic growth and differentiation in *Pisum sativum* root nodules. Can J Bot 54:2163–2186

Newcomb W, Sippel D, Peterson RL (1979) The early morphogenesis of *Glycine max* and *Pisum sativum* root nodules. Can J Bot 57:2603–2616

Newman MEJ (2003) The structure and function of complex networks. SIAM Rev 45(2):167–256

Orth JD, Palsson BØ (2010) Systematizing the generation of missing metabolic knowledge. Biotechnol Bioeng 107(3):403–412

Orth JD, Thiele I, Palsson BØ (2010) What is flux balance analysis? Nat Biotechnol 28 (3):245–248

Orth JD, Conrad TM, Na J, Lerman JA, Nam H, Feist AM, Palsson BØ (2011) A comprehensive genome-scale reconstruction of *Escherichia coli* metabolism. Mol Syst Biol 7:535

Palsson BØ (2006) Systems biology: properties of reconstructed networks. Cambridge University Press, Cambridge

Palsson BØ (2011) Systems biology: simulation of dynamic network states. Cambridge University Press, Cambridge

Patriarca EJ, Tate R, Iaccarino M (2002) Key role of bacterial NH4+ metabolism in Rhizobium-plant symbiosis. Microbiol Mol Biol Rev 66:203–222

Prell J, Poole P (2006) Metabolic changes of rhizobia in legume nodules. Trends Microbiol 14:161–168

Price ND, Reed JL, Palsson BØ (2004) Genome-scale models of microbial cells: evaluating the consequences of constraints. Nature 2(11):886–897

Ravasz E, Somera AL, Mongru DA, Oltvai ZN, Barabasi AL (2002) Hierarchical organization of modularity in metabolic networks. Science 297(5586):1551–1555

Redmond JW, Batley M, Djordjevic MA, Innes RW, Kuempel PL, Rolfe BG (1986) Flavones induce expression of nodulation genes in Rhizobium. Nature 323:632–635

Resendis-Antonio O (2009) Filling kinetic gaps: dynamic modeling of metabolism where detailed kinetic information is lacking. PLoS One 4(3):e4967

Resendis-Antonio O, Freyre-Gonzalez JA, Menchaca-Mendez R, Gutierrez-Rios RM, Martinez-Antonio A, Avila-Sanchez C, Collado-Vides J (2005) Modular analysis of the transcriptional regulatory network of *E. coli*. Trends Genet 21(1):16–20

Resendis-Antonio O, Reed JL, Encarnacion S, Collado-Vides J, Palsson BØ (2007) Metabolic reconstruction and modeling of nitrogen fixation in *Rhizobium etli*. PLoS Comput Biol 3:1887–1895

Resendis-Antonio O, Hernandez M, Salazar E, Contreras S, Martinez-Batallar G, Mora Y, Encarnacion S (2011) Systems biology of bacterial nitrogen fixation: high-throughput technology and its integrative description with constraint-based modeling. BMC Syst Biol 5:120

Resendis-Antonio O, Hernandez M, Mora Y, Encarnacion S (2012) Functional modules, structural topology, and optimal activity in metabolic networks. PLoS Comput Biol 8(10):e1002720

Rodriguez-Plaza JG, Villalon-Rojas A, Herrera S, Garza-Ramos G, Torres Larios A, Amero C, Zarraga Granados G, Gutierrez Aguilar M, Lara Ortiz MT, Polanco Gonzalez C, Uribe Carvajal S, Coria R, Peña Diaz A, Bredesen DE, Castro-Obregon S, del Rio G (2012) Moonlighting peptides with emergent function. PLoS One 7(7):e40125

Rolfe BG, Gresshoff PM (1988) Genetic analysis of legume nodule initiation. Annu Rev Plant Physiol Plant Mol Biol 39:297–319

Sarma AD, Emerich DW (2005) Global protein expression pattern of *Bradyrhizobium japonicum* bacteroids: a prelude to functional proteomics. Proteomics 5:4170–4184

Sarma AD, Emerich DW (2006) A comparative proteomic evaluation of culture grown versus nodule isolated *Bradyrhizobium japonicum*. Proteomics 6:3008–3028

Schwmtzer CR, Tjepkema JD (1990) The biology of Frankia and actinorhizal plants. Academic, San Diego, CA

Stumpf MPH, Porter MA (2012) Critical truths about power laws. Science 335:665

Trainer MA, Charles TC (2006) The role of PHB metabolism in the symbiosis of rhizobia with legumes. Appl Microbiol Biotechnol 71:377–386

Turgean BG, Bauer WD (1982) Early events in the infection of soybean by *Rhizobium japonicum*. Time course and cytology of the initial infection process. Can J Bot 60:152–161

Udvardi M, Day D (1997) Metabolite transport across symbiotic membranes of legume nodules. Annu Rev Plant Physiol Plant Mol Biol 48:493–523

Varma A, Palsson BO (1994) Stoichiometric flux balance models quantitatively predict growth and metabolic byproduct secretion in wild-type *Escherichia coli* W3110. Appl Environ Microbiol 60:3724–3731

Vasse J, de Billy F, Camut S, Truchet G (1990) Correlation between ultrastructural differentiation of bacteroids and nitrogen fixation in alfalfa nodules. J Bacteriol 172(8):4295–4306

Wagner A, Fell DA (2001) The small world inside large metabolic networks. Proc R Soc Lond B Biol Sci 268:1803–1810

Watts DJ, Strogatz SH (1998) Collective dynamics of 'small-world' networks. Nature 393 (6684):409–410

Chapter 3
Carbon Metabolism During Symbiotic Nitrogen Fixation

Emmanouil Flemetakis and Trevor L. Wang

3.1 Introduction

Legumes have the ability to grow in low nitrogen soils because of their capacity to incorporate atmospheric nitrogen into amino acids. This ability has been gained via the sequestering of the pre-existing mycorrhizal developmental pathway (Parniske 2008) and through the formation of a specialised organ, the root nodule, in which there is a symbiotic relationship with rhizobium bacteria of the soil. The plant supplies carbon (C) to the symbiotic partner, and in return, the bacterium (in modified form known as a bacteroid) fixes atmospheric nitrogen (N_2) to generate amino acids that are supplied to the plant. Legumes pay a price for this relationship. In seminal studies from Pate's laboratory (Pate and Herridge 1977; Pate et al. 1979a, b), it was reported that in the annual lupin (*Lupinus albus* L.) approximately 51 % of C from photosynthesis was sequestered by the roots and 4.0–6.5 g C utilised by the nodule to fix every gram of nitrogen (N). However, 34 % of the C supplied to the nodule returned to the shoot in the form of symbiotic nitrogen fixation (SNF) products. This large commitment of C being translocated to nodules was supported by later studies on the same species (Layzell et al. 1981) and in other species (see Gordon 1992), which gave values between 40 % and 50 %. In a study on *Trifolium repens* L. cv. Blanca (white clover), Gordon et al. (1987) showed that half of the 45 % processed by the nodule was respired. Here, we review the importance of some recent developments and key enzymes involved in the metabolic pathway of carbon to the bacteroid.

E. Flemetakis (✉)
Laboratory of Molecular Biology, Department of Agricultural Biotechnology,
Agricultural University of Athens, Iera Odos 75, Athens 11855, Greece
e-mail: mflem@aua.gr

T.L. Wang
Metabolic Biology, John Innes Centre, Norwich Research Park, Colney, Norwich NR4 7UH, UK
e-mail: trevor.wang@jic.ac.uk

The metabolic pathway of C supplied from the leaves of a legume plant to its root nodules can be divided into three phases: the catabolism to hexoses of photosynthesis-derived sucrose entering infected cells, the metabolism of hexoses to phosphoenolpyruvate (PEP) via glycolysis and the conversion of PEP to dicarboxylic acids (DCA) for transport into bacteroids. The upregulation of genes during nodulation (nodule enhanced, NE) as exemplified by the nodulin genes (Verma et al. 1986) has often been used as prime facie evidence for the specific involvement of their products in SNF. However, as we shall see later, this is not always the case as it may reflect another facet of the nodulation process not associated with nodule function per se, for example, cell division. The use of the model legumes *Lotus japonicus* and *Medicago truncatula* has greatly enhanced our understanding of nodulation and C delivery since one can investigate enzymes in the pathway through the use of specific mutants. Moreover, large datasets for transcripts and metabolites (e.g. http://mtgea.noble.org/v2/) are now available together with mutants for several genes encoding enzymes of the pathway (e.g. Horst et al. 2007). By interrogating such databases, one can clearly identify that a large set of genes concerned with carbohydrate metabolism is altered in expression during nodulation, and this is a reflection of both metabolism (Fig. 3.1) and the creation of a nodule.

3.2 Sucrose Catabolism

When sucrose from the aerial parts of the plant is delivered via the phloem to the root nodule, there are two routes by which it can be utilised by a cell, both involving enzymatic cleavage of the disaccharide to its individual monomers. The two enzymes involved are sucrose synthase (SUS, EC 2.4.1.13) and invertase (INV, EC 3.2.1.26), each using a different mechanism. The former catalyses the cleavage of sucrose to UDP-glucose and fructose in a reversible reaction, whereas the latter generates the individual hexoses, glucose and fructose, in an irreversible reaction. It was not until mutants of SUS were isolated in *Pisum sativum* (pea; Wang et al. 1998; Craig et al. 1999) that it became clear that this enzyme was the only one essential for SNF and assimilation.

3.2.1 Sucrose Synthase

Sucrose is believed to move to the infected core of the nodule via the uninfected cells. This is where the highest content of SUS can be found in both *Glycine max* (soybean) and white clover (Gordon et al. 1992, 1995). The first defined plant mutants shown to affect N assimilation were isolated from a forward screen of pea seeds for a wrinkled phenotype (Wang et al. 1990, 1998). The mutants were at the *rug4* locus of pea, and their seeds had a lowered starch content, which generated the

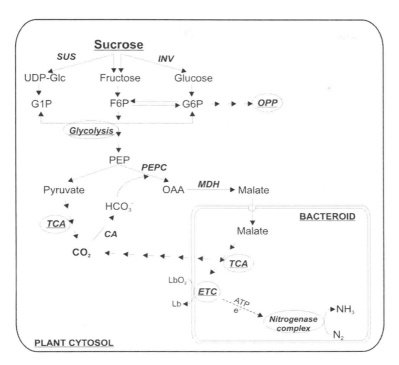

Fig. 3.1 Schematic representation of the possible sucrose catabolising metabolic pathways in nitrogen-fixing nodules. For simplicity, only enzymes discussed in this chapter are shown. *SUS* sucrose synthase, *INV* invertase, *PEPC* phospho*enol*pyruvate carboxylase, *CA* carbonic anhydrase, *MDH* malate dehydrogenase, *OPP* oxidative pentose phosphate pathway, *TCA* tricarboxylic acid cycle, *ETC* electron transport chain

wrinkled phenotype (Wang and Hedley 1991). Subsequently, it was found that the locus encoded a sucrose synthase (SUS) whose activity was clearly required for starch synthesis in the seed (Craig et al. 1999). The *rug4* mutant plants when grown in the presence of rhizobium in N-free composts or poor soil in the field showed severe symptoms of N starvation. The plants had low chlorophyll contents and a low N content indicating that they derived little of their N from fixation. Craig et al. (1999) concluded therefore that this SUS isoform was needed to supply C for SNF. Furthermore, the *rug4* nodules showed the early senescence phenotype that is characteristic of this class of mutants (e.g. Novak et al. 1995). Since the plants grew normally when supplied with N, it was concluded that normal growth and development was supported by other SUS isoforms. In a detailed analysis of these mutants, Gordon et al. (1999) showed that several changes occurred in the enzymes of N assimilation, some decreasing in activity (phosphofructokinase, PEP carboxylase, Gln synthase, Gln oxoglutarate aminotransferase and Asp aminotransferase), others increasing their activities (PPi-dependent, Fru-6-P phosphotransferase, pyruvate decarboxylase and alcohol dehydrogenase). INV also decreased and there was substantially less leghaemoglobin in the mutants. Neither nodules nor isolated bacteroids of *rug4* mutants showed apparent nitrogenase activity despite the presence of normal levels of nitrogenase protein. Furthermore, mutant plants showed little of no N accumulation. Gordon et al. (1999) hypothesised that SUS was needed to maintain nitrogenase activity and they concluded that the *Rug4* isoform of SUS was essential for SNF, INV being unable to compensate for its loss in the mutant.

The *Rug4* isoform of SUS represents the NE form (nodulin-100; Thummler and Verma 1987). Studies by Barratt and co-workers showed that SUS was encoded by a small gene family in both pea (Barratt et al. 2001) and Arabidopsis (Bieniawska et al. 2007). In pea, there are six cytosolic isoforms of SUS, two of which are present in the nodule (SUS1 and SUS3; Barratt et al. 2001). It is in legumes where we see the most dramatic effect of reduced SUS activity, since a quadruple knockout mutant of the four main isoforms in Arabidopsis showed no phenotype indicating a high level of redundancy within the gene family in this species (Barratt et al. 2009). The use of *L. japonicus* permitted a further analysis of SUS isoforms. Like pea and Arabidopsis, there are six isoforms, and two are present in the nodule, LjSUS1 and LjSUS3 (Horst et al. 2007); LjSUS3 is the main and the NE isoform. However, when the NE isoform was knocked out through the isolation of a TILLING mutant creating a premature stop codon in the coding sequence, the plants could grow and fix nitrogen although with reduced capacity, the plants showing symptoms of N starvation. It was only when the activity of both isoforms was removed by creating a double mutant that the plant became severely impaired and unable to fix N, indicating that LjSUS1 must also contribute to the maintenance of N assimilation. The wild type (WT) phenotype could be restored, however, by application of N. The different contributions of the NE isoform in pea (*rug4*) and in Lotus (LjSUS3) to the total nodule SUS activity (90 % vs. 70 %) may account for the different phenotypes (Horst et al. 2007). In *M. truncatula*, when the gene encoding the main nodule-enhanced isoform (Hohnjec et al. 1999) was targeted by using an antisense construct (Baier et al. 2007), a severe decrease in protein content of up to 90 % in some transgenic lines was observed using western blotting. The *M. truncatula* downregulated plants showed a similar growth and development phenotype to those in the pea *rug4* and *L. japonicus* double mutants albeit less severe. However, enzyme activity was not measured in the *M. truncatula* plants and so it was not clear how much activity was present due to the remaining protein. Only the NE isoform was detected in nodules by western blotting, but an additional isoform was observed in roots. Interestingly, the antisense construct decreased the transcript and protein levels of *both* isoforms.

One further point can be made concerning the role of starch accumulation in nodule function from the detailed analysis of the *rug4* mutant by Gordon et al. (1999). The mutant nodules contained substantial amounts of starch albeit lower than in the WT, which indicates that nodule starch cannot be mobilised to substitute for the loss of UDP-glucose delivered by SUS. This is also supported by the fact that the starchless plastidial phosphoglucomutase mutants, *Psrug3* and *Ljpgm1*, have perfectly functional nodules. In contrast, *Ljgwd1* mutants that cannot break down starch and accumulate very large amounts in all starch-storing organs, including nodules, have impaired nodule function (Vriet et al. 2010). Put together, these data indicate a rather counterintuitive concept that starch accumulation in nodules is in competition with the delivery of C to the bacteroids for SNF rather than acting as a source of C. However, different species have adopted different strategies to maintain a supply of C to nodules especially during darkness (Gordon 1992), and so this may not be the case for all legumes.

3.2.2 Invertase

INV are a large family of proteins, divided by their pH optima. The acid forms are present in the cell wall or vacuole whereas the others are cytosolic with a pH optimum that is neutral or alkaline (N/A). Early findings in soybean showed that both N/A INV and SUS were present in nodules (Morell and Copeland 1984, 1985) although it was not clear which catalysed the hydrolysis of sucrose. Soluble acid INV were not detected in legume nodules initially (Gordon et al. 1999), but low activity was subsequently found in those of *L. japonicus* (Flemetakis et al. 2006). Despite there being much information on the acidic forms, there was a paucity of information about the N/A isoforms until recently. In this respect, the breakthrough was made through research on legumes.

Genomic studies in rice, Arabidopsis and poplar revealed families of genes encoding the different isoforms (Ji et al. 2005; Qi et al. 2007; Bocock et al. 2008). The gene sequences encoding the different classes of INV, however, are so different that they cannot be combined into a single phylogenetic tree. Map-based cloning of a gene in rice (*Oscyt-inv1*) and Arabidopsis (*Atcytinv1*) gave the first indication that N/A INV could have a specific effect on plant growth and development, but due to redundancy of genes in these species (Qi et al. 2007; Jia et al. 2008), the phenotype was unclear and the precise role was not found until studies were initiated on *L. japonicus*. Flemetakis et al. (2006) identified two genes expressed in *L. japonicus* roots encoding N/A isoforms that accounted for most of the activity present; only one, LjINV1, was expressed in the nodule. Subsequently, seven genes encoding N/A INV were described in *L. japonicus* (Welham et al. 2009), *LjINV1* being the most highly expressed in all organs examined. Because the main isoform, LjINV1, was NE, Flemetakis et al. (2006) proposed that it had a role in supplying hexose phosphates for nodule metabolism and SNF. Subsequently, the characterization of TILLING mutants indicated that LjINV1 was not essential for nodule formation or function but rather had a general role in growth and cell development of the whole plant (Welham et al. 2009).

Hence, SUS activity is sufficient to supply C for SNF. Since more ATP molecules are generated for each sucrose molecule converted to malate via the SUS reaction, this reaction may be favoured over the N/A INV route, the products being less dependent on ATP for further metabolism (Gordon 1992). The SUS route also supports the hypothesis that regulation of nodule activity could be achieved via a regulation of SUS activity (Gordon et al. 1997; Galvez et al. 2005).

From the data on SUS and INV alone, it is clear that NE expression can be very misleading although it is used as a rule of thumb for involving a specific gene product in nodulation, SNF and assimilation. However, if one considers that the nodule is also an organ with its own meristem, either initially as in determinate nodules (such as Lotus and soybean) or persistently as a distinct zone within indeterminate nodules (such as pea and Medicago), it is hardly surprising that genes involved with meristem development and function would not also be highly expressed when compared to whole roots, stems or leaves. Thus the true measure of nodule enhancement may be best made by comparison with meristems such as the shoot or root apex, a comparison that has never been carried out in transcript studies involving nodulation.

3.3 PEP to Malate and Transport into the Bacteroid

3.3.1 PEP Carboxylase

Phospho*enol*pyruvate carboxylase (PEPC, EC 4.1.1.31) is a widespread enzyme present in multiple isoforms in bacteria, cyanobacteria, green algae and higher plants. In photosynthetic higher plants, different PEPC isoforms fulfil distinct physiological functions. During C_4 photosynthesis and Crassulacean acid metabolism (CAM), PEPC is responsible for the initial fixation of inorganic C in the form of bicarbonate (HCO_3^-; Izui et al. 2004). In C_3 leaves and non-photosynthetic sink organs, including seeds, roots and nodules, specific PEPC isoforms represent key players during dark CO_2 fixation serving various biochemical and physiological functions (Vidal and Chollet 1997). PEPC accounts for about 0.5–2.0 % of the soluble protein in *M. sativa* (alfalfa) and soybean nodules (Vance et al. 1994). The significance of dark CO_2 fixation during SNF will be discussed later (see Sect. 3.4). In vascular plants, PEPC is typically regulated post-translationally by metabolite allosteric effectors, including the activators glucose-6-phosphate and triose phosphates, and the inhibitors malate and aspartate (Zhang and Chollet 1997). Furthermore, PEPC activity is regulated by the reversible phosphorylation by a specific Ca^{2+}-insensitive Ser/Thr kinase, phospho*enol*pyruvate carboxylate kinase (PEPCK), also including NE isoforms (Nimmo et al. 2001; Xu et al. 2007). In *L. japonicus*, antisense-induced suppression of the NE *Ljpepc1* transcript resulted in plants showing typical nitrogen-deficiency symptoms and reduced nitrogenase activity. Furthermore, the decrease of nodule PEPC activity resulted in significant changes in SUS and asparagine aminotransferase activities coupled to lower contents for sucrose, succinate, asparagine, aspartate and glutamate (Nakagawa et al. 2003; Nomura et al. 2006) implicating PEPC in the regulation of C/N metabolic fluxes in *L. japonicus* nodules. Interestingly, the observed low efficiency of SNF in *M. truncatula*, when compared to that of alfalfa, was correlated to a significantly lower PEPC activity and organic acid content in this model legume (Sulieman and Schulze 2010).

3.3.2 Carbonic Anhydrase

One of the final products of sucrose metabolism is CO_2, which accumulates to a high level in nodules. Carbonic anhydrase (CA; carbonate dehydratase, EC 4.2.1.1) is a zinc-containing enzyme that catalyses the reversible hydration of carbon dioxide to form bicarbonate, which is the required substrate for PEP carboxylase. Several CA families (α-, β-, γ-, δ- and ζ- CAs) are present in plants, animals and microorganisms, suggesting that this simple conversion of a membrane-permeable gas into a membrane-impermeable ionic product is vital to many important biological functions. Currently, only β-CAs have been extensively studied in higher plants. In contrast, plant α-CA isoforms are biochemically and physiologically poorly characterised, including dioscorin-like nectarine (Nectarine III, NEC3) from tobacco, reported to possess both monodehydroascorbate reductase and CA activity (Carter and Thornburg 2004), and storage proteins (dioscorins, DB2 and DB3) from the yam tuber *Dioscorea batatas* which lacks CA activity (Gaidamashvili et al. 2004). Recently, γ-CA isoforms have been localised to the mitochondrial complex I of *Arabidopsis thaliana* (Sunderhaus et al. 2006).

The presence of high CA activity in nodules was reported quite early during the study of SNF (Atkins 1974). Nodule CA activity was later measured in a number of crop legumes including pea, bean and *Lupinus angustifolius* (Lane et al. 2005). The first nodule-specific CA transcripts belonging to the β-class CAs were identified in the nodules of soybean (Kavroulakis et al. 2000), alfalfa (de la Pena et al. 1997) and *L. japonicus* (Flemetakis et al. 2003), but the exact biochemical and physiological role of this CA isoform during symbiosis remains unclear. Based on transcript accumulation and localization patterns, it has been proposed that β-CAs may fulfil distinct physiological roles during nodule development. In *L. japonicus* and soybean at early stages of nodule development, the expression of *CA1* precedes the expression of nitrogenase. In addition, both LjCA1 protein and CA activity are present in all cell types, co-localising with PEPC. Thus the expression of *LjCA1* has been correlated with the need to convert most of the CO_2 produced by respiration to bicarbonate, which in turn can be channelled to various biosynthetic processes including production of oxaloacetate for amino acid biosynthesis, lipogenesis, pyrimidine biosynthesis and gluconeogenesis (Kavroulakis et al. 2003; Hoang and Chapman 2002; Flemetakis et al. 2003). At later stages of nodule development, all nodule β-CAs studied so far have been found to localise to a few layers of inner cortical cells in both determinate and indeterminate nodules (de la Pena et al. 1997; Kavroulakis et al. 2000; Flemetakis et al. 2003). These cell layers appear more compact with less obvious intercellular gas spaces, compared to the immediately adjacent outer cortical cells and to the cells forming the boundary layer, and they are proposed to participate in the formation of a physical barrier to the diffusion of gases (Parsons and Day 1990). The strictly regulated spatial accumulation of β-CA transcripts has been implicated in an osmoregulatory mechanism involving the localised production of malate through the combined activities of CA and PEPC and is analogous to the process that occurs in stomatal guard cells (de la Pena et al. 1997). Alternatively, the localization of β-CA transcripts and protein in the nodule inner cells has been proposed to facilitate the diffusion of the excess CO_2 towards the rhizosphere, in a mechanism similar to that proposed for the facilitated diffusion of atmospheric CO_2 towards the chloroplasts in photosynthetic tissues (Badger and Price 1994).

However, in all the aforementioned studies, the strictly localised expression of the nodule-specific β-CAs could not explain the high levels of CA activity detected in both the infected and uninfected cells of the central tissue in mature nitrogen-fixing nodules (Atkins et al. 2001; Flemetakis et al. 2003). This raised the possibility that alternative CA isoforms, of either plant or microbial origin and differing enough at the nucleotide level compared to β-CAs, are responsible for the observed CA activity. Recently, the availability of genome and EST sequence databases from the model legumes and their microsymbionts revealed the presence of several CA-like sequences, belonging to α- and γ- classes. In *L. japonicus*, two genes encoding NE α-CA isoforms have been indentified (Tsikou et al. 2011). Both CO_2 hydration and bicarbonate dehydration activities of the full-length proteins were demonstrated by heterologous expression. Temporal and spatial expression analysis of *LjCAA1* and *LjCAA2* revealed that both genes are induced early during nodule development and remain active in nodule inner cortical cells, vascular bundles and the central tissue during all stages of nodule development. Interestingly, both genes were slightly to moderately downregulated in ineffective nodules formed by mutant *Mesorhizobium loti* strains, indicating that these genes may also be involved in biochemical and physiological processes not directly linked to SNF and assimilation. In addition, it was recently demonstrated that also *M. loti* harbours an active periplasmic α-CA on the symbiosis island (Kalloniati et al. 2009). The *MlCaa1* gene was found to be expressed in both nitrogen-fixing bacteroids and free-living bacteria. Interestingly, gene expression in batch cultures was induced by increasing the pH of the medium. Nodulation of *L. japonicus* with a *MlCaa1* deletion mutant strain showed no differences in shoot traits and nutritional status, but the plants consistently formed more nodules and exhibited a higher fresh weight, N content, nitrogenase activity and ^{13}C content. It was proposed that although the deletion mutant does not abolish the ability to form nitrogen-fixing nodules, this α-CA may participate in an auxiliary ATP-independent mechanism that could buffer the bacteroid periplasm by providing an alternative source of protons, thus creating an environment favourable for NH_3 protonation and facilitating its diffusion and transport to the plant (Day et al. 2001; White et al. 2007). An interesting analogous mechanism to the proposed role of *MlCaa1* during symbiosis comes from the physiological role of the *Helicobacter pylori* periplasmic α-CA homologue (Fig. 3.2).

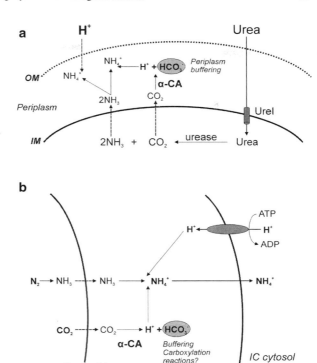

Fig. 3.2 (**a**) The role of urease and α-CA in acid acclimation of *Helicobacter pylori* (Marcus et al. 2005). *IM* inner membrane, *OM* outer membrane. (**b**) Schematic representation of the proposed role of *Ml*Caa1 during SNF. The CA catalysed hydration of the CO_2 produced by bacteroid respiration provides protons which can be used for the formation of NH_4^+, thus facilitating the diffusion of further NH_3, while the produced bicarbonate can be used for the buffering of the bacteroid periplasm. *IM* inner membrane, *PBS* peribacteroid membrane, *IC* infected cell

3.3.3 Malate Dehydrogenase

There is substantial evidence that the dicarboxylic acid (DCA) imported into the nodule is malate (reviewed by Udvardi and Day 1997; White et al. 2007). Although rhizobia do take up and can grow on sucrose and other sugars, bacteroids cannot use them, preferring malate or succinate to be most effective but also utilising formate and oxaloacetate (Herrada et al. 1989; Ou Yang et al. 1990). A plant DCA transporter has been identified in *Alnus glutinosa* that is expressed in nodules at the interface between plant cells and bacteria (Jeong et al. 2004), but in the absence of studies on specific plant mutants for DCA transporters, the exact specificity remains unknown. Furthermore, bacteroid membranes contain DCA transporters, and defects in bacteria transporters lead to ineffective nodules (Ronson et al. 1981; Udvardi and Day 1997). Malate is a key player in plant metabolism; it is a product of starch breakdown in plastids; it is utilised by mitochondria, glyoxisomes and peroxisomes; and it is the predominant DCA present in nodules (Rosendahl et al. 1990). Its level is also sensitive to SNF activity and decreases markedly in ineffective nodules (Schulze et al. 2002). There are numerous isoforms of malate dehydrogenase (MDH) in the cell, which makes analysis of enzyme activities difficult. Miller et al. (1998) characterised several isoforms in alfalfa and detected a unique protein in the nodules whose expression was enhanced during nodulation. The gene encoding this isoform has no introns, unlike all other isoforms. This group also detected a similar protein in pea (Fedorova et al. 1999). The enzyme's K_m favoured very markedly the formation of malate from oxaloacetate. The authors therefore regarded this enzyme as a key player in the manipulation of SNF, and preliminary investigations of transgenic lines showed that some had an increase in transcript and protein levels of the enzyme and improved SNF, although this has not been followed up (Schulze et al. 2002).

In the absence of specific mutants for NE MDH, however, the essential nature of malate for SNF remains unproven. In our preliminary studies with a TILLING mutant of Lj*NEMDH* predicted to have a deleterious effect on the activity of this isoform, there was a whole plant and root cell phenotype, but nodule function was unaffected (Welham, Edwards, and Wang, unpublished). This situation is similar to that observed for *LjINV1* (see Sect. 3.2.2) where there was no effect on SNF but a profound effect on plant development (Welham et al. 2009). In alfalfa, the NE MDH isoform represents 50 % of the total activity for this enzyme in the nodule (Miller et al. 1998), but in Lotus, when using the same antibodies for immunoprecipitation (Miller et al. 1998), this isoform only appears to represent ca. 20 % of the activity (Welham, Edwards and Wang, unpublished). Hence, as for SUS activity, other MDH isoforms may also be important for nodule function, at least in *L. japonicus*.

3.4 Dark CO_2 Fixation in Nitrogen-Fixing Nodules

Several studies have demonstrated that N_2-fixing nodules are capable of fixing CO_2, while the oxaloacetate produced can be utilised either for the synthesis of malate or as a source of C skeletons for the assimilation of the symbiotically reduced N (Chollet et al. 1996). Studies using $^{14}CO_2$ have demonstrated that dark CO_2 fixation in nodules supplies the bacteroids with C skeletons, mainly in the form of organic acids, for respiration and other biochemical processes (Rosendahl et al. 1990; Vance and Heichel 1991). In addition, organic acids containing dark-fixed CO_2 can be transported from nodules to shoots of legumes (Minchin and Pate 1973; Maxwell et al. 1984; Vance et al. 1985). Recently, $Na_2^{13}CO_3$ labelling of *L. japonicus* roots showed that N_2-fixing nodules and both nodulated and non-nodulated roots could incorporate $^{13}CO_2$. However, nodulated plants exported significantly higher amounts of ^{13}C from roots to the shoots (Fotelli et al. 2011). In addition, dark CO_2 fixation in nodules was found to be directly linked with SNF as ineffective nodules formed by $\Delta nifA$ and $\Delta nifH$ mutant *M. loti* strains incorporated significantly less label when compared to the N_2-fixing nodules. Fix^- plants also exhibited a significant decrease in ^{13}C accumulation in leaves, an observation possibly reflecting the lack of amide amino acid export from the fix^- nodules. In contrast, ^{13}C labelling in the stems of fix^- plants remained constant, indicating that compounds other than amides may account for the ^{13}C label in stems, with organic acids being the most prominent candidates (Fotelli et al. 2011), since significant amounts of these compounds are present in the xylem sap of amide-transporting legumes (Vance et al. 1985). Gene expression studies also revealed that transcripts for both NAD- and NADP-malic enzymes are significantly upregulated in the stems of rhizobium-inoculated plants when compared to the stems of non-inoculated plants (Fotelli et al. 2011). Interestingly, transcript accumulation in stems was only dependent on the presence of nodules rather than on SNF and export, as a similar effect was observed in plants harbouring ineffective nodules. These findings suggest that dark CO_2 fixation may fulfil multiple biochemical and physiological roles during SNF. It has been well established that CO_2 concentration in the rhizosphere can influence SNF, and prolonged elevated CO_2 supply to the root/nodule system of alfalfa led to a significant increase in nodule CO_2 fixation, SNF and growth (Grobbelaar et al. 1971; Yamakawa et al. 2004; Fischinger et al. 2010). These beneficial effects of nodule dark CO_2 fixation are traditionally attributed to the production of malate, the principal source of energy and C skeletons for the bacteroids (Rosendahl et al. 1990; Vance and Heichel 1991). However, previous studies have demonstrated that organic acids translocated from roots to the shoots can be decarboxylated with the released CO_2 recycled through photosynthesis to form carbohydrates (Cramer and Richards 1999; Hibberd and Quick 2002). The observation that nodulation induces the expression of genes encoding malate decarboxylating enzymes in the stems of *L. japonicus* points to the possible refixation of CO_2 respired or taken up by nodules in the stems through a C_4-type mechanism. The existence of this mechanism could have a positive impact on the C budget of the plant as a whole, by at least partially reducing the C costs of nodules while at the same time maintaining a constant pool of malate in nodules, as organic acid accumulation is suggested to have a negative regulatory impact on nitrogenase activity (Le Roux et al. 2008). Future work should aim towards the quantification of the CO_2 recycled through such a mechanism and this relative contribution to the whole plant C budget.

3.5 Conclusions and Perspectives

Considerable progress has been made over the last 10 years towards our understanding of the mechanisms operating to set up a nodule, so much so that it is now timely to attempt to transfer the ability to nodulate to non-legumes (http://www.gatesfoundation.org/How-We-Work/Quick-Links/Grants-Database/Grants/2012/06/OPP1028264). Creating a nodule to receive bacteria is only part of the story, however, since many metabolic processes require adjustments before the nodule can function correctly (Udvardi and Poole 2013). The use of model legumes has helped us to determine, for example, which is the key enzyme for delivering carbohydrate to the nodule, but we still have little understanding of its regulation or which plant processes are key to nodule function since the lesions in only a few non-fixing mutants have been determined (Udvardi and Poole 2013). Many of the enzyme isoforms required for correct functioning of carbon metabolism exist in non-legumes, in common with some of the components of signalling (Antolín-Llovera et al. 2012; Zhu et al. 2006), but their regulation will be different. As mentioned, legumes pay a significant price, which can be viewed as a 'carbon tax' on the plant, for receiving fixed nitrogen from bacteroids. This requires modifications to their metabolism and to specific isoforms involved in the process. Until we fully understand both the steps in the pathway of carbon to the nodule and its regulation, our ability to levy a tax on non-legumes or even manipulate the level of tax in legumes will be severely limited. In this regard, future attention needs to be refocussed on nodule fixation and especially those plant genes concerned with taxation.

References

Antolín-Llovera M, Ried MK, Binder A, Parniske M (2012) Receptor kinase signaling pathways in plant-microbe interactions. Annu Rev Phytopathol 50:451–473
Atkins CA (1974) Occurrence and some properties of carbonic anhydrase from legume root nodules. Phytochemistry 13:93–98
Atkins C, Smith P, Mann A, Thumfort P (2001) Localization of carbonic anhydrase in legume nodules. Plant Cell Environ 24:317–326
Badger MR, Price GD (1994) The role of carbonic-anhydrase in photosynthesis. Annu Rev Plant Physiol Plant Mol Biol 45:369–392
Baier MC, Barsch A, Kuster H, Hohnjec N (2007) Antisense repression of the *Medicago truncatula* nodule-enhanced sucrose synthase leads to a handicapped nitrogen fixation mirrored by specific alterations in the symbiotic transcriptome and metabolome. Plant Physiol 145:1600–1618
Barratt DHP, Barber L, Kruger NJ, Smith AM, Wang TL, Martin C (2001) Multiple, distinct isoforms of sucrose synthase in pea. Plant Physiol 127:655–664
Barratt DHP, Derbyshire P, Findlay K, Pike M, Wellner N, Lunn J, Feil R, Simpson C, Maule A, Smith AM (2009) Sucrose catabolism in Arabidopsis requires cytosolic invertase but not sucrose synthase. Proc Natl Acad Sci USA 106:13124–13129
Bieniawska Z, Barratt DHP, Garlick AP, Thole V, Kruger NJ, Martin C, Zrenner R, Smith AM (2007) Analysis of the sucrose synthase gene family in Arabidopsis. Plant J 49:810–828
Bocock PN, Morse AM, Dervinis C, Davis JM (2008) Evolution and diversity of invertase genes in *Populus trichocarpa*. Planta 227:565–576
Carter CJ, Thornburg RW (2004) Tobacco Nectarin III is a bifunctional enzyme with monodehydroascorbate reductase and carbonic anhydrase activities. Plant Mol Biol 54:415–425
Chollet R, Vidal J, O'Leary MH (1996) Phospho*enol*pyruvate carboxylase, a ubiquitous highly regulated enzyme in plants. Annu Rev Plant Physiol Plant Mol Biol 47:273–298
Craig J, Barratt P, Tatge H, Dejardin A, Handley L, Gardner CD, Barber L, Wang T, Hedley C, Martin C, Smith AM (1999) Mutations at the *rug4* locus alter the carbon and nitrogen metabolism of pea plants through an effect on sucrose synthase. Plant J 17:353–362
Cramer MD, Richards MB (1999) The effect of rhizosphere dissolved inorganic carbon on gas exchange characteristics and growth rates of tomato seedlings. J Exp Bot 50:79–87
Day DA, Poole PA, Tyerman SD, Rosendahl L (2001) Ammonia and amino acid transport across symbiotic membranes in nitrogen-fixing legume nodules. Cell Mol Life Sci 58:61–71

de la Pena TC, Frugier F, McKhann HI, Bauer P, Brown S, Kondorosi A, Crespi M (1997) A carbonic anhydrase gene is induced in the nodule primordium and its cell-specific expression is controlled by the presence of Rhizobium during development. Plant J 11:407–420

Fedorova M, Tikhonovich IA, Vance CP (1999) Expression of C-assimilating enzymes in pea (*Pisum sativum* L.) root nodules: in situ localization in effective nodules. Plant Cell Environ 22:1249–1262

Fischinger SA, Hristozkova M, Mainassara Z-A, Schulze J (2010) Elevated CO_2 concentration around alfalfa nodules increases N_2 fixation. J Exp Bot 61:121–130

Flemetakis E, Dimou M, Cotzur D, Aivalakis G, Efrose RC, Kenoutis C, Udvardi M, Katinakis P (2003) A *Lotus japonicus* β-type carbonic anhydrase gene expression pattern suggests distinct physiological roles during nodule development. Biochim Biophys Acta 1628:186–194

Flemetakis E, Efrose RC, Ott T, Stedel C, Aivalakis G, Udvardi MK, Katinakis P (2006) Spatial and temporal organisation of sucrose metabolism in *Lotus japonicus* nitrogen-fixing nodules suggests a role for the elusive alkaline/neutral invertase. Plant Mol Biol 62:53–69

Fotelli MN, Tsikou D, Kolliopoulou A, Aivalakis G, Katinakis P, Udvardi MK, Rennenberg H, Flemetakis E (2011) Nodulation enhances dark CO_2 fixation and recycling in the model legume *Lotus japonicus*. J Exp Bot 62:2959–2971

Gaidamashvili M, Yuki O, Iijima S, Takayama T, Ogawa T, Muramoto K (2004) Characterization of the yam tuber storage proteins from *Dioscorea batatas* exhibiting unique lectin activities. J Biol Chem 279:26028–26035

Galvez L, Gonzalez EM, Arrese-Igor C (2005) Evidence for carbon flux shortage and strong carbon/nitrogen interactions in pea nodules at early stages of water stress. J Exp Bot 56:2551–2561

Gordon AJ (1992) Carbon metabolism in the legume nodule. In: Pollock CJ, Farrar JF, Gordon AJ (eds) Carbon partitioning within and between organisms. Bios Scientific, Oxford

Gordon AJ, Mitchell DF, Ryle GJA, Powell CE (1987) Diurnal production and utilization of photosynthate in nodulated white clover. J Exp Bot 38:84–98

Gordon AJ, Thomas BJ, Reynolds PHS (1992) Localization of sucrose synthase in soybean root nodules. New Phytol 122:35–44

Gordon AJ, Thomas BJ, James CL (1995) The location of sucrose synthase in root nodules of white clover. New Phytol 130:523–530

Gordon AJ, Minchin FR, Skøt L, James CL (1997) Stress-induced declines in soybean N_2 fixation are related to nodule sucrose synthase activity. Plant Physiol 114:937–946

Gordon AJ, Minchin FR, James CL, Komina O (1999) Sucrose synthase in legume nodules is essential for nitrogen fixation. Plant Physiol 120:867–878

Grobbelaar N, Hough MC, Clarke B (1971) Nodulation and nitrogen fixation of isolated roots of *Phaseolus vulgaris* L. 3. Effect of carbon dioxide and ethylene. Plant Soil 35:215–278

Herrada G, Puppo A, Rigaud J (1989) Uptake of metabolites by bacteriod containing vesicles and by free bacteroids from French bean nodules. J Gen Microbiol 135:3165–3177

Hibberd JM, Quick PW (2002) Characteristics of C_4 photosynthesis in stems and petioles of C_3 flowering plants. Nature 415:451–454

Hoang CV, Chapman KD (2002) Biochemical and molecular inhibition of plastidial carbonic anhydrase reduces the incorporation of acetate into lipids in cotton embryos and tobacco cell suspensions and leaves. Plant Physiol 128:1417–1427

Hohnjec N, Becker JD, Puhler A, Perlick AM, Kuster H (1999) Genomic organization and expression properties of the *MtSucS1* gene, which encodes a nodule-enhanced sucrose synthase in the model legume *Medicago truncatula*. Mol Gen Genet 261:514–522

Horst I, Welham T, Kelly S, Kaneko T, Sato S, Tabata S, Parniske M, Wang TL (2007) TILLING mutants *of Lotus japonicus* reveal that nitrogen assimilation and fixation can occur in the absence of nodule-enhanced sucrose synthase. Plant Physiol 144:806–820

Izui K, Matsumura H, Furumoto T, Kai Y (2004) Phospho*enol*pyruvate carboxylase: a new era of structural biology. Annu Rev Plant Biol 55:69–84

Jeong J, Suh S, Guan C, Tsay Y-F, Moran N, Oh CJ, An CS, Demchenko KN, Pawlowski K, Lee Y (2004) A nodule-specific dicarboxylate transporter from alder is a member of the peptide transporter family1. Plant Physiol 134:969–978

Ji XM, van den Ende W, van Laere A, Cheng SH (2005) Structure, evolution, and expression of the two invertase gene families of rice. J Mol Evol 60:615–634

Jia LQ, Zhang BT, Mao CZ, Li JH, Wu YR, Wu P, Wu ZC (2008) OsCYT-INV1 for alkaline/neutral invertase is involved in root cell development and reproductivity in rice (*Oryza sativa* L.). Planta 228:51–59

Kalloniati C, Tsikou D, Lampiri V, Fotelli MN, Rennenberg H, Chatzipavlidis I, Fasseas C, Katinakis P, Flemetakis E (2009) Characterization of a *Mesorhizobium loti* α-type carbonic anhydrase and its role in symbiotic nitrogen fixation. J Bacteriol 191:2593–2600

Kavroulakis N, Flemetakis E, Aivalakis G, Katinakis P (2000) Carbon metabolism in developing soybean nodules: the role of carbonic anhydrase. Mol Plant Microbe Interact 13:14–22

Kavroulakis N, Flemetakis E, Aivalakis G, Dahiya P, Brewin NJ, Fasseas K, Hatzopoulos P, Katinakis P (2003) Tissue distribution and subcellular localization of carbonic anhydrase in mature soybean root nodules indicates a role in CO_2 diffusion. Plant Physiol Biochem 41:479–484

Lane TW, Saito MA, George GN, Pickeing IJ, Prince RC, Morel FM (2005) Biochemistry: a cadmium enzyme from a marine diatom. Nature 435:42

Layzell DB, Pate JS, Atkins CA, Canvin DT (1981) Partitioning of carbon and nitrogen and nutrition of root and shoot apex in a nodulated legume. Plant Physiol 67:30–36

Le Roux MR, Khan S, Valentine AJ (2008) Organic acid accumulation may inhibit N_2 fixation in phosphorus-stressed lupin nodules. New Phytol 177:956–962

Marcus EA, Moshfegh AP, Sachs G, Scott DR (2005) The periplasmic alpha-carbonic anhydrase activity of *Helicobacter pylori* is essential for acid acclimation. J Bacteriol 187:729–738

Maxwell CA, Vance CP, Heichel GH, Stade S (1984) CO_2 fixation in alfalfa and birdsfoot trefoil root nodules and partitioning of ^{14}C to the plant. Crop Sci 24:257–264

Miller SS, Driscoll BT, Gregerson RG, Gantt JS, Vance CP (1998) Alfalfa malate dehydrogenase (MDH): molecular cloning and characterization of five different forms reveals a unique nodule-enhanced MDH. Plant J 15:173–184

Minchin FR, Pate JS (1973) The carbon balance of a legume and the functional economy of its root nodules. J Exp Bot 24:259–270

Morell M, Copeland L (1984) Enzymes of sucrose breakdown in soybean nodules. Plant Physiol 74:1030–1034

Morell M, Copeland L (1985) Sucrose synthase of soybean nodules. Plant Physiol 78:149–154

Nakagawa T, Izumi T, Banba M, Umehara Y, Kouchi H, Izui K, Shingo H (2003) Characterization and expression analysis of genes encoding phospho*enol*pyruvate carboxylase and phospho*enol*pyruvate carboxylase kinase of *Lotus japonicus*, a model legume. Mol Plant Microbe Interact 16:281–288

Nimmo GA, Wilkins MB, Nimmo HG (2001) Partial purification and characterization of a protein inhibitor of phospho*enol*pyruvate carboxylase kinase. Planta 213:250–257

Nomura M, Thu Mai H, Fujii M, Hata S, Izui K, Tajima S (2006) Phospho*enol*pyruvate carboxylase plays a crucial role in limiting nitrogen fixation in *Lotus japonicus* nodules. Plant Cell Physiol 47:613–621

Novak K, Pesina K, Ncbcsarova J, Skrdleta V, Lisa L, Nasincc V (1995) Symbiotic tissue degradation pattern in the ineffective nodules of three nodulation mutants of pea. Ann Bot 76:301–313

Ou Yang L, Udvardi MK, Day D (1990) Specificity and regulation of the dicarboxylate carrier on the peribacteroid membrane of soybean nodules. Planta 182:437–444

Parniske M (2008) Arbuscular mycorrhiza: the mother of plant root endosymbioses. Nat Rev Microbiol 6:763–775

Parsons R, Day DA (1990) Mechanism of soybean nodule adaptation to different oxygen pressures. Plant Cell Environ 13:501–512

Pate JS, Herridge DF (1977) Partitioning and utilization of net photosynthate in a nodulated annual legume. J Exp Bot 29:401–412

Pate JS, Layzell DB, Atkins CA (1979a) Economy of carbon and nitrogen in a nodulated and non-nodulated (NO_3-grown) legume. Plant Physiol 64:1008–1083

Pate JS, Layzell DB, McNeil DL (1979b) Modeling the transport and utilization of carbon and nitrogen in a nodulated legume. Plant Physiol 63:730–737

Qi XP, Wu ZC, Li JH, Mo XR, Wu SH, Chu J, Wu P (2007) AtCYT-INV1, a neutral invertase, is involved in osmotic stress-induced inhibition on lateral root growth in Arabidopsis. Plant Mol Biol 64:575–587

Ronson CW, Lyttleton P, Robertson JG (1981) C4-dicarboxylate transport mutants of *Rhizobium trifolii* form ineffective nodules on *Trifolium repens*. Proc Natl Acad Sci USA 78:4284–4288

Rosendahl L, Vance CP, Pedersen WB (1990) Products of dark CO_2 fixation in pea root nodules support bacteroid metabolism. Plant Physiol 93:12–19

Schulze J, Tesfaye M, Litjens RHMG, Bucciarelli B, Trepp G, Miller S, Samac D, Allan D, Vance CP (2002) Malate plays a central role in plant nutrition. Plant Soil 247:133–139

Sulieman S, Schulze J (2010) The efficiency of nitrogen fixation of the model legume *Medicago truncatula* (Jemalog A17) is low compared to *Medicago sativa*. J Plant Physiol 167:683–692

Sunderhaus S, Dudkina NV, Jänsch L, Klodmann J, Heinemeyer J, Perales M, Zabaleta E, Boekema EJ, Braun HP (2006) Carbonic anhydrase subunits form a matrix-exposed domain attached to the membrane arm of mitochondrial complex I in plants. J Biol Chem 281:6482–6488

Thummler F, Verma DPS (1987) Nodulin-100 of soybean is the subunit of sucrose synthase regulated by the availability of free heme in nodules. J Biol Chem 262:14730–14736

Tsikou D, Stedel C, Kouri ED, Udvardi MK, Wang TL, Katinakis P, Labrou NE, Flemetakis E (2011) Characterization of two novel nodule-enhanced α-type carbonic anhydrases from *Lotus japonicus*. Biochim Biophys Acta 1814:496–504

Udvardi MK, Day DA (1997) Metabolite transport across symbiotic membranes of legume nodules. Annu Rev Plant Physiol Plant Mol Biol 48:493–523

Udvardi M, Poole PS (2013) Transport and metabolism in legume-rhizobia symbioses. Annu Rev Plant Biol 64:781–805

Vance CP, Heichel GH (1991) Carbon in N_2 fixation, limitation or exquisite adaptation. Annu Rev Plant Physiol Plant Mol Biol 42:373–392

Vance CP, Boylan KLM, Maxwell CA, Heichel GH, Hardman LL (1985) Transport and partitioning of CO_2 fixed by root nodules of ureide and amide producing legumes. Plant Physiol 78:774–778

Vance CP, Gergerson RG, Robinson DL, Miller SS, Gantt JS (1994) Primary assimilation of nitrogen in alfalfa nodules: molecular features of enzymes involved. Plant Sci 101:51–64

Verma DPS, Fortin MG, Stanley J, Mauro VP, Purohit S, Morrison N (1986) Nodulins and nodulin genes of *Glycine max*. Plant Mol Biol 7:51–61

Vidal J, Chollet R (1997) Regulatory phosphorylation of C4 PEP carboxylase. Trends Plant Sci 2:230–237

Vriet C, Welham T, Brachmann A, Pike J, Perry J, Parniske M, Sato S, Tabata S, Smith AM, Wang TL (2010) A suite of *Lotus japonicus* starch mutants reveals both conserved and novel features of starch metabolism. Plant Physiol 154:643–655

Wang TL, Hedley CL (1991) Seed development in peas: knowing your three 'r's' (or four, or five). Seed Sci Res 1:3–14

Wang TL, Hadavizideh A, Harwood A, Welham TJ, Harwood WA, Faulks R, Hedley CL (1990) An analysis of seed development in *Pisum sativum*. XIII. The chemical induction of storage product mutants. Plant Breed 105:311–320

Wang TL, Bogracheva TY, Hedley CL (1998) Starch: as simple as A, B, C? J Exp Bot 49:481–502

Welham T, Pike P, Horst I, Flemetakis E, Katinakis P, Kaneko T, Sato S, Tabata S, Perry J, Parniske P, Wang TL (2009) A cytosolic invertase is required for normal growth and cell development in the model legume, *Lotus japonicus*. J Exp Bot 60:3353–3365

White J, Prell J, James K, Poole P (2007) Nutrient sharing between symbionts. Plant Physiol 144:604–614

Xu W, Sato SJ, Clemente TE, Chollet R (2007) The PEP-carboxylase kinase gene family in *Glycine max* (GmPpcK1-4): an in-depth molecular analysis with nodulated, non-transgenic and transgenic plants. Plant J 49:910–923

Yamakawa T, Ikeda T, Ishizuka J (2004) Effects of CO_2 concentration in rhizosphere on nodulation and N_2 fixation of soybean and cowpea. Soil Sci Plant Nutr 50:713–720

Zhang XQ, Chollet R (1997) Phospho*enol*pyruvate carboxylase protein kinase from soybean root nodules: partial purification, characterization, and up/down-regulation by photosynthate supply from the shoots. Arch Biochem Biophys 343:260–268

Zhu H, Riely BK, Burns NJ, Ané J-M (2006) Tracing nonlegume orthologs of legume genes required for nodulation and arbuscular mycorrhizal symbioses. Genetics 172:2491–2499

Chapter 4
Genomic and Functional Diversity of the Sinorhizobial Model Group

Alessio Mengoni, Marco Bazzicalupo, Elisa Giuntini, Francesco Pini, and Emanuele G. Biondi

4.1 Introducing Sinorhizobia and the Model *Sinorhizobium meliloti*

Bacteria called rhizobia, which are associated with pasture legumes, contribute to a substantial proportion of the nitrogen fixed into plants from the atmosphere (Lugtenberg and Kamilova 2009). Among the rhizobia that belong to the *Alphaproteobacteria* class, the genus *Sinorhizobium* (syn. *Ensifer*) is one of the most investigated, accounting for 1,847 records in PubMed (http://www.ncbi.nlm.nih.gov/) and 4,143 in ISI Web of Knowledge (http://apps.webofknowledge.com) databases (6 June 2012). According to the Bergeys' Manual of Systematic Bacteriology (Brenner et al. 2005), cells of the genus *Sinorhizobium* are small rods (0.5–1.0 × 1.2–3.0 μm in size), which are motile by one polar or subpolar flagellum or 1–6 peritrichous flagella. The optimal growth temperature is 25–30 °C, but sinorhizobia can tolerate a wide range of temperatures. The growth rate is quite high (fast-growing rhizobia) and colonies appear within 2–5 days of growth on yeast mannitol-mineral salts agar media. In the 2nd Edition of Bergeys' Manual of Systematic Bacteriology (Brenner et al. 2005), the genus includes eight species: *S. meliloti*, *S. medicae*, *S. arboris*, *S. fredii*, *S. kostiense*, *S. saheli*, *S. xinjiangense*, and *S. terangae* (Table 4.1). Additionally to these eight species, other nine have been described in the literature: *S. chiapanecum*, *S. americanum*, *S. abri*, *S. indiaense*, *S. kummerowiae*, *S. morelense* (syn. *Ensifer*

A. Mengoni (✉) • M. Bazzicalupo
Department of Biology, Univeristy of Florence, Sesto Fiorentino 50019, Italy
e-mail: alessio.mengoni@unifi.it; marco.bazzicalupo@unifi.it

E. Giuntini
Brighton, UK
e-mail: elisa.giuntini@gmail.com

F. Pini • E.G. Biondi
Institut de Recherche Interdisciplinaire – IRI CNRS USR3078, Villeneuve d'Ascq, France
e-mail: alessio.mengoni@unifi.it; emanuele.biondi@iri.univ-lille1.fr

Table 4.1 Members of the genus *Sinorhizobium* (*Ensifer*) and their host range

Species	Host plant	Reference
S. abri	Abrus precatorius	Ogasawara et al. (2003)
S. americanum	Acacia spp. (A. farnesiana, A. acatlensis, A. pennatula, A. macilenta, and A. cochliacantha)	Toledo et al. (2003)
S. arboris	Acacia senegal, Prosopis chilensis	Brenner et al. (2005)
S. chiapanecum	Acaciella angustissima	Rincón-Rosales et al. (2009)
S. fredii	Medicago sativa, Cajanus cajan, Glycine max cv. Peking, Glycine soja, and Vigna unguiculata	Brenner et al. (2005)
S. garamanticus	Argyrolobium uniflorum and Medicago sativa	Merabet et al. (2010)
S. indiaense	Sesbania rostrata	Ogasawara et al. (2003)
S. kostiense	Acacia senegal, Prosopis chilensis	Brenner et al. (2005)
S. kummerowiae	Kummerowia stipulacea	Wei et al. (2002)
S. medicae	Medicago orbicularis, M. polymorpha, M. rugosa, and M. truncatula	Brenner et al. (2005)
S. meliloti	Species of genera Melilotus, Medicago, and Trigonella	Brenner et al. (2005)
S. mexicanus	Acacia angustissima	Lloret et al. (2007)
S. morelense	Leucaena leucocephala	Wang et al. (2002)
S. numidicus	Argyrolobium uniflorum and Lotus creticus	Merabet et al. (2010)
S. saheli	Sesbania species (S. cannabina, S. grandiflora, S. rostrata, S. pachycarpa), Acacia seyal, Leucaena leucocephala, and Neptunia natans	Brenner et al. (2005)
S. terangae	Acacia spp. (A. senegal, A. laeta, A. tortilis subsp. raddiana, A. horrida, A. mollissima) and Sesbania spp. (S. rostrata, S. cannabina, S. aculeata, S. sesban)	Brenner et al. (2005)
S. xinjiangense	Glycine max	Brenner et al. (2005)

adhaerens), *Ensifer mexicanus*, *Ensifer garamanticus*, and *Ensifer numidicus* (Table 4.1). In the same order (*Rhizobiales*) of sinorhizobia, there are important human pathogens such as *Bartonella* and *Brucella*, and several plant-associated bacteria of major agricultural importance, such as *Agrobacterium*, *Ochrobactrum*, *Bradyrhizobium*, *Mesorhizobium*, and *Rhizobium* (Alexandre et al. 2008). In this context, the study of sinorhizobia also allows to infer some of the biological properties of other animal–bacteria associations. A phylogenetic tree of sinorhizobia (*Ensifer*) and related genera has been produced by multilocus sequence typing (Martens et al. 2007).

The species *S. meliloti* has been deeply investigated, in leguminous plants of the genus *Medicago* (Sprent 2001), which have high agronomic value worldwide (as *M. sativa* L., alfalfa) and includes one of the most important model legume for molecular biology of plant–microbe interaction and symbiotic nitrogen fixation (*M. truncatula* Gaertn.). *S. meliloti* is distributed worldwide in many soils at temperate latitudes, both in association with legumes or in a free-living form (Denison and Kiers 2004; Brenner et al. 2005). This species is a model for studying plant–bacteria interactions and in particular legume–rhizobia symbiosis and symbiotic nitrogen fixation at the molecular details (Gibson et al. 2008). To date (June 2012), the genomes of strains belonging to the species *S. meliloti* (Galardini et al. 2011; Galibert et al. 2001; Schneiker-Bekel et al. 2011), *S. medicae* (Bailly et al. 2011; Reeve et al. 2010), and *S. fredii* (Schmeisser et al. 2009; Weidner et al. 2012) have been determined and genome sequencing of additional strains and species has been announced under the Community Sequencing Programs of Joint Genome Institute (CSP2010-JGI) and GEBA-JGI initiatives (http://genome.jgi.doe.gov/genome-projects/). Moreover, a large collection of *S. meliloti* nodulation mutants is available (Mao et al. 2005), as well as genetic manipulation techniques and dedicated databases (Gibson et al. 2008; Krol et al. 2011). Because of this body of information on *S. meliloti*, this chapter will discuss mainly *S. meliloti*'s genomic and functional diversity. A paragraph (4.6) will be dedicated to *S. medicae*, for which the genomes of several strains have also been recently sequenced (Bailly et al. 2011).

4.2 Soil, Nodules, and Plant Tissues: Alternative Lifestyles of Sinorhizobia

Rhizobia can live as free bacteria in soil or plants, but, when conditions are suitable, may form symbiotic associations with leguminous plants (van Rhijn and Vanderleyden 1995).

Nodulation of legumes by rhizobia has probably evolved from the ancient symbiosis between arbuscular mycorrhiza (AM) fungi and land plants (Sprent and James 2007). Rhizobial symbiosis is related to bacterial invasion of the host roots through breaks in the epidermis, due to the emergence of lateral roots (often referred to as "crack infection") (Sprent 2008). Then, a tightly organized developmental cascade in the host plant and in the bacterial guest is initiated and leads to the formation of the nodule and the terminal differentiation of the guest (Gibson et al. 2008). The guest, now defined as "symbiont," could then be interpreted as a "domesticated" pathogen (Gage 2004; Gibson et al. 2008).

Temperate legumes (e.g., *Medicago sativa*, *Pisum sativum*, *Vicia hirsuta*) form indeterminate nodules that are characterized by persistent meristematic activity, which causes the continuous elongation of nodules and allows the bacteria inside the nodule to continue to infect plant cells (Gage 2004); the internal tissue of such nodules consists of a number of distinct zones containing invaded plant cells at different stages of differentiation, in which also bacteria show a progressive differentiation (Patriarca et al. 2002; Pawlowski and Bisseling 1996). Indeed, inside this type of nodules, bacterial cells become elongated and polyploid and are called bacteroids (Mergaert et al. 2006). These dramatic changes are induced by specific plant peptides (Mergaert et al. 2006; Van de Velde et al. 2010; Wang et al. 2010). However, an indeterminate nodule still contains, within the zone of invasion, around 10^5–10^{10} undifferentiated bacteria (Gibson et al. 2008). It is assumed that a single symbiotic rhizobium has a greater fitness if it successfully colonizes a nodule, compared with a non-symbiont strain that resides in soil, where nutrient availability limitation could impair growth (Gibson et al. 2008). An acceptable evolutionary model implies that on the one hand rhizobia have advantages nodulating legumes while on the other hand plants have developed mechanisms to prevent the parasitism by non-fixing rhizobia sanctioning inefficient nodules (Gibson et al. 2008; Kiers et al. 2003; West et al. 2002).

Recently rhizobia have been found also in nonsymbiotic association inside nontarget plant species such as rice and maize (Chi et al. 2005), colonizing both the intercellular and intracellular spaces of epidermis, cortex, and vascular system (Chi et al. 2010), suggesting the existence of different life strategies other than symbiosis and free life in soil. These data show that ecological niches of rhizobia are wider than previously expected (soil and root nodules of legumes), including the endosphere (root, stem, and leaf tissues) of potentially all higher plants. A study, using a quantitative PCR approach (Trabelsi et al. 2009), was recently published reporting the presence of *S. meliloti* in the leaf tissue of *M. sativa* (Pini et al. 2012). Since the bacteria were detected only by their genetic relationship with corresponding nodulating strains still undetermined (are they coming from root nodules, or they constitute an independent population?), however it is still difficult to understand their genetic relationship with corresponding nodulating strains. The capability to colonize all plant compartments suggests the occurrence of high genetic variability within *S. meliloti* populations, and potential new ecological and functional roles for this species, not investigated so far. Moreover soil and nodules *S. meliloti* populations showed similar values of genetic diversity (Pini et al. 2012); additionally, *S. meliloti* populations of soils and nodules are also strongly different in terms of haplotypes as previously found for other rhizobial species in chickpea (Sarita et al. 2005) and clover (Zézé et al. 2001).

4.2.1 Relationships Between Genetic Diversity and Symbiotic Preferences

Symbiotic nitrogen fixation through the association between legumes and rhizobia provides about 90 millions of tons of fixed nitrogen per year worldwide (Bouton 2007). The amount of nitrogen fixed annually by the *Sinorhizobium–Medicago* (mainly alfalfa, *Medicago sativa*) symbioses is estimated to be worth more than 200 million dollars saved in comparison with chemical fertilization, and the strict economic value of alfalfa cultivation in the USA is estimated to US$8.1 billion per year (Bouton 2007). Although several essential features of the nodulation process have been elucidated and nowadays scientists are able to explain most of the major steps of nodule formation, many aspects are still not understood (Gibson et al. 2008). One of the less considered aspects of *Sinorhizobium*–legume symbiosis concerns the effect of genetic variation of natural strains on symbiosis establishment, performance, and host specificity. Despite the fact that many natural isolates keep the ability to infect *Medicago* plants in laboratory conditions, the survival on soil and the symbiotic efficiency of different strains might indeed depend on features unrelated or poorly related to the strict nodulation process (Sprent 2001). The understanding of this symbiotic variability and the possibility to select the best strain–cultivar association for crop yield improvement has stirred the attention of investigators and also of crop companies (see, for instance, http://www.continentalsemences.com/). In the Community Sequencing Programs 2008 and 2010 of the Joint Genome Institute (JGI), the genome sequencing of *S. meliloti* AK83, BL225C (Galardini et al. 2011), and then of BO21CC and AK58 strains (still ongoing, http://genome.jgi.doe.gov) have been motivated by the necessity to understand the genetic basis of the symbiotic performances displayed by these strains toward *M. sativa*.

In the past decades, several efforts have been made to address the specific issue of genotype–genotype interaction in *Sinorhizobium*–*Medicago* symbiosis and also on the symbiosis of *Sinorhizobium* strains with other legumes (as, for instance, *Cicer*, *Lotus*, etc.). Early studies showed indeed that a large genetic polymorphism exists in natural populations of *S. meliloti* and that different *S. meliloti* strains have different host preferences with respect to plant genotype (Paffetti et al. 1996, 1998), indicating a sort of plant variety effect. In particular, on the basis of the analysis of DNA sequences of *nodC* gene, four biovars have been identified, each with a different plant host species specificity (bv. mediterranense, bv. lancerottense, bv. medicaginis, bv. meliloti) (Leòn-Barrios et al. 2009; Mnasri et al. 2007; Rogel et al. 2011; Villegas et al. 2006). Additionally, *nifH* gene sequences identify also a biovar ciceri, specific of chickpea (Maâtallah et al. 2002). For these biovars, since they are defined by both genetic methods and symbiotic capabilities toward host plants, the term symbiovar has been recently proposed (Rogel et al. 2011). The symbiovar would also reflect an assembly of genes suitable for host specificity, thus providing the basis for the identification of genetic determinants of symbiotic specificity.

Concerning the *S. meliloti* bv. meliloti strains isolated from root nodules of *M. sativa* in Italy, a study on a large panel of strains (Carelli et al. 2000) has been carried out, monitoring the changes of *S. meliloti* population nodulating three cultivars of alfalfa over a 4-year period. The aims of the study were to establish whether and to what extent the nodulating populations changed during the experiment and to evaluate the influence of environmental factors, such as soil, cultivar, and plant genotype. Results obtained in this study showed that the genetic structure of the nodulating *S. meliloti* population was apparently not affected by the cultivar or soil in the first 2 years. Here interestingly, more than plant genotype effect, there was a strong single host plant effect (Table 4.2), leaving to hypothesize that the

Table 4.2 Analysis of molecular variance on RAPD profiles from nodulating populations of *S. meliloti* data from Carelli et al. (2000) during 3 years of cultivation

Source of variation	1st year	2nd year	3rd year
Among cultivars	19.5	17.1	−1.4 (n.s.)
Plant within cultivars	9.9.	14.0	19.3
Nodules within plants	70.7	68.9	82.1

Most of the molecular diversity is present in strains isolated from nodules of the same plant. On the first 2 years, a plant cultivar effect was detected but then disappeared in favor of a single-plant effect (Data report the distribution of variance of RAPD profiles obtained from *S. meliloti* isolated from nodules of plants of three different cultivars grown in pot. All values are significant at $P < 0.0001$. *n.s.* not significant)

strains best adapted to the relationship with each single plant progressively emerged in all soils for all cultivars. Consequently, the effect of individual plants on the genetic variability of the symbiotic population appeared to increase from the beginning to the end of the trial. It should be mentioned here that the host plant species used in this study (*M. sativa*) is highly polymorphic, even within the same cultivar, due to its autotetraploidy and to the outcrossing mating behavior (Mengoni et al. 2000a, b). Each individual plant has indeed a peculiar genotype, only partially shared with members of the same cultivar, which could explain an individual-based more than a cultivar-based selection, at least in the experimental conditions tested (single pots for each individual plant).

In most cases, the genetic diversity of *S. meliloti* strains "selected" by root nodules was found to be very high. Indeed, analyses done with different molecular techniques (RAPD, BOX-PCR, PCR-RFLP, AFLP, MLST, etc.) have shown that the diversity of strains isolated from single plants or small plant populations is so high that each isolate is often characterized by a peculiar molecular fingerprint (especially when using the high-resolution AFLP technique), thus constituting a single strain (Bailly et al. 2006; Biondi et al. 2003b; Paffetti et al. 1996; Talebi Bedaf et al. 2008). Moreover, several strains of *S. meliloti* have been shown to harbor a consistent number of mobile genetic elements (as transposons and mobile introns), which contribute in generating new genetic diversity (Biondi et al. 1999, 2003a, 2011). Consequently, even the sampling of hundreds of isolates will not include the whole genetic diversity present in the nodulating population of *S. meliloti*. Nonetheless, these studies allowed to shed some light on the diversity of sinorhizobia associated with wild populations of *Medicago* plants, showing evidence of selection in the symbiotic genes (Bailly et al. 2006), and a possible phylogeographic structuring of populations (Talebi Bedaf et al. 2008). For instance, within *S. meliloti*, homologous recombination in the *nod* genes region has been detected, which limits the divergence of the different strains (Bailly et al. 2007). The same study additionally showed reproductive isolation between *S. meliloti* and *S. medicae* even though these bacteria can co-occur in the same individual host plants.

The recent development of a cultivation-independent approach for the analysis of the genetic diversity of *S. meliloti* populations (Trabelsi et al. 2010) could spur

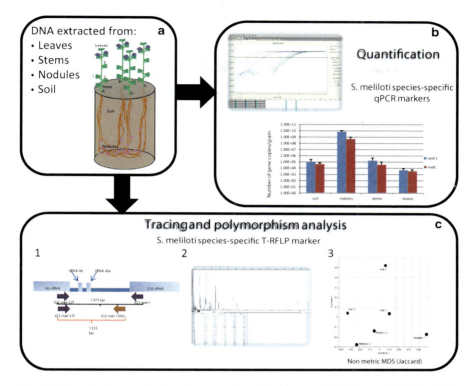

Fig. 4.1 *Sinorhizobium meliloti* population analysis. Workflow used in Pini et al. (2012). (**a**) DNA has been extracted from one pooled soil sample, one pooled sample from all the nodules and four plants (one stem and 2–3 pools of leaves *per* plant). (**b**) Population size was estimated by qPCR using two species-specific primer pairs which amplify, respectively, chromosomal (*rpo*E1) and megaplasmidic loci (*nod*C on pSymA) (Trabelsi et al. 2009). (**c**) The genetic diversity of *S. meliloti* populations was measured using (1) the 1.3 kbp long 16S–23S ribosomal intergenic spacer (IGS) (Trabelsi et al. 2010). (2) IGS-T-RFs (16S–23S ribosomal intergenic spacer terminal-restriction fragments) were detected after separate digestion with four restriction enzymes (*Alu*I, *Msp*I, *Hinf*I, *Hha*I). (**c**) A non-metric MDS plot of similarities of IGS-T-RFLP profiles was drawn (Pini et al. 2012)

more in-depth exploration of the relationships between sinorhizobial diversity and environmental variables (soil, geographic origin, plant, etc.) as recently shown (Pini et al. 2012) (Fig. 4.1).

4.2.2 Phenomic (Symbiotic and Nonsymbiotic) Diversity

The large genetic diversity present within sinorhizobial species, notably on *S. meliloti*, is also reflected by the wide range of phenotypes observed, both in symbiotic functions and in other nonsymbiotic-associated functions. This huge

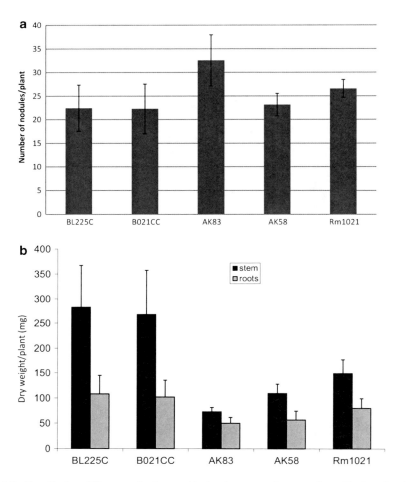

Fig. 4.2 Quantitative differences in the symbiotic phenotypes between *S. meliloti* strains. The number of nodules (**a**) and plant dry weight (**b**) formed on *M. sativa* plants by five different *S. meliloti* strains (1021, AK83, AK58, BL225C, BO21CC) are shown. Data from Biondi et al. (2009). Bars represent one standard deviation

phenotypic diversity is stimulating more and more attention due to its possible biotechnological exploitation on the selection of inoculants and on the search for genes responsible for the symbiotic efficiency (Fig. 4.2). The nodulation-mutant database (http://nodmutdb.vbi.vt.edu/) (Mao et al. 2005) actually provides a very important comprehensive resource for depositing, organizing, and retrieving information on symbiosis-related genes, mutants, and published literature to facilitate the genetic understanding of symbiotic processes in both rhizobia and their host plants. However, in spite of the large body of data already present, several aspects related to the quantitative differences among natural rhizobial strains are poorly understood. For instance, Fig. 4.2 reports a comparison of the number of nodules and plant dry weight after inoculation with five strains, the laboratory strain 1021 and four "natural" strains (AK83, AK58, BL225C, BO21CC). Even though all strains are able to nodulate the plant, it is clear that differences are present among strains, some of them promoting plant growth (as dry weight) better than others (i.e., BL225C vs. AK83).

In the perspective of understanding the molecular bases of symbiotic variability, very recently a study on the evaluation of transcriptomic responses of both plant (*M. trucantula*) and rhizobial partner (*S. meliloti*) has been performed (Heath et al. 2012). Results of this study have shown that the expression of nearly a quarter of rhizobial genome, including nodulation genes (ca. 1,600 genes), vary according to the host plant genotype. However, which genes may play major roles in quantitative variability of symbiotic phenotypes remains elusive.

Regarding the nonsymbiosis-related phenotypes, other lines of research are using *S. meliloti* as a model system for investigating metabolic diversity at a genomic scale, with the aim to clarify the functional and metabolic role of the genetic polymorphism observed and to explore the metabolic versatility and diversity of a model soil bacterium. In this view, the direct high-throughput assessment of phenotypes (phenome) by Phenotype Microarray™ (PM) technique (Biolog Inc.) (Bochner 2009) has drawn much of the attention of molecular biologists and microbiologists. Recently, an analysis has been performed comparing *S. meliloti* strains using "Biolog" (Biondi et al. 2009) to explore the "panphenome" of the species, that is, the extent of the phenotypic diversity shown by different strains of the species. This study showed the presence of high metabolic diversity among strains, in particular in carbon substrates utilization and resistance to osmolytes and pH gradients, which may be related to the ability to grow and persist in soil and to use different carbon sources derived from plant material degradation in the rhizosphere. In comparison with other bacterial species, especially those of clinical importance, *S. meliloti* strains seem to be more versatile in the use carbon sources, as, for instance, galactosides, which are abundant in the plant rhizosphere. Intriguingly, when the PM diversity was compared to the genomic diversity of the same strains, only one phenotypic difference could be directly related to a genetic diversity, the great majority of differences in phenotypes being not obviously connected to genes diversity as it is shown by the comparison of patterns of Fig. 4.3, which show different groupings of sample for the genomic vs. phenomic data.

4.2.3 The Pangenome of S. meliloti: *Clues into Phenotypic Variability*

Recently, the core and accessory genome of the species *S. meliloti* have begun to be elucidated comparing three genomes, the type strain Rm1021, AK83, and BL225C (Galardini et al. 2011). This study identified a very large set of genes, about 37 % of

Fig. 4.3 Phenomic-genomic relationships in *S. meliloti*. Principal component analysis of comparative genomic hybridization data (CGH) (**a**) and phenotype microarray data (**b**) *Numbers* on axes report the percentage of variance explained by the first (PC1) and second (PC2) principal component. Data from Biondi et al. (2009) and Giuntini et al. (2005)

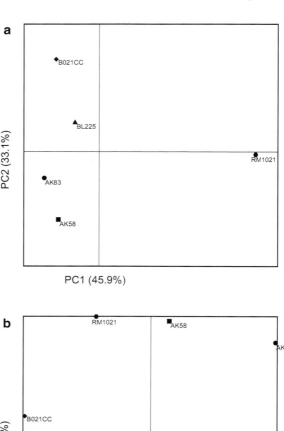

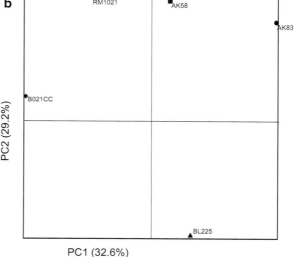

the total number of genes annotated, belonging only to one or two of the strains analyzed, which is referred to as accessory genome; this proportion is similar to the pangenomic content of *Escherichia coli*, with about 42 % of the genes belonging to the accessory genome, while for other species such as *Bacillus anthracis* and *Streptococcus pneumoniae*, the size of the accessory genome is slightly larger, 60 % and 77 %, respectively (Tettelin et al. 2008). This large number of accessory

Table 4.3 *S. meliloti* genes present in the accessory genome fraction related to symbiotic interaction (data from Galardini et al. 2011)

Gene or protein name	Strain(s)
fixK-like	Rm1021/BL225C
fixNOQP$_3$	Rm1021/BL225C
norBCDEQ	Rm1021/BL225C
nosRZDFYLX	Rm1021/BL225C
nirKV	Rm1021/BL225C
nnrRSU	Rm1021/BL225C
nrtABC	Rm1021/BL225C
cycB$_2$	Rm1021
hemN	Rm1021/BL225C
Symbiosis-related SDR	Rm1021/BL225C
nwsB	BL225C
rhbABCDE, rhtAX, rhrA	Rm1021/BL225C
acdS	AK83/BL225C
fixK weak homolog	AK83/BL225C
nodQ$_1$	Rm1021/AK83
fixT$_3$	Rm1021
nodP$_2$	Rm1021
fixT$_2$	Rm1021
fixK$_2$	Rm1021
C P450	AK83/BL225C
hupE	AK83
hemA homolog	AK83/BL225C
pcs distant homolog	BL225C
CTP:phosphocholine cytidylyl transferase	AK83
Cadherin-like protein	AK83
cgmB	Rm1021
expR (fragment)	Rm1021
Sugar isomerase	Rm1021/BL225C
nodM (AK83)	AK83
napC/nirT-like	BL225C
glnA-like	AK83
fixS2	Rm1021/BL225C
fixO-like	AK83/BL225C
fixT1-like	AK83/BL225C
fixT-like	BL225C
fixL-related	Rm1021

genes supported the observation of the vast phenotypic diversity of these strains and of the species in general (Biondi et al. 2009), as illustrated in the previous section.

Symbiotic-related genes have been shown to be highly variable among different rhizobial species (Amadou et al. 2008) (Table 4.3). Indeed, the analysis of *S. meliloti* pangenome revealed that also the symbiotic accessory genome in this species is highly variable, representing about 22 % of all the symbiotic genes. The most notable feature found was a large variability among the "microaerophilic" genes, which includes the transcriptional regulator annotated as FixK-like, a third copy of electron transport chain (*fixNOQP*), and several genes related to nitrogen metabolism (*nos*, *nor*, *nir*, *nnr*, and *nrt*). The pangenome results were consistent with previous data made by comparative genomic hybridization (Giuntini et al. 2005) and by phenotypic microarray on different metabolic activity of these strains in different nitrogen sources (Biondi et al. 2009).

More recently several other *S. meliloti* genomes are becoming available (Li et al. 2012; Schneiker-Bekel et al. 2011) and many genomes of *S. meliloti* and other sinorhizobia are under sequencing at JGI, both under CSP2010 (strains AK58 and BO21CC) and under GEBA initiatives (*S. arbores*, *S. terangae*, plus other strains of both *S. meliloti* and *S. medicae*), allowing researchers to expand this pangenome analysis in the near future and to relate pangenome with panphenome diversity on a large panel of strains.

4.3 Sinorhizobia Population Genomics: The *S. medicae* Case

In spite of the fact that most evolutionary and functional studies have used *S. meliloti* as a model bacterium because the genome of the strain *S. meliloti* 1021 was available since more than 10 year ago (Galibert et al. 2001), another sinorhizobial species (*S. medicae*) is stirring the attention of researchers. The reason for most of the studies on *S. medicae* resides on the ability of strains of this species to be superior symbiont than *S. meliloti* 1021 of annual medics, in particular of the model legume *M. truncatula* (Terpolilli et al. 2008). Recently the genome of one *S. medicae* strain (WSM 419) has been released (Reeve et al. 2010) allowing genomic studies to be performed also on this species. *S. meliloti* and *S. medicae* are closely related species that form a tight phylogenetic clade together with *S. arboris* (Martens et al. 2007). The genomes of *S. meliloti* 1021 and *S. medicae* WSM 419 share similar architectures. Both include three main replication units: (1) a chromosome which harbors most of the housekeeping genes; (2) a chromid (Harrison et al. 2010), where many genes involved in polysaccharide synthesis are clustered (pSMED01 for *S. medicae* WSM 419 and pSymB for *S. meliloti* 1021); and (3) a megaplasmid, where *nod* genes and genes involved in nitrogen fixation are located (pSMED02 for WSM 419 and pSymA for 1021). The genome of *S. medicae* WSM 419 also includes a 219-kb plasmid called pSMED03. Based on sequencing of several loci, *S. medicae* is less diverse than *S. meliloti*, especially at chromosomal loci (Bailly et al. 2006; van Berkum et al. 2006). Indications of directional and balancing selection have been identified for the *nod* genes region on pSMED02/pSymA and the region involved in polysaccharide synthesis on pSMED01/pSymB, respectively, in sympatric populations of *S. meliloti* and *S. medicae* (Bailly et al. 2006), suggesting that recombination has an important role in shaping the diversity of both species. Within each species, linkage disequilibrium analyses suggest that homologous recombination occurs preferentially at loci located on the chromid and megaplasmid, rather than on the chromosome.

Recently, a comparison of partial genome sequences of 12 isolates of *S. medicae* (isolated in 1 m^2 of soil at York, UK) with the reference strain WSM 419 (originally isolated in Sardinia, Italy) was done. The mapping of the sequence reads against the reference genome of *S. medicae* WSM 419 showed that most indels (sequences containing genes gained by horizontal transfer and gene losses) have affected pSMED02 and pSMED03. This is in agreement with the observation that most indel events affecting the *S. meliloti* genome occur on pSymA (Galardini et al. 2011; Giuntini et al. 2005) and with the low proportion of homologous genes observed between pSMED02 of *S. medicae* and pSymA of *S. meliloti* when compared with the strong synteny between pSMED01 and pSymB and between the chromosomes of *S. meliloti* 1021 and that of *S. medicae* WSM 419. Very interestingly was the evidence that the relative divergence of the Sardinian strain WSM 419 from the York population was lower for pSMED01 and pSMED02 than for the chromosome and that the majority of York strains were more similar to WSM 419 than to some other highly diverged York strains. In other words, strains within the population in one square meter of soil were shown to be more diverged from each other in these replicons than from a strain originating very far away in strongly different climatic conditions. Another intriguing result coming from *S. medicae* population genomic analysis is the discovery of genes involved in the synthesis of rhizobitoxine (*rtx*), previously known in *Bradyrhizobium* only (Yasuta et al. 2001), which enhances nodulation capabilities of the bacteria by inhibiting ethylene synthesis in plant tissues. Interestingly, these genes are not present in the *S. meliloti* strains analyzed so far (Duan et al. 2009; Galardini et al. 2011), which, on the contrary, seem to cope with plant ethylene production by using a different pathway based on the ACC deaminase gene (*acdS*), which led to an increase of nodulation competitiveness by metabolizing a precursor of ethylene (Glick 2005). This divergence between *S. meliloti* and *S. medicae* pangenomes raises questions about the ecological factors that could limit the spread of such functions among rhizobia, as a potential trade-off between the fitness of symbiotic partners (Ratcliff and Denison 2009). The high frequency of genes influencing the ethylene pathways in natural populations illustrates the outcome of a conflict of interest between rhizobia and their host regarding the investment of legumes in symbiotic functions. The existence of two alternative bacterial systems in so similar species confirms that defining ecologically important functions is not straightforward, as different pathways can be involved in similar phenotypes. More generally, the analysis of *S. medicae* pangenome has highlighted that studies of accessory genome composition are important in defining the selective pressures that describe the ecological niche and drive the evolution of bacterial species. Indeed, one could argue that, while the core genome defines the taxonomy of bacteria, the accessory genome has an equal or greater importance in defining their ecological niche.

4.4 Concluding Remarks and Perspectives

The use of next-generation sequencing technologies and of high-throughput systems for substrate utilization phenotype analysis (Phenotype Microarray, Biolog, Inc.) is offering an exciting opportunity to researchers interested in the genetic and phenotypic diversity. It is now feasible to produce draft genome sequences of virtually every interesting strain and to compare it with phenotypic features, from symbiosis to the large panel of biochemical conditions (nearly 2,000) tested on multiwell plates. This huge amount of data, if well organized and interpreted, could improve our understanding of rhizobial genome functionality and ecology, as well as provide an exceptional tool for the biotechnological use of phenotypic and genotypic diversity of rhizobia. In this perspective, one of the most critical challenges of the near future will be to develop computational tools and improve or create new databases, specifically aimed at deciphering the genomic basis of phenotypic diversity.

References

Alexandre A, Laranjo M, Young JPW, Oliveira S (2008) dnaJ is a useful phylogenetic marker for alphaproteobacteria. Int J Syst Evol Microbiol 58:2839–2849

Amadou C, Pascal G, Mangenot S et al (2008) Genome sequence of the β-rhizobium *Cupriavidus taiwanensis* and comparative genomics of rhizobia. Genome Res 18(9):1472–1483

Bailly X, Olivieri I, De Mita S, Cleyet-Marel JC, Bena G (2006) Recombination and selection shape the molecular diversity pattern of nitrogen-fixing *Sinorhizobium* sp. associated to *Medicago*. Mol Ecol 15:2719–2734

Bailly X, Olivieri I, Brunel B, Cleyet-Marel JC, Bena G (2007) Horizontal gene transfer and homologous recombination drive the evolution of the nitrogen-fixing symbionts of Medicago species. J Bacteriol 189:5223–5236

Bailly X, Giuntini E, Sexton MC et al (2011) Population genomics of *Sinorhizobium medicae* based on low-coverage sequencing of sympatric isolates. ISME J 5(11):1722–1734

Biondi EG, Fancelli S, Bazzicalupo M (1999) ISRm10: a new insertion sequence of *Sinorhizobium meliloti*: nucleotide sequence and geographic distribution. FEMS Microbiol Lett 181:171–176

Biondi EG, Femia AP, Favilli F, Bazzicalupo M (2003a) IS Rm31, a new insertion sequence of the IS 66 family in Sinorhizobium meliloti. Arch Microbiol 180:118–126

Biondi EG, Pilli E, Giuntini E et al (2003b) Genetic relationship of *Sinorhizobium meliloti* and *Sinorhizobium medicae* strains isolated from Caucasian region. FEMS Microbiol Lett 220:207–213

Biondi EG, Tatti E, Comparini D et al (2009) Metabolic capacity of *Sinorhizobium* (*Ensifer*) *meliloti* strains as determined by phenotype microarray analysis. Appl Environ Microbiol 75:5396–5404

Biondi EG, Ns T, Bazzicalupo M, Martinez-Abarca F (2011) Spread of the group II intron RmInt1 and its insertion sequence target sites in the plant endosymbiont *Sinorhizobium meliloti*. Mob Genet Elements 1:2–7

Bochner BR (2009) Global phenotypic characterization of bacteria. FEMS Microbiol Rev 33:191–205

Bouton J (2007) The economic benefits of forage improvement in the United States. Euphytica 154:263–270

Brenner DJ, Krieg NR, Staley JT (2005) Bergeys' manual of systematic bacteriology, vol 2, 2nd edn, The proteobacteria. Part C: The alpha-, beta-, delta-, and epsilonproteobacteria. Springer, Berlin

Carelli M, Gnocchi S, Fancelli S et al (2000) Genetic diversity and dynamics of *Sinorhizobium meliloti* populations nodulating different alfalfa varieties in Italian soils. Appl Environ Microbiol 66:4785–4789

Chi F, Shen SH, Cheng HP et al (2005) Ascending migration of endophytic rhizobia, from roots to leaves, inside rice plants and assessment of benefits to rice growth physiology. Appl Environ Microbiol 71:7271–7278

Chi F, Yang P, Han F, Jing Y, Shen S (2010) Proteomic analysis of rice seedlings infected by Sinorhizobium meliloti 1021. Proteomics 10:1861–1874

Denison RF, Kiers ET (2004) Lifestyle alternatives for rhizobia: mutualism, parasitism, and forgoing symbiosis. FEMS Microbiol Lett 237:187–193

Duan J, Müller K, Charles T, Vesely S, Glick B (2009) 1-aminocyclopropane-1-carboxylate (ACC) deaminase genes in rhizobia from southern Saskatchewan. Microb Ecol 57:423–436

Gage DJ (2004) Infection and invasion of roots by symbiotic, nitrogen-fixing rhizobia during nodulation of temperate legumes. Microbiol Mol Biol Rev 68:280–300

Galardini M, Mengoni A, Brilli M et al (2011) Exploring the symbiotic pangenome of the nitrogen-fixing bacterium *Sinorhizobium meliloti*. BMC Genomics 12:235

Galibert F, Finan TM, Long SR et al (2001) The composite genome of the legume symbiont *Sinorhizobium meliloti*. Science 293:668–672

Gibson KE, Kobayashi H, Walker GC (2008) Molecular determinants of a symbiotic chronic infection. Annu Rev Genet 42:413–441

Giuntini E, Mengoni A, De Filippo C et al (2005) Large-scale genetic variation of the symbiosis-required megaplasmid pSymA revealed by comparative genomic analysis of *Sinorhizobium meliloti* natural strains. BMC Genomics 6:158

Glick BR (2005) Modulation of plant ethylene levels by the bacterial enzyme ACC deaminase. FEMS Microbiol Lett 251:1–7

Harrison PW, Lower RPJ, Kim NKD, Young JPW (2010) Introducing the bacterial 'chromid': not a chromosome, not a plasmid. Trends Microbiol 18:141–148

Heath KD, Burke PV, Stinchcombe JR (2012) Coevolutionary genetic variation in the legume-rhizobium transcriptome. Mol Ecol 21:4735–4747

Kiers ET, Rousseau RA, West SA, Denison RF (2003) Host sanctions and the legume-rhizobium mutualism. Nature 425:78–81

Krol E, Blom J, Winnebald J et al (2011) RhizoRegNet-a database of rhizobial transcription factors and regulatory networks. J Biotechnol 155:127–134

Leòn-Barrios M, Lorite MJ, Donate-Correa J, Sanjuan J (2009) *Ensifer meliloti* bv. lancerottense establishes nitrogen-fixing symbiosis with Lotus endemic to the Canary Islands and shows distinctive symbiotic genotypes and host range. Syst Appl Microbiol 32:413–420

Li Z, Ma Z, Hao X, Wei G (2012) Draft genome sequence of *Sinorhizobium meliloti* CCNWSX0020, a nitrogen-fixing symbiont with copper tolerance capability isolated from lead-zinc mine tailings. J Bacteriol 194:1267–1268

Lloret L, Ormeno-Orrillo E, Rincon R et al (2007) *Ensifer mexicanus* sp. nov. a new species nodulating Acacia angustissima (Mill.) Kuntze in Mexico. Syst Appl Microbiol 30:280–290

Lugtenberg B, Kamilova F (2009) Plant-growth-promoting rhizobacteria. Annu Rev Microbiol 63:541–556

Maâtallah J, Berraho EB, Muñoz S, Sanjuan J, Lluch C (2002) Phenotypic and molecular characterization of chickpea rhizobia isolated from different areas of Morocco. J Appl Microbiol 93:531–540

Mao C, Qiu J, Wang C, Charles TC, Sobral BWS (2005) NodMutDB: a database for genes and mutants involved in symbiosis. Bioinformatics 21:2927–2929

Martens M, Delaere M, Coopman R et al (2007) Multilocus sequence analysis of *Ensifer* and related taxa. Int J Syst Evol Microbiol 57:489–503

Mengoni A, Gori A, Bazzicalupo M (2000a) Use of RAPD and microsatellite (SSR) variation to assess genetic relationships among populations of tetraploid alfalfa, *Medicago sativa*. Plant Breed 119:311–317

Mengoni A, Ruggini C, Vendramin GG, Bazzicalupo M (2000b) Chloroplast microsatellite variations in tetraploid alfalfa. Plant Breed 119:509–512

Merabet C, Martens M, Mahdhi M et al (2010) Multilocus sequence analysis of root nodule isolates from *Lotus arabicus* (Senegal), *Lotus creticus*, *Argyrolobium uniflorum* and *Medicago sativa* (Tunisia) and description of *Ensifer numidicus* sp. nov. and *Ensifer garamanticus* sp. nov. Int J Syst Evol Microbiol 60:664–674

Mergaert P, Uchiumi T, Alunni B et al (2006) Eukaryotic control on bacterial cell cycle and differentiation in the Rhizobium-legume symbiosis. Proc Natl Acad Sci USA 103:5230–5235

Mnasri B, Mrabet M, Laguerre G, Aouani M, Mhamdi R (2007) Salt-tolerant rhizobia isolated from a Tunisian oasis that are highly effective for symbiotic N2-fixation with *Phaseolus vulgaris* constitute a novel biovar (bv. *mediterranense*) of *Sinorhizobium meliloti*. Arch Microbiol 187:79–85

Ogasawara M, Suzuki T, Mutoh I et al (2003) *Sinorhizobium indiaense* sp. nov. and *Sinorhizobium abri* sp. nov. isolated from tropical legumes, *Sesbania rostrata* and *Abrus precatorius*, respectively. Symbiosis 34:53–68

Paffetti D, Scotti C, Gnocchi S, Fancelli S, Bazzicalupo M (1996) Genetic diversity of an Italian Rhizobium meliloti population from different *Medicago sativa* varieties. Appl Environ Microbiol 62:2279–2285

Paffetti D, Daguin F, Fancelli S et al (1998) Influence of plant genotype on the selection of nodulating *Sinorhizobium meliloti* strains by *Medicago sativa*. Antonie Van Leeuwenhoek 73:3–8

Patriarca EJ, Tate R, Iaccarino M (2002) Key role of bacterial NH(4)(+) metabolism in Rhizobium-plant symbiosis. Microbiol Mol Biol Rev 66:203–222

Pawlowski K, Bisseling T (1996) Rhizobial and actinorhizal symbioses: what are the shared features? Plant Cell 8:1899–1913

Pini F, Frascella A, Santopolo L et al (2012) Exploring the plant-associated bacterial communities in *Medicago sativa* L. BMC Microbiol 12:78

Ratcliff WC, Denison RF (2009) Rhizobitoxine producers gain more poly-3-hydroxybutyrate in symbiosis than do competing rhizobia, but reduce plant growth. ISME J 3:870–872

Reeve WG, Chain P, O'Hara G et al (2010) Complete genome sequence of the *Medicago* microsymbiont *Ensifer* (*Sinorhizobium*) *medicae* strain WSM419. Stand Genomic Sci 2:1

Rincón-Rosales R, Lloret L, Ponce E, Martínez-Romero E (2009) Rhizobia with different symbiotic efficiencies nodulate *Acaciella angustissima* in Mexico, including *Sinorhizobium chiapanecum* sp. nov. which has common symbiotic genes with *Sinorhizobium mexicanum*. FEMS Microbiol Ecol 67:103–117

Rogel MA, Ormeno-Orrillo E, Martinez Romero E (2011) Symbiovars in rhizobia reflect bacterial adaptation to legumes. Syst Appl Microbiol 34:96–104

Sarita S, Sharma PK, Priefer UB, Prell J (2005) Direct amplification of rhizobial *nodC* sequences from soil total DNA and comparison to *nodC* diversity of root nodule isolates. FEMS Microbiol Ecol 54:1–11

Schmeisser C, Liesegang H, Krysciak D et al (2009) *Rhizobium* sp. strain NGR234 possesses a remarkable number of secretion systems. Appl Environ Microbiol 75:4035–4045

Schneiker-Bekel S, Wibberg D, Bekel T et al (2011) The complete genome sequence of the dominant *Sinorhizobium meliloti* field isolate SM11 extends the *S. meliloti* pan-genome. J Biotechnol 155:20–33

Sprent JI (2001) Nodulation in legumes. Royal Botanic Gardens Kew, London

Sprent JI (2008) 60Ma of legume nodulation: what's new? What's changing? J Exp Bot 59 (5):1081–1084

Sprent JI, James EK (2007) Legume evolution: where do nodules and mycorrhizas fit in? Plant Physiol 144:575–581

Talebi Bedaf M, Bahar M, Saeidi G, Mengoni A, Bazzicalupo M (2008) Diversity of *Sinorhizobium* strains nodulating *Medicago sativa* from different Iranian regions. FEMS Microbiol Lett 288:40–46

Terpolilli JJ, O'Hara GW, Tiwari RP, Dilworth MJ, Howieson JG (2008) The model legume *Medicago truncatula* A17 is poorly matched for N_2 fixation with the sequenced microsymbiont *Sinorhizobium meliloti* 1021. New Phytol 179:62–66

Tettelin H, Riley D, Cattuto C, Medini D (2008) Comparative genomics: the bacterial pan-genome. Curr Opin Microbiol 11:472–477

Toledo I, Lloret L, Martìnez-Romero E (2003) *Sinorhizobium americanus* sp. nov., a new *Sinorhizobium* species nodulating native *Acacia* spp. in Mexico. Syst Appl Microbiol 26:54–64

Trabelsi D, Pini F, Aouani ME, Bazzicalupo M, Mengoni A (2009) Development of real-time PCR assay for detection and quantification of *Sinorhizobium meliloti* in soil and plant tissue. Lett Appl Microbiol 48:355–361

Trabelsi D, Pini F, Bazzicalupo M et al (2010) Development of a cultivation-independent approach for the study of genetic diversity of *Sinorhizobium meliloti* populations. Mol Ecol Resour 10:170–172

van Berkum P, Elia P, Eardly BD (2006) Multilocus sequence typing as an approach for population analysis of *Medicago*-nodulating rhizobia. J Bacteriol 188:5570–5577

Van de Velde W, Zehirov G, Szatmari A et al (2010) Plant peptides govern terminal differentiation of bacteria in symbiosis. Science 327:1122–1126

van Rhijn P, Vanderleyden J (1995) The Rhizobium-plant symbiosis. Microbiol Rev 59:124–142

Villegas MDC, Rome S, Maure L et al (2006) Nitrogen-fixing sinorhizobia with *Medicago laciniata* constitute a novel biovar (bv. *medicaginis*) of *S. meliloti*. Syst Appl Microbiol 29:526–538

Wang ET, Tan ZY, Willems A et al (2002) *Sinorhizobium morelense* sp. nov., a *Leucaena leucocephala*-associated bacterium that is highly resistant to multiple antibiotics. Int J Syst Evol Microbiol 52:1687–1693

Wang D, Griffitts J, Starker C et al (2010) A nodule-specific protein secretory pathway required for nitrogen-fixing symbiosis. Science 327:1126–1129

Wei GH, Wang ET, Tan ZY, Zhu ME, Chen WX (2002) *Rhizobium indigoferae* sp. nov. and *Sinorhizobium kummerowiae* sp. nov., respectively isolated from *Indigofera* spp. and *Kummerowia stipulacea*. Int J Syst Evol Microbiol 52:2231–2239

Weidner S, Becker A, Bonilla I et al (2012) Genome sequence of the soybean symbiont *Sinorhizobium fredii* HH103. J Bacteriol 194:1617–1618

West SA, Kiers ET, Simms EL, Denison RF (2002) Sanctions and mutualism stability: why do rhizobia fix nitrogen? Proc Biol Sci 269:685–694

Yasuta T, Okazaki S, Mitsui H et al (2001) DNA sequence and mutational analysis of Rhizobitoxine biosynthesis genes in *Bradyrhizobium elkanii*. Appl Environ Microbiol 67:4999–5009

Zézé A, Mutch LA, Young JPW (2001) Direct amplification of *nodD* from community DNA reveals the genetic diversity of *Rhizobium leguminosarum* in soil. Environ Microbiol 3:363–370

Part II
Actinorhizal Symbiosis

Chapter 5
Establishment of Actinorhizal Symbioses

Alexandre Tromas, Nathalie Diagne, Issa Diedhiou, Hermann Prodjinoto, Maïmouna Cissoko, Amandine Crabos, Diaga Diouf, Mame Ourèye Sy, Antony Champion, and Laurent Laplaze

5.1 Introduction

To cope with nitrogen limitations, some plants have developed the ability to fix atmospheric nitrogen through symbiotic interaction with soil bacteria. The most efficient and intimate of such associations lead to the formation of new root organs called nodules were bacteria are hosted intracellularly and fix atmospheric nitrogen in optimal conditions. Two types of nitrogen-fixing root nodule symbioses (RNS) have been described: the well-studied legume–rhizobial symbiosis (see Part I) and actinorhizal symbioses (Perrine-Walker et al. 2011). The latter involves filamentous bacteria of the genus *Frankia* that interact with more than 200 plant species from eight different families, collectively called actinorhizal plants (Baker and Mullin 1992). Besides *Datisca* these plants are woody shrubs or trees and are present on every continent except Antarctica (Baker and Schwintzer 1990). The efficiency of nitrogen fixation in actinorhizal symbioses is comparable to the one in

---

A. Tromas • A. Crabos • A. Champion • L. Laplaze (✉)
Laboratoire Commun de Microbiologie IRD/ISRA/UCAD, Centre de Recherche de Bel Air, BP 1386 Dakar, Senegal

Institut de Recherche pour le Développement (IRD), UMR DIADE, 911 Avenue Agropolis, 34394 Montpellier cedex 5, France
e-mail: alexandre.tromas@ird.fr; amandine.crabos@ird.fr; antony.champion@ird.fr; laurent.laplaze@ird.fr

N. Diagne • I. Diedhiou • H. Prodjinoto • M. Cissoko
Laboratoire Commun de Microbiologie IRD/ISRA/UCAD, Centre de Recherche de Bel Air, BP 1386 Dakar, Senegal
e-mail: nathalie.diagne@ird.fr; diedhiouissa@yahoo.fr; lheureux018@yahoo.fr; Maimouna.Cissoko@ird.fr

D. Diouf • M.O. Sy
Laboratoire Campus de Biotechnologies Végétales, Département de Biologie Végétale, Faculté des Sciences & Techniques, Université Cheikh Anta Diop, BP 5005 Dakar-Fann, Senegal
e-mail: diaga.diouf@ucad.edu.sn; oureyesy1@yahoo.fr

legumes with an estimated rate of 240–350 kg ha^{-1} year^{-1} (Wall 2000). Because of this, actinorhizal plants play very important roles in various ecosystems, as pioneer species able to colonise poor or degraded soils and to improve their fertility. These plants are used in soil fixation, agroforestry, reforestation, to build windbreaks or as source of timber or firewood (Diem and Dommergues 1990; Zhong et al. 2011).

Interestingly, phylogenetic analyses have demonstrated that all plant-forming RNS belong to the same clade (Rosid I). It was suggested that this common ancestor possessed some kind of predisposition towards RNS formation (Soltis et al. 1995; Doyle 2011). Studies suggest that actinorhizal symbioses appeared 3–4 times independently during evolution (Swensen 1996).

Despite their ecological importance, the establishment and functioning of actinorhizal symbioses are still poorly understood. Progresses have been hindered by the lack of genetic transformation system in *Frankia* and the fact that actinorhizal plants are mostly trees and shrubs thus making genetic approaches difficult. Moreover, no model system to study actinorhizal symbioses as emerged so far and different groups work with different experimental systems. Nevertheless, technical breakthroughs in the last decade have opened new avenues to the study of actinorhizal symbioses. First of all, the genomes of several *Frankia* strains have been sequenced (Normand et al. 2007; Persson et al. 2011) opening the way for comparative genomic studies (e.g. Bickhart et al. 2009; Udwary et al. 2011) and the analysis of global gene expression during symbiosis (Alloisio et al. 2010). Similarly, genomic resources are now available for some actinorhizal plants including ESTs database and microarrays (Hocher et al. 2006, 2011a). Finally, stable and hairy root transformation has now been achieved for actinorhizal plants of the *Casuarinaceae* (Diouf et al. 1995; Franche et al. 1997), *Datiscaceae* (Markmann et al. 2008) and *Rhamnaceae* (Imanishi et al. 2011) family paving the way for functional plant gene studies through RNAi (Gherbi et al. 2008a, b; Markmann et al. 2008). These new tools have led to advances in our understanding of the molecular mechanisms controlling actinorhizal symbioses formation. In this chapter, we give an overview of the current knowledge on the events leading to actinorhizal symbioses establishment. Multiple reviews addressing specific aspects of actinorhizal symbioses are available for further reading (Péret et al. 2009; Perrine-Walker et al. 2011; Hocher et al. 2011b; Abdel-Lateif et al. 2012; Pawlowski and Demchenko 2012).

5.2 Pre-infection, a Molecular Dialogue Between Symbiotic Partners

The first step towards symbiosis is the recognition of compatible symbionts. Host specificity in actinorhizal symbiosis is not as stringent as in legume–rhizobial symbiosis (Pawlowski and Sprent 2008; Pawlowski and Demchenko 2012). Host specificity originates from both partners. Some *Frankia* subgroups like the

"Casuarina" strains evolved high levels of specificity and are able to nodulate only two *Casuarinaceae* genera, *Casuarina* and *Allocasuarina*. On the plant side, some genera like *Gymnostoma* (*Casuarinaceae*) accept a wide range of *Frankia* strains. In legumes, host specificity derives from co-evolution of the bacterial signal molecule and the plant receptor. For the bacterial signal, specificity would result from chemical substitutions on the same chemical backbone (Wall 2000). We can hypothesise that actinorhizal plants with a broad range of bacterial host would recognise a common feature of the signal molecule whereas actinorhizal plants with restricted bacterial hosts would recognise a specific decoration of the signal molecule. On the *Frankia* side, strains able to nodulate divergent plant species would be able to synthesise multiple molecules corresponding to receptors of various plant species (Pueppke and Broughton 1999).

Nevertheless, little is known about the molecular interactions between *Frankia* and host plants in the rhizosphere prior to infection. It was shown recently that aqueous root exudates from the actinorhizal trees *Casuarina glauca* and *C. cunninghamiana* changed *Frankia* physiology and symbiotic properties (Beauchemin et al. 2012). Root exudates increased the growth of *Frankia* and caused hyphal curling, suggesting a chemotrophic response and/or surface property changes. Interestingly, Beauchemin et al. (2012) showed that root exudates altered the bacterial surface properties at the fatty acid and carbohydrate level. More importantly, *Frankia* cells treated with root exudates formed nodules significantly earlier than controls (Beauchemin et al. 2012). These data support the hypothesis of early chemical signalling between actinorhizal host plants and *Frankia* in the rhizosphere. However, the signals involved have not been identified yet. In legumes, flavonoids have been demonstrated to be the symbiotic plant signal attracting rhizobia and initiating the production of bacterial Nod factors (Ferguson et al. 2010). Flavonoids are secondary metabolites derived from the phenylpropanoid pathway. They are widely distributed in plants and fulfil many functions from pigmentation to cell cycle regulation (Abdel-Lateif et al. 2012; Hassan and Mathesius 2012). The impact of root exudates from *C. cunninghamiana* on *Frankia* could not be mimicked by some flavonoids that were shown to be active in the legume–rhizobia symbiosis (Beauchemin et al. 2012). In *Myrica gale*, flavonoids extracted from fruits changed *Frankia* growth and nitrogen fixation according to the symbiotic specificity of strains, inducing compatible and inhibiting incompatible strains (Popovici et al. 2010). This suggests that flavonoids might be plant signals involved in defining symbiotic specificity. In order to analyse the role of flavonoids in actinorhizal nodule formation, *C. glauca* plants with reduced flavonoids biosynthesis were produced by downregulation of *CgCHS1* (Laplaze et al. 1999) using RNA interference. *CgCHS1* encodes a chalcone synthase, the first enzyme of the flavonoid biosynthetic pathway. In these plants, the level of flavonoids in the roots was drastically reduced (Abdel Lateif and Hocher, in preparation). This led to a delay in nodulation and a reduction of the percentage of nodulation (Abdel Lateif and Hocher, in preparation). These results suggest that flavonoids are important for actinorhizal symbiosis formation and might represent symbiotic signals emitted by the root.

Flavonoids might also be involved in actinorhizal prenodule or nodule development by inhibiting auxin transport as demonstrated in legumes (Hassan and Mathesius 2012).

In the legume–rhizobial symbiosis, flavonoids trigger the expression of the bacterial *nod* genes (Chap. 1). These genes encode proteins involved in the biosynthesis of lipo-chitooligosaccharides molecules named Nod factors that act as bacterial symbiotic signals (Chap. 1). Many aspects of the molecular dialogue between rhizobia and legumes have been elucidated from synthesis to perception and transduction of the symbiotic signals (Chap. 1, Jones et al. 2007). In actinorhizal symbioses, the bacterial signal is not yet identified. The recent sequencing of *Frankia* genomes revealed a lack of canonical *nod* genes essential for Nod factors biosynthesis and that symbiotic genes (such as *nif* genes, *hup1*, *hup2* and *shc*) are not organised in symbiotic island and are not induced under symbiotic conditions in *Frankia* (Alloisio et al. 2010; Normand et al. 2007). Altogether, this suggests that *Frankia* might synthesise chemically distinct signalling molecules. Previous attempts to identify symbiotic signals secreted by *Frankia* led to the isolation of an unknown compound found to be heat stable, hydrophobic, resistant to chitinase and smaller than lipo-chitooligosaccharides (Cérémonie et al. 1999). Besides, *Frankia* is known to produce some auxins such as phenylacetic acid (PAA; Wheeler et al. 1984). Treatment of *Alnus glutinosa* plants with PAA was reported to induce nodule-like structure thus suggesting that these molecules might also be involved in pre-infection signalling (Hammad et al. 2003).

The mechanisms of perception and transduction of the bacterial symbiotic signal in actinorhizal plants are poorly known. Genetic studies in model legumes have revealed that legume-rhizobial and the more ancient arbuscular mycorhizal (AM) symbioses (Part III) share part of their genetic programme leading to endosymbiosis including part of the symbiotic signal transduction pathway. This is in accordance with studies that suggest that the legume–rhizobial symbiosis would be derived from the more ancient AM symbiosis (Parniske 2008). Homologues of the genes involved in the common symbiotic transduction pathway have been identified in EST databases from the actinorhizal plants *C. glauca* and *A. glutinosa* (Hocher et al. 2011a, b). Transcriptome analyses showed that some of these genes are more expressed in actinorhizal nodules compared to non-inoculated roots (Hocher et al. 2011a). The receptor kinase SYMRK is part of the common Nod and Myc signalling pathway. A homologue was recently isolated from two actinorhizal species *C. glauca* and *Datisca glomerata* (Gherbi et al. 2008a; Markmann et al. 2008). Knockdown of *SYMRK* by RNA interference led to inhibition of nodulation and mycorrhisation in these plants when inoculated with compatible *Frankia* bacteria and AM fungi, respectively. Moreover, *C. glauca SYMRK* complemented the *Lotus japonicus ljsymrk* mutant for both nodulation and mycorrhisation (Gherbi et al. 2008a; Markmann et al. 2008). This demonstrated that *SYMRK* is a common signalling element shared between AM, legume–rhizobia and actinorhizal symbioses, supporting the hypothesis that the capacity to accommodate N_2-fixing bacteria evolved at least partly from the more ancient AM genetic programme. Interestingly, complementation of *ljsymrk* using *SYMRK* genes isolated from nodulating and non-nodulating species showed that all the genes tested were able to complement the lack of mycorrhisation but only genes with three LRRs (found in *Tropaeolum majus* and all nodulating plants) were able to complement the nodulation defect (Markmann et al. 2008). The appearance of an additional LRR motif in SYMRK might therefore be one of the evolutionary events that led to nitrogen-fixing RNS apparition in the Fabid clade.

In conclusion, while chemical signalling between *Frankia* and its host plants seems to involve molecules different from Nod factors, the same common symbiotic transduction pathway was recycled from the more ancient AM symbiosis to perceive symbiotic signals in legume–rhizobial and actinorhizal symbioses.

5.3 Infection

After pre-infection events, *Frankia* can enter the plant root either intracellularly via root hairs or intercellularly via the middle lamellas of cell epidermal (Obertello et al. 2003; Wall and Berry 2008). The same *Frankia* strains can induce both types of infection in different host plants indicating that the type of infection is controlled by the host plant (Racette and Torrey 1989; Pawlowski and Demchenko 2012).

5.3.1 Intracellular Infection

Intracellular infection is found in the *Myrica*, *Comptonia*, *Alnus* and *Casuarina* genera (Berry and Sunnel 1990). *Frankia* bacteria secrete factors that induce root hair deformation (Fig. 5.1a; Callaham and Torrey 1977). Studies carried out by Torrey (1976) demonstrate that all root hairs are deformed during *Casuarina* infection while just some root hairs are deformed in *Comptonia* (Callaham et al. 1979). Sugar-binding lectins produced by *Frankia* might help the bacteria to bind the root hairs in some actinorhizal plants such as *A. glutinosa* (Pujic et al. 2012). *Frankia* hyphae become entrapped by plant cell polysaccharides at the tip of some deformed root hairs and a local hydrolysis of primary cell occurs at the site of *Frankia* penetration (Berry et al. 1986). Surprisingly, recent analyses of *Frankia* genomes did not find any conserved secreted polysaccharide-degrading enzymes that might be responsible for this degradation (Mastronunzio et al. 2008). This might suggest that *Frankia* secretes effector-like molecules to communicate with its host and trigger the local loosening to the cell wall necessary for infection to occur (Mastronunzio et al. 2008). In some deformed root hair, the plasma membrane invaginates and forms an infection thread structure (Fig. 5.1b). Within this structure, growing *Frankia* hyphae are encapsulated by a cell wall-like matrix made of xylan, hemicellulose, cellulose and pectin (Berg 1990).

Root hair deformation occurs 24–28 h after inoculation. However, only growing root hairs are infected by *Frankia* (Callaham et al. 1979). In these infected root

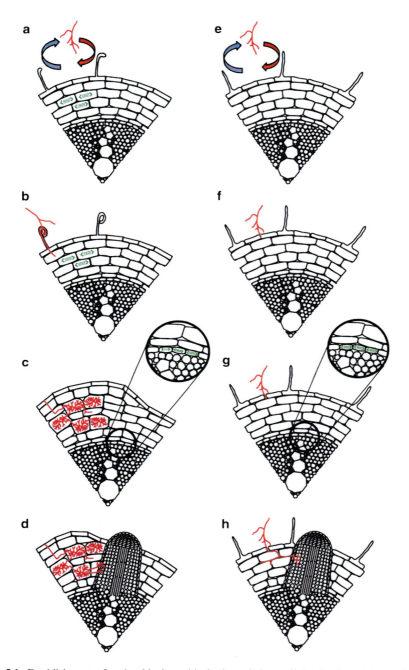

Fig. 5.1 Establishment of actinorhizal symbiosis through intracellular (**a–d**) and intercellular infection (**e–h**). *Frankia* hyphae are shown in *red* and dividing cells in *green*. (**a, e**). Exchange of symbiotic signals between the two partners with, only in (**a**), deformation of root hairs and metabolic modifications in cortical cells close to the site of perception of the bacterial signal. (**b**) Penetration of *Frankia* within a curled root hair and initiation of division in cortical cells.

hairs, a high metabolic activity is observed (Berry et al. 1986; Berry and Sunnel 1990). Simultaneously with infection, cell divisions occur in cortical cells adjacent to the infected root hair inducing the formation of a protuberance called the prenodule (Fig. 5.1b). Infected threads grow towards the prenodule and invade some of its cells that become hypertrophied and both the plant cell and bacteria differentiate to fix nitrogen (Fig. 5.1c; Laplaze et al. 2000a). While the prenodule is an obligatory step of the infection process, it is not the precursor of a nodule lobe. As the prenodule develops, cell divisions occur in the pericycle opposite to a protoxylem pole giving rise to a nodule primordium (Fig. 5.1c). The nodule primordium develops into a nodule lobe that is infected by *Frankia* hyphae coming from the prenodule (Fig. 5.1d).

A gene encoding a protease of the subtilase family called *Cg12* or *Ag12* in *C. glauca* and *A. glutinosa,* respectively, is specifically expressed in plant cells infected by *Frankia* (Laplaze et al. 2000b; Ribeiro et al. 1995; Svistoonoff et al. 2003). No expression was found during intracellular symbiosis with the AM fungus *Glomus intraradices* (Svistoonoff et al. 2003). It has been proposed that these proteases might be involved in cell wall remodelling or the processing of peptidic signals during symbiotic infection. A recent comparative transcriptome analysis of genes induced during actinorhizal, rhizobial and AM symbioses indicates that protease-encoding genes are among the core genes that are induced in all three endosymbioses (Tromas et al. 2012). This suggests that proteases are important component for setting endosymbioses and that some of the genes encoding proteases involved in the ancient AM symbiosis have been recycled to form RNS. Interestingly, the infection-specific induction of a *ProCG12::GUS* construct was retained in the model legume *M. truncatula* (Svistoonoff et al. 2004) indicating that gene regulation during infection in legume–rhizobial and actinorhizal symbioses might use conserved regulators.

Several studies suggest a role of the phytohormone auxin during infection by *Frankia*. A gene named *CgAUX1* encoding a functional auxin influx carrier is expressed in plant cells infected by *Frankia* but not by the AM fungi *G. intraradices* in *C. glauca* (Péret et al. 2007, 2008). Inhibition of auxin influx using 1-naphtoxy acetic acid (1-NOA) delays actinorhizal nodule formation and leads to the formation of small nodules in *C. glauca* (Péret et al. 2007). *Frankia* produces auxins, indole-3-acetic acid (IAA) and phenylacetic acid (PAA), in vitro (Wheeler and Henson 1979; Hammad et al. 2003; Perrine-Walker et al. 2010). Recent immunolocalisation experiments showed specific accumulation of both IAA and PAA in plant cells infected by *Frankia* in *C. glauca* (Perrine-Walker et al. 2010). Gene expression, immunolocalisation and modelling experiments suggest that this specific accumulation is due to auxin production by *Frankia in planta*

Fig. 5.1 (continued) (**f**) Penetration of *Frankia* in between epidermal cells. (**c**) Branching of *Frankia* hyphae within the prenodule and divisions of pericyle cells at the site of initiation of the nodule, in front of a xylem pole. (**g**) Initiation of the nodule without formation of a prenodule. (**d, h**) emergence of the nodule colonised by *Frankia*

and the specific localisation of auxin influx and efflux carriers in *C. glauca* nodules (Perrine-Walker et al. 2010). Altogether, these studies link symbiotic infection to auxin accumulation in *C. glauca*. However, we do not know if this is a common feature of actinorhizal symbioses. Moreover, the role, if any, of auxin in those cells infected by *Frankia* is still unknown. We are currently addressing this question by inhibiting auxin signalling specifically during *Frankia* infection in *C. glauca* (Laplaze, unpublished data).

5.3.2 *Intercellular Infection*

Intercellular infection occurs in some actinorhizal plant genera such as *Elæagnus*, *Ceanothus*, *Cercocarpus*, *Hippophae*, *Shepherdia* and *Discaria* (Miller and Baker 1985; Berry and Sunnel 1990; Valverde and Wall 1999; Imanishi et al. 2011). During intercellular infection, some signal exchange must occur between the two partners but no root hair deformation is observed (Fig. 5.1e). Instead, *Frankia* enters through the middle lamella between adjacent epidermal cells (Fig. 5.1f) and then progresses intercellularly in the root cortex (Fig. 5.1g). As in intracellular infection, this is associated with pectolytic activity that might be of plant rather than bacterial origin (Mastronunzio et al. 2008). During intercellular infection, prenodule formation has not been reported. However, some cortical cell divisions occur in *Ceanothus* but these new cells are not infected by *Frankia* (Berry and Sunnel 1990). Nodule primordium formation occurs through cell divisions in the pericycle in front of a xylem pole. The nodule lobe primordium is then colonised by intercellular hyphae (Fig. 5.1g). *Frankia* hyphae become intracellular when they invade cortical cells of the young nodule primordium.

5.4 Actinorhizal Nodule Formation

Actinorhizal nodule lobes are formed from cell divisions occurring in the pericycle in front of a xylem pole. New nodule lobes are formed by branching, giving rise to a coralloid actinorhizal nodule formed of multiple lobes. Each lobe contains a meristem at its apex, a central vascular bundle and a periderm. In the nodule, four zones have been defined: (1) the meristematic zone, (2) the infection zone, (3) the fixation zone and (4) the senescence zone. The *meristematic zone* is localised at the apex and produces new cells responsible for the indeterminate growth of actinorhizal nodules. The *infected zone* is adjacent to the apical zone. In this zone, *Frankia* hyphae infect some of the new cells. The *fixation zone* is composed of infected and uninfected cells. Infected cells are filled with *Frankia* hyphae and hypertrophied. Vesicles differentiate and nitrogen fixation occurs. Assimilation of the fixed N probably occurs in uninfected cells (Wall 2000). The *senescence zone* is localised at the base of old nodules. In this zone, the host cytoplasm and the bacteria degenerate.

Because of its origin, i.e. cells divisions in the pericycle in front of xylem poles and its structure, the actinorhizal nodule lobe has been considered as a modified lateral root (Pawlowski and Bisseling 1996). Moreover, in some actinorhizal plants such as *C. glauca*, a structure called nodular root is formed. The nodular root is a very specialised root showing negative geotropism (growing upward) and cortical aerenchyma and lacks a root cap and root hairs. It has been suggested that it plays an important role under flooding or waterlogged conditions by increasing gas exchange between the nodule and the atmosphere (Silvester et al. 1990). Furthermore, study carried out by Schwintzer et al. (1982) shows a correlation between the external oxygen tension and the length of the modified root. The formation of lenticels is noted at the nodule periderm of certain actinorhizal genera (*Alnus*, *Coriaria* and *Datisca*). This structure is also involved in nodule aeration (Silvester et al. 1990). The presence of both structures, lenticel and root nodule in some Datisca species, has been reported by Pawlowski et al. (2007).

5.5 Conclusions

Nitrogen availability is one of the major limiting factors of crop production worldwide. The high price of nitrogen fertilisers and their environmental impact has recently renewed the interest of the plant science community and funding charities for research on the transfer of biological nitrogen fixation to crops such as cereals (Den Herder et al. 2010; Beatty and Good 2011). Two groups of plants evolved the ability to form nitrogen-fixing RNS: the rhizobial symbiosis is restricted to the *Fabaceae* with the notable exception of the genus *Parasponia* (*Cannabaceae* family), while actinorhizal symbioses occur in eight angiosperm families. Interestingly, molecular phylogenies studies showed that (1) all these plants belong to a single N_2-fixing clade (Soltis et al. 1995) and (2) that nodulation appeared several times independently in both groups (Doyle 2011). This led to the suggestion that the common ancestor of the N_2-fixing clade had a yet unknown genetic predisposition to form RNS. Recent results have shown that actinorhizal and rhizobial symbioses rely on similar molecular components that were recycled from the ancient and widespread AM symbiosis (reviewed in Geurts et al. 2012). More studies on actinorhizal and rhizobial symbioses (including the atypical *Parasponia–Rhizobium* symbiosis) are needed to understand how during evolution RNS appeared in the nitrogen-fixing clade. Moreover, because actinorhizal (and *Parasponia*) nodules are simpler structure (modified lateral roots) that appeared independently in different plant families, it might represent a better model for the transfer of nodulation to other plants. New strategies are being developed to characterise the molecular mechanisms of actinorhizal symbioses formation and functioning. This should shed a new light on the evolution of root endosymbioses and hopefully pave the way to the transfer of nitrogen fixation to important crops.

References

Abdel-Lateif K, Bogusz D, Hocher V (2012) The role of flavonoids in the establishment of plant roots endosymbioses with arbuscular mycorrhiza fungi, rhizobia and *Frankia* bacteria. Plant Signal Behav 7(6):636–641

Alloisio N, Queiroux C, Fournier P, Pujic P, Normand P, Vallenet D, Médigue C, Yamaura M, Kakoi K, Kucho K (2010) The *Frankia alni* symbiotic transcriptome. Mol Plant Microbe Interact 23:593–607

Baker DD, Mullin BC (1992) Actinorhizal symbioses. In: Stacey G, Burris RH, Evans HJ (eds) Biological nitrogen fixation. Chapman and Hall, New York, NY, pp 259–292

Baker DD, Schwintzer CR (1990) Introduction. In: Schwintzer CR, Tjepkema JD (eds) The biology of *Frankia* and Actinorhizal plants. Academic, New York, NY, pp 157–176

Beatty PH, Good AG (2011) Plant science. Future prospects for cereals that fix nitrogen. Science 333:416–417

Beauchemin NJ, Furnholm T, Lavenus J, Svistoonoff S, Doumas P, Bogusz D, Laplaze L, Tisa LS (2012) Casuarina root exudates alter the physiology, surface properties, and plant infectivity of *Frankia* sp. strain CcI3. Appl Environ Microbiol 78:575–580

Berg RH (1990) Cellulose and xylans in the interface capsule in symbiotic cells of actinorhizae. Protoplasma 159:35–43

Berry AL, Sunnel LA (1990) The infection process and nodule development. In: Schwintzer CR, Tjepkema JD (eds) The biology of *Frankia* and actinorhizal plants. Academic, New York, NY, pp 61–81

Berry AM, McIntyre L, McCully M (1986) Fine structure of root hair infection leading to nodulation in the *Frankia-Alnus* symbiosis. Can J Bot 64:292–305

Bickhart DM, Gogarten JP, Lapierre P, Tisa LS, Normand P, Benson DR (2009) Insertion sequence content reflects genome plasticity in strains of the root nodule actinobacterium *Frankia*. BMC Genomics 10:468

Callaham D, Torrey JG (1977) Prenodule formation and primary nodule development in roots of *Comptonia* (Myricaceae). Can J Bot 55:2306–2318

Callaham D, Newcomb W, Torrey JG, Peterson RL (1979) Root hair infection in actinomycete-induced root nodule initiation in *Casuarina*, *Myrica* and *Comptonia*. Bot Gaz 140:S1–S9

Cérémonie H, Debellé F, Fernandez MP (1999) Structural and functional comparison of *Frankia* root hair deforming factor and rhizobia Nod factor. Can J Bot 77:1293–1301

Den Herder G, Van Isterdael G, Beeckman T, De Smet I (2010) The roots of a new green revolution. Trends Plant Sci 15:600–607

Diem HG, Dommergues YR (1990) Current and potential uses and management of Casuarinaceae in the tropics and subtropics. In: Schwintzer CR, Tjepkema JD (eds) The biology of *Frankia* and actinorhizal plants. Academic, San Diego, CA, pp 317–342

Diouf D, Gherbi H, Prin Y, Franche C, Duhoux E, Bogusz D (1995) Hairy root nodulation of *Casuarina glauca*: a system for the study of symbiotic gene expression in an actinorhizal tree. Mol Plant Microbe Interact 8:532–537

Doyle JJ (2011) Phylogenetic perspectives on the origins of nodulation. Mol Plant Microbe Interact 24:1289–1295

Ferguson BJ, Indrasumunar A, Hayashi S, Lin M-H, Lin Y-H, Reid DE, Gresshoff PM (2010) Molecular analysis of legume nodule development and autoregulation. J Integr Plant Biol 52:61–76

Franche C, Diouf D, Le QV, Bogusz D, N'Diaye A, Gherbi H, Gobé C, Duhoux E (1997) Genetic transformation of the actinorhizal tree *Allocasuarina verticillata* by *Agrobacterium tumefaciens*. Plant J 11:897–904

Geurts R, Lillo A, Bisseling T (2012) Exploiting an ancient signalling machinery to enjoy a nitrogen fixing symbiosis. Curr Opin Plant Biol 15:1–6

Gherbi H, Markmann K, Svistoonoff S, Estevan J, Autran D, Giczey G, Auguy F, Péret B, Laplaze L, Franche C et al (2008a) SymRK defines a common genetic basis for plant root endosymbioses with arbuscular mycorrhiza fungi, rhizobia, and *Frankia* bacteria. Proc Natl Acad Sci 105:4928–4932

Gherbi H, Nambiar-Veetil M, Chonglu Z, Félix J, Autran D, Girardin R, Vaissayre V, Auguy F, Bogusz D, Franche C (2008b) Post-transcriptional gene silencing in the root system of the actinorhizal tree *Allocasuarina verticillata*. Mol Plant Microbe Interact 21:518–524

Hammad Y, Nalin R, Marechal J, Fiasson K, Pepin R, Berry AM, Normand P, Domenach A-M (2003) A possible role for phenyl acetic acid (PAA) on *Alnus glutinosa* nodulation by *Frankia*. Plant Soil 254:193–205

Hassan S, Mathesius U (2012) The role of flavonoids in root–rhizosphere signalling: opportunities and challenges for improving plant–microbe interactions. J Exp Bot 63(9):3429–3444

Hocher V, Auguy F, Argout X, Laplaze L, Franche C, Bogusz D (2006) Expressed sequence-tag analysis in *Casuarina glauca* actinorhizal nodule and root. New Phytol 169:681–688

Hocher V, Alloisio N, Florence A, Fournier P, Doumas P, Pujic P, Gherbi H, Queiroux C, Da Silva C, Wincker P et al (2011a) Transcriptomics of actinorhizal symbioses reveals homologs of the whole common symbiotic signaling cascade. Plant Physiol 156:700–711

Hocher V, Alloisio N, Bogusz D, Normand P (2011b) Early signaling in actinorhizal symbioses. Plant Signal Behav 6:1377–1379

Imanishi L, Vayssières A, Franche C, Bogusz D, Wall L, Svistoonoff S (2011) Transformed hairy roots of the actinorhizal shrub *Discaria trinervis*: a valuable tool for studying actinorhizal symbiosis in the context of intercellular infection. BMC Proc 5:P85

Jones KM, Kobayashi H, Davies BW, Taga ME, Walker GC (2007) How rhizobial symbionts invade plants: the *Sinorhizobium-Medicago* model. Nat Rev Microbiol 5:619–633

Laplaze L, Gherbi H, Frutz T, Pawlowski K, Franche C, Macheix JJ, Auguy F, Bogusz D, Duhoux E (1999) Flavan-containing cells delimit *Frankia*-infected compartments in *Casuarina glauca* nodules. Plant Physiol 121:113–122

Laplaze L, Duhoux E, Franche C, Frutz T, Svistoonoff S, Bisseling T, Bogusz D, Pawlowski K (2000a) *Casuarina glauca* prenodule cells display the same differentiation as the corresponding nodule cells. Mol Plant Microbe Interact 13:107–112

Laplaze L, Ribeiro A, Franche C, Duhoux E, Auguy F, Bogusz D, Pawlowski K (2000b) Characterization of a *Casuarina glauca* nodule-specific subtilisin-like protease gene, a homolog of *Alnus glutinosa* ag12. Mol Plant Microbe Interact 13:113–117

Markmann K, Gábor G, Parniske M (2008) Functional adaptation of a plant receptor-kinase paved the way for the evolution of intracellular root symbioses with bacteria. PLoS Biol 6:e68

Mastronunzio JE, Tisa LS, Normand P, Benson DR (2008) Comparative secretome analysis suggests low plant cell wall degrading capacity in *Frankia* symbionts. BMC Genomics 9:47

Miller IM, Baker DD (1985) The initiation, development and structure of root nodules in *Elaeagnus angustifolia* L. (Elaeagnaceae). Protoplasma 128:107–119

Normand P, Lapierre P, Tisa LS, Gogarten Johann P, Alloisio N, Bagnarol E, Bassi CA, Berry AM, Bickhart DM, Choisne N et al (2007) Genome characteristics of facultatively symbiotic *Frankia* sp. strains reflect host range and host plant biogeography. Genome Res 17:7–15

Obertello M, Sy MO, Laplaze L et al (2003) Actinorhizal nitrogen fixing nodules: infection process, molecular biology and genomics. Afr J Biotechnol 2:528–538

Parniske M (2008) Arbuscular mycorrhiza: the mother of plant root endosymbioses. Nat Rev Microbiol 6:763–775

Pawlowski K, Bisseling T (1996) Rhizobial and actinorhizal symbioses: what are the shared features? Plant Cell 8:1899

Pawlowski K, Demchenko KN (2012) The diversity of actinorhizal symbiosis. Protoplasma 249(4):967–979

Pawlowski K, Sprent JI (2008) Comparison between actinorhizal and legume symbiosis. In: Pawlowski K, Sprent JI, Pawlowski K, Newton WE, Dilworth MJ, James EK, Sprent Janet I, Newton WE (eds) Nitrogen fixation: origins, applications, and research progress, Nitrogen-fixing actinorhizal symbioses. Springer, Dordrecht, pp 261–288

Pawlowski K, Jacobsen KR, Alloisio N, Ford Denison R, Klein M, Tjepkema JD, Winzer T, Sirrenberg A, Guan C, Berry AM (2007) Truncated hemoglobins in actinorhizal nodules of *Datisca glomerata*. Plant Biol (Stuttg) 9:776–785

Péret B, Swarup R, Jansen L, Devos G, Auguy F, Collin M, Santi C, Hocher V, Franche C, Bogusz D et al (2007) Auxin influx activity is associated with *Frankia* infection during actinorhizal nodule formation in *Casuarina glauca*. Plant Physiol 144:1852–1862

Péret B, Svistoonoff S, Lahouze B, Auguy F, Santi C, Doumas P, Laplaze L (2008) A role for auxin during actinorhizal symbioses formation? Plant Signal Behav 3:34–35

Péret B, Svistoonoff S, Laplaze L (2009) When plants socialize: symbioses and root development. In: Beeckman T (ed) Root development, vol 37, Annual plant reviews. Wiley-Blackwell, Hoboken, NJ, pp 209–238

Perrine-Walker F, Doumas P, Lucas M, Vaissayre V, Beauchemin NJ, Band LR, Chopard J, Crabos A, Conejero G, Péret B et al (2010) Auxin carriers localization drives auxin accumulation in plant cells infected by *Frankia* in *Casuarina glauca* actinorhizal nodules. Plant Physiol 154:1372–1380

Perrine-Walker F, Gherbi H, Imanishi L, Hocher V, Ghodhbane-Gtari F, Lavenus J, Benabdoun FM, Nambiar-Veeti M, Svistoonoff S, Laplaze L (2011) Symbiotic signaling in actinorhizal symbioses. Curr Protein Pept Sci 12:156–164

Persson T, Benson DR, Normand P, Vanden Heuvel B, Pujic P, Chertkov O, Teshima H, Bruce DC, Detter C, Tapia R et al (2011) Genome sequence of 'Candidatus Frankia datiscae' Dg1, the uncultured microsymbiont from nitrogen-fixing root nodules of the Dicot *Datisca glomerata*. J Bacteriol 193:7017–7018

Pujic P, Fournier P, Alloisio N, Hay A-E, Maréchal J, Anchisi S, Normand P (2012) Lectin genes in the *Frankia alni* genome. Arch Microbiol 194:47–56

Popovici J, Comte G, Bagnarol E, Alloisio N, Fournier P, Bellvert F, Bertrand C, Fernandez MP (2010) Differential effects of rare specific flavonoids on compatible and incompatible strains in the *Myrica gale-Frankia* actinorhizal symbiosis. Appl Environ Microbiol 76(8):2451–2460

Pueppke SG, Broughton WJ (1999) *Rhizobium* sp. strain NGR234 and *R. fredii* USDA257 share exceptionally broad, nested host ranges. Mol Plant Microbe Interact 12:293–318

Racette S, Torrey JG (1989) Root nodule initiation in *Gymnostoma* (Casuarinaceae) and *Shepherdia* (Elaeagnaceae) induced by *Frankia* strain HFPGpI1. Can J Bot 67:2873–2879

Ribeiro A, Akkermans AD, van Kammen A, Bisseling T, Pawlowski K (1995) A nodule-specific gene encoding a subtilisin-like protease is expressed in early stages of actinorhizal nodule development. Plant Cell 7:785–794

Schwintzer CR, Berry AM, Disney LD (1982) Seasonal patterns of root nodule growth, endophyte morphology, nitrogenase activity, and shoot development in *Myrica-Gale*. Can J Bot 60:746–757

Silvester WB, Harris SL, Tjepkema JD (1990) Oxygen regulation and hemoglobin. In: Schwintzer CR, Tjepkema JD (eds) The biology of *Frankia* and actinorhizal plants. Academic, San Diego, CA, pp 157–176

Soltis DE, Soltis PS, Morgan DR, Swensen SM, Mullin BC, Dowd JM, Martin PG (1995) Chloroplast gene sequence data suggest a single origin of the predisposition for symbiotic nitrogen fixation in angiosperms. Proc Natl Acad Sci USA 92:2647–2651

Svistoonoff S, Laplaze L, Auguy F, Runions J, Duponnois R, Haseloff J, Franche C, Bogusz D (2003) cg12 expression is specifically linked to infection of root hairs and cortical cells during *Casuarina glauca* and *Allocasuarina verticillata* actinorhizal nodule development. Mol Plant Microbe Interact 16:600–607

Svistoonoff S, Laplaze L, Liang J, Ribeiro A, Gouveia MC, Auguy F, Fevereiro P, Franche C, Bogusz D (2004) Infection-related activation of the cg12 promoter is conserved between actinorhizal and legume-rhizobia root nodule symbiosis. Plant Physiol 136:3191–3197

Swensen SM (1996) The evolution of actinorhizal symbioses: evidence for multiple origins of the symbiotic association. Am J Bot 83:1503–1512

Torrey JG (1976) Initiation and development of root nodules of *Casuarina* (*Casuarinaceae*). Am J Bot 63:335–345

Tromas A, Parizot B, Diagne N, Champion A, Hocher V, Cissoko M, Crabos A, Prodjinoto H, Lahouze B, Bogusz D, Laplaze L, Svistoonoff S (2012) Heart of endosymbioses: transcriptomics reveals a conserved genetic program among arbuscular mycorrhizal, actinorhizal and legume-rhizobial symbioses. PLoS One 7:e44742

Udwary DW, Gontang EA, Jones AC, Jones CS, Schultz AW, Winter JM, Yang JY, Beauchemin N, Capson TL, Clark BR et al (2011) Comparative genomic and proteomic analysis of the actinorhizal symbiont *Frankia* reveals significant natural product biosynthetic potential. Appl Environ Microbiol 77(110):3617–3625

Valverde C, Wall LG (1999) Time course of nodule development in the *Discaria trinervis* (Rhamnaceae)—*Frankia* symbiosis. New Phytol 141:345–354

Wall LG (2000) The actinorhizal symbiosis. J Plant Growth Regul 19:167–182

Wall LG, Berry AM (2008) Early interactions, infection and nodulation in actinorhizal symbiosis. In: Pawlowski K, Newton WE, Dilworth MJ, James EK, Sprent Janet I, Newton WE (eds) Nitrogen fixation: origins, applications, and research progress, Nitrogen-fixing actinorhizal symbioses. Springer, Dordrecht, pp 147–166

Wheeler CT, Henson IE (1979) Hormones in plants bearing actinomycete nodules. Bot Gaz 140:52–57

Wheeler C, Crozier A, Sandberg G (1984) The biosynthesis of indole-3-acetic acid by *Frankia*. Plant Soil 78:99–104

Zhong C, Pinyopusarek K, Kalinganire A, Franche C (2011) Improving smallholder livelihoods through improved casuarina productivity: proceedings of the 4th international casuarina workshop. China Forestry Publishing House, Haikou

Chapter 6
Abiotic Factors Influencing Nitrogen-Fixing Actinorhizal Symbioses

Hiroyuki Tobita, Ken-ichi Kucho, and Takashi Yamanaka

6.1 Introduction

6.1.1 Frankia

The genus *Frankia* is a gram-positive soil bacterium that forms N_2-fixing symbioses in the root nodules of angiosperm species in 25 genera distributed among eight families and known collectively as actinorhizal plants (Huss-Danell 1997). *Frankia* is an actinomycete that grows as multicellular hyphae and develops spores, but it has distinctive abilities to form symbioses with plants and fix atmospheric N_2. This symbiotic N_2-fixation capability is analogous to legume symbiont rhizobia (comprising mostly gram-negative α-proteobacteria), but *Frankia* is phylogenetically distinct from the bacterial partners in legume. Genome analysis shows that *Frankia* strains do not have canonical *nod* genes found in rhizobia (Normand et al. 2007), indicating different molecular bases of symbiosis between these two groups of bacteria. *Frankia* differentiates spherical cells called "vesicles" that are specialized for N_2-fixation. The vesicles are surrounded by multiple layers of hopanoid lipids that function as barriers to oxygen diffusion that provide protection for an oxygen-labile nitrogenase (Berry et al. 1993). Two different modes of infection occur in actinorhizal symbioses (Huss-Danell 1997); *Frankia* invades host plants either intracellularly via root hairs or intercellularly via gaps between the epidermal cells of roots. Root hairs deform when contacted by *Frankia* cells during intracellular infection. The infection thread forms through a

---

H. Tobita (✉) • T. Yamanaka
Forestry and Forest Products Research Institute, Tsukuba 305-8687, Japan
e-mail: tobi@ffpri.affrc.go.jp; yamanaka@affrc.go.jp

K.-i. Kucho
Department of Chemistry and Bioscience, Graduate School of Science and Engineering, Kagoshima University, Korimoto 1-21-35, Kagoshima 890-0065, Japan
e-mail: kkucho@sci.kagoshima-u.ac.jp

root hair acting as an entry route for *Frankia* hyphae that invade plant cells. Cell division is subsequently initiated at the root cortex, leading to formation of a prenodule. A genuine nodule develops from the pericycle; it accepts *Frankia* hyphae from the prenodule. Neither root hair deformation nor prenodule formation occurs when infection is intercellular.

6.1.2 Actinorhizal Plants

There are eight angiosperm families in symbioses with *Frankia* that are known to fix atmospheric N_2 in root nodules. Three families (Casuarinaceae, Elaeagnaceae, Coriariaceae) among the eight are completely actinorhizal, but the other five are not (Swensen and Benson 2008). For example, *Alnus* is the only genus of actinorhizal plants among the Betulaceae. Currently, >200 plant species are known to be actinorhizal (Gardner and Barrueco 1999). Actinorhizal species are distributed on every continent except Antarctica. North and South America have the greatest diversity of native actinorhizal plants (Baker and Schwintzer 1990). Actinorhizal plants occur primarily in the temperate zone, but the Casuarinaceae and Myricaceae are truly tropical. Although some species of *Alnus* and *Elaeagnus* occur in tropical latitudes, they are confined to high mountainous regions where the climate is actually temperate. Most actinorhizal plants are found in nitrogen (N)-poor soils, such as in sites disturbed by natural disasters, sand dunes, gravel, and along the shores of streams and lakes. The success of these pioneer plants as colonizers of diverse soil types is a consequence of N_2-fixation by their microsymbiont partner *Frankia*. Actinorhizal plants, such as *A. japonica* and *Myrica gale* var. *tomentosa* in Japan, are also able to grow in wetlands (Yamanaka and Okabe 2008), suggesting a tolerance to anaerobiosis. Actinorhizal plants have been used as timber, fuel wood, vegetation recovery, and for amenity planting. Alder species have been used for timber in Japan's northern region, for land stabilization in watersheds, and for vegetation recovery after volcanic eruptions or landslides (Yamanaka and Okabe 2008). *Casuarina* species were introduced as windbreaks and for stabilizing dunes against wind erosion on some southeastern Japanese islands and on the Bonin Islands. *Elaeagnus* and *M. rubra* have been used for amenity planting in parks and along roadsides. *M. rubra* is cultivated for its fruits in Japan's southern region; *Hippophae rhamnoides* has been introduced into the country's northern region (Ishii 2003).

A phylogenetic study using chloroplast-gene sequence data (*rbcL*) showed that all N_2-fixing and nodulated plants (actinorhizal/leguminous plants) are in a single clade interspersed with non-nodulated plants, suggesting that N_2-fixing nodules evolved only once during the evolution of angiosperms (Doyle 1998). However, later analyses demonstrated the existence of plant clades that diverged earlier than the *Frankia* clade, suggesting that *Frankia*–actinorhizal symbioses evolved independently at least three or four times (Swensen 1996; Jeong et al. 1999).

6.2 Free-Living Growth, Nodulation, and N_2-Fixation by *Frankia*

Numerous abiotic factors affect the free-living and symbiotic properties of *Frankia*, including soil depth, drought, moisture, aeration, temperature, pH, organic matter, and inorganic chemicals; excellent reviews have summarized these effects (Huss-Danell 1997; Dawson 2008; Valdes 2008). This section focuses mainly on details of the effects of inorganic chemicals. Some of these chemicals act positively on growth and symbiotic activity of *Frankia*, while others act negatively. Importantly, chemicals with positive effects act negatively at high doses. A cross talk of effects between positive and negative agents may occur, for example, between phosphorus (P) and N.

6.2.1 Iron (Fe)

Fe and molybdenum (Mo) are essential cofactors for nitrogenase, and their availability limits N_2-fixing activity. Although Fe is abundant in soil, availability to organisms is made difficult under aerobic conditions at biological pH by the presence of Fe in the form of insoluble Fe oxyhydroxide polymers (Singh et al. 2008). Fe in an insoluble form is taken up into cells as a complex chelated with siderophores. *Frankia* strains isolated from *Hippophae salicifolia* produce hydroxamate- and catecholate-type siderophores (Singh et al. 2008). The level of siderophore production increases as the concentration of Fe decreases or the concentration of ethylenediaminetetraacetic acid (EDTA) increases, suggesting that siderophore production is regulated in response to the availability of metals.

6.2.2 Nickel (Ni)

Ni is a cofactor of uptake hydrogenase. This enzyme is used to recover the reducing power from hydrogen (H_2), an inevitable byproduct of N_2-fixation. Growth of *Frankia* in N-free medium is stimulated by addition of up to 2.25 mM Ni (Wheeler et al. 2001), which suggests that additional Ni enhances the activity of uptake hydrogenase, thus stimulating N_2-fixation (Mattsson and Sellstedt 2002). The growth and nodulation of the host plant *Alnus glutinosa* are more sensitive to Ni than *Frankia* (Wheeler et al. 2001). *Alnus* plants treated with 0.5 mM of Ni develop severe chlorosis and reduced growth, nodulation, and nitrogenase activity (NA). Interestingly, addition of 0.025 mM Ni increases NA in comparison with a concentration of 0.12 µM Ni; this difference between effects at different Ni concentrations may be due to stimulated uptake hydrogenase activity. Ni accumulates in both roots and nodules.

6.2.3 Aluminum (Al)

Al, which is solubilized at acidic pH values, generally has inhibitory effects on plant growth. Igual et al. (1997) demonstrated that this is not always the case for the actinorhizal plant *Casuarina cunninghamiana*. Although increasing the concentration of Al reduces the number of nodulated plants, nodulation does occur in the presence of Al at a concentration of 880 µM. Surprisingly, NA is highest at an intermediate concentration of Al (220 µM), which produces better plant growth than control conditions without Al. Consistent with these effects, the free-living growth of *Frankia* strains is stimulated by addition of 125–500 µM Al (Igual and Dawson 1999). Al may ameliorate the inhibitory effect of low pH on growth by modifying the electrical potential of the cell surface.

6.2.4 Boron (B)

B is widely recognized as a micronutrient essential for higher plants, but it also affects the physiology of *Frankia*. Bolaños et al. (2002) tested the *Discaria trinervis–Frankia* sp. strain BCU110501 symbiosis for B requirement and found that nodulation is reduced when the plant or bacterium is cultivated without B. Depletion of B inhibits the free-living growth of this *Frankia* strain in both the presence and absence of a N source, arresting growth completely when N is absent. B-deficient *Frankia* cells have a highly disorganized structure with weakly packed cell walls and almost empty cytoplasm. N_2-fixation activity is not detectable in such cells because nitrogenase expression is inhibited and vesicle structure is impaired. These experiments indicate that B is an important nutrient for both nodulation and N_2-fixation.

6.2.5 Sodium Chloride (Salinity, NaCl)

Some actinorhizal plants, such as *Casuarina* spp., are salinity tolerant and often planted in coastal areas as components of windbreaks. Tani and Sasakawa (2003) reported that *C. equisetifolia* plants are nodulated under salinities as high as 300 mM NaCl, although NA is severely reduced. The growth of a *Frankia* strain Ceq1 isolated from this plant species is suppressed almost linearly with increasing NaCl concentration, but growth still occurs at 500 mM NaCl. Hyphae are morphologically changed under high salinity, e.g., they become thickened and shortened. The intracellular concentration of NaCl is significantly lower than that in the medium, which suggests the existence of mechanisms that function to reduce influx of salt into *Frankia* cells. In contrast, *Frankia* isolated from a less salt-tolerant plant species (*Elaeagnus macrophylla*) is less tolerant to NaCl in both its nodulation and growth processes (Tani and Sasakawa 2000). The simple prediction that *Frankia* strains from salt-tolerant plants will also be tolerant to high salinity is not, however, always true; salt tolerance is very variable among *Frankia* strains, e.g., even when the strains are isolated from the same plant species (Hafeez et al. 1999).

6.2.6 Zinc (Zn)

As actinorhizal plants are often used to restore fertility in polluted soils like mine tailings, the response of *Frankia* to heavy metals is a pressing issue for investigation. Cusato et al. (2007) reported the effect of Zn on nodulation of *Discaria americana*. Concentrations of 500–1,500 mg kg^{-1} Zn in dry soil delayed both shoot and root growth and the development of nodules, but ultimately plants grew to sizes similar to controls without Zn. Hence, *D. americana* is likely resistant to high concentrations of Zn. *D. americana* plants absorb Zn, which mainly accumulates in the roots. The number of nodules decreases as the concentration of Zn increases, but the total weight of nodules is constant (Cusato et al. 2007). Thus, Zn affects *Frankia* infections, but not nodule development, and the large nodules satisfy plant growth demands for N.

6.2.7 Other Heavy Metals

Two previous studies investigated the tolerance of *Frankia* strains to a wide variety of heavy metals (Richards et al. 2002; Bélanger et al. 2011a). Although some outcomes were inconsistent between the studies due to different growth conditions, strains, and evaluations of cell viability, several consensus conclusions can be identified. Most *Frankia* strains tested are resistant to lead (Pb) while being sensitive to cadmium (Cd). The spectra of sensitivity to metals vary depending on strains. For example, strain CcI3 is more tolerant of Cd, chromium (Cr), Zn, and Ni than the other strains tested (Bélanger et al. 2011a). Many *Frankia* strains are resistant to arsenate (AsO_4^{3-}), Cr, and selenium (Se) (Richards et al. 2002), but sensitive to silver (Ag), AsO_2^-, antimony (Sb) (Richards et al. 2002), cobalt (Co), copper (Cu), and Cr (Bélanger et al. 2011a).

6.2.8 Reactive Oxygen Species

During nodule formation in legumes, the rhizobia are initially recognized as intruders and subjected to the host plant's defense responses, such as reactive oxygen species (ROS) generation (Tavares et al. 2007). Nitric oxide (NO) and hydrogen peroxide (H_2O_2) play key roles in the defense system. Despite the lack of any direct evidence for ROS generation in actinorhizal plants, *Frankia* apparently has mechanisms for coping with such responses. *Frankia* strain CcI3 has two truncated hemoglobin genes (*hboN* and *hboO*); NO induces the expression of *hboN*, while *hboO* is induced by hypoxia (Niemann and Tisa 2008). Hemoglobin is able to convert NO to nontoxic nitrate. The activity and transcription of two catalase genes (*katA* and *katB*) in *Frankia* strain ACN14a are induced when cells are challenged with H_2O_2 (Santos et al. 2007). ROS also affects N_2-fixation in the nodule. NA of *Alnus firma* nodules increases following treatment with an NO scavenger, suggesting that NO is generated in the nodules of actinorhizal plants and inhibits N_2-fixation (Sasakura et al. 2006). *Alnus firma* induces expression of the hemoglobin gene (*AfHb1*) in nodules, probably to detoxify NO (Sasakura et al. 2006).

6.2.9 N and P

N occurs in two biologically distinguishable chemical forms (N_2 and combined N) and both affect actinorhizal symbiotic N_2-fixation. Combined N is in the form of ammonium, which is a direct product of N_2-fixation, and biologically converted derivatives, such as nitrate. Nodulation and N_2-fixation are suppressed by the presence of combined N (Huss-Danell 1997). This form of biological regulation also operates in legume symbioses, in which nodulation is systemically repressed through a signaling pathway involving the receptor-like kinase HAR1 and the small polypeptide LjCLE-RS2 (Okamoto et al. 2009). In *Alnus incana*, which is infected by *Frankia* via root hairs, nodulation is regulated at a stage preceding cortical cell division (Gentili et al. 2006). In contrast to combined N, P has positive effects on symbiotic N_2-fixation. Symbiotic N_2-fixation is a P-consuming activity that accompanies de novo synthesis of DNA and plasma membranes for cell divisions, which are required for nodule development, and ATP synthesis for reduction of N_2. Addition of P thus increases NA and the numbers and sizes of nodules (Wall et al. 2000; Valverde et al. 2002). Stimulation by P is specific to nodulation and is not simply mediated via plant growth (Gentili and Huss-Danell 2002, 2003; Gentili 2006). The effects of combined N and P on nodulation are interactive in that P alleviates inhibitory effects of combined N. Such alleviation only occurs when the amount of combined N relative to P is below a critical level. In the case of *Alnus incana*, a N/P ratio of 7 or lower allows alleviation (Wall et al. 2000). Split root experiments demonstrate whether the regulation of symbiotic N_2-fixation by combined N and P is local or systemic. In most cases, the suppression of nodulation and NA by combined N is systemic (Arnone et al. 1994; Gentili and Huss-Danell 2002, 2003), i.e., applying combined N to a root inhibits nodulation and N_2-fixation in other roots. The stimulation of nodulation by P is also systemic (Gentili and Huss-Danell 2003).

Wall et al. (2003) showed that N_2 affects actinorhizal symbiosis. Depletion of N_2 (by replacement with argon gas) increases nodule biomass, but the process varies among plant species with different infection pathways. Intracellularly infected *A. incana* increases nodule biomass by developing new nodules, whereas intercellularly infected *D. trinervis* invests in increasing the sizes of existing nodules rather than forming new ones.

Combined N and P affect free-living properties of *Frankia*. An abundance of combined N compounds represses free-living N_2-fixation, while their depletion induces it. These physiological changes accompany the orchestrated modification of gene expression. Proteome analysis using two-dimensional gel electrophoresis has identified 11 proteins whose expressions change between N_2-fixing and N-replete conditions (Alloisio et al. 2007). Transcriptome analysis using DNA microarrays has identified many more (96) differentially expressed genes (Alloisio et al. 2010), including those related to N_2-fixation, vesicle formation, energy metabolism, transcription, and translation, which are upregulated under N_2-fixing conditions. Most of these genes are upregulated in nodules as well, but the induction rates are elevated in this case because the proportion of N_2-fixing cells (vesicles) is much higher in nodules than under free-living conditions. Limiting combined N or P conditions induces the formation of spores in *Frankia* strain CcI3 but not in two other strains under the same conditions (Burleigh and Dawson 1991).

6.3 Actinorhizal Symbiosis

N_2-fixation by the nodules of actinorhizal plants depends on the transport of newly formed photosynthates from the leaves of host plants to the nodules (Valverde and Wall 2003). A large amount of assimilated carbon is required for the N_2-fixation process, and substantially more is required for N assimilation from the soil (Vitousek et al. 2002). Therefore, many abiotic and biotic factors that affect the photosynthesis of actinorhizal plants also have effects on *Frankia* symbiosis, as previously reviewed by Huss-Danell (1997) and Dawson (2008).

6.3.1 Light Availability and Seasonality

In general, actinorhizal plants tend to be light-demanding, early successional, and shade-intolerant species (Claessens et al. 2010). Light-saturated photosynthetic rates (A_{sat}) of *Alnus hirsuta* are very different between sun and shade leaves within the crown (Koike et al. 2001), which indicates a typical photosynthetic advantage under high light intensity. High rates of both nodule growth and N_2-fixation are maintained even in early autumn, as observed in *Alnus* species (Kaelke and Dawson 2003). The N resorption rate from senescing leaves in deciduous actinorhizal plants is usually lower than in other woody species (Chaia and Vobis 2000). Since leaf N content is highly correlated with photosynthetic activity, prolonged N retention in the autumn is associated with prolonged photosynthesis (Tateno 2003), and this may maintain high rates of NA and N_2-fixation (Dawson 2008).

The interaction of various abiotic and biotic factors affects seasonal variation in nodule NA (Huss-Danell 1990). No NA is detectable before budbreak, even at soil temperatures above 10 °C, because the supply of photosynthate is limited (Ekblad et al. 1994). During summer when leaves are fully developed, there are NA variations that may be due to fluctuating weather conditions (Huss-Danell 1990). For example, NA decreases in response to drought stress (Sharma et al. 2010); conversely, high NA occurs in the rainy season (Uliassi and Ruess 2002). The N_2-fixation rate estimated when several major assumptions are incorporated, such as the conversion rate in the acetylene reduction method, reportedly ranges from a few kgN ha^{-1} year^{-1} up to 320 kgN ha^{-1} year^{-1} (Hibbs and Cromack 1990; Son et al. 2007; Sharma et al. 2010). Hurd et al. (2001) showed that N_2-fixation may account for an estimated 85–100 % of foliar N in an *A. incana* ssp. *rugosa* stand.

6.3.2 Rising Carbon Dioxide Concentration ([CO_2]) in the Atmosphere

The increasing level of atmospheric [CO_2] is an important factor in global climate change (IPCC 2007). Many plant species grown under elevated [CO_2] have enhanced A_{sat} and growth when other environmental resources do not limit productivity (Körner 2006), but these enhancements are often inhibited by low N supply and insufficient sink capacity for the photosynthate (Ainsworth and Rogers 2007). As N availability commonly limits forest productivity, a sustained stimulation of productivity at elevated [CO_2] requires an enhanced N supply to match the increase in carbon acquisition (Luo et al. 2004). Symbiotic N_2-fixation may be a source of N under elevated [CO_2] (Vitousek et al. 2002). As N_2-fixers are largely independent of soil N, they often respond more directly in photosynthetic activity and growth to increased [CO_2] than non-N_2-fixers (Vogel et al. 1997; Eguchi et al. 2008; Watanabe et al. 2008). In *Alnus* species, elevated [CO_2] increases the total amount of N_2-fixation per plant through promoting formation of a greater nodule mass (Tobita et al. 2005, 2010), increasing NA (Temperton et al. 2003), or both (Arnone and Gordon 1990; Vogel et al. 1997). Under free-air CO_2 enrichment (FACE) conditions, the ratio of N taken up from the soil and by N_2-fixation in *Alnus* is unaffected by elevated [CO_2] (Hoosbeek et al. 2011).

6.3.2.1 N and P Availability

A_{sat} and the growth of *Alnus* species increase at elevated [CO_2], even under N deficiency (Tobita et al. 2005, 2011; Watanabe et al. 2008); this is not the case for plants that do not fix molecular N. Thomas et al. (2000) demonstrated that in

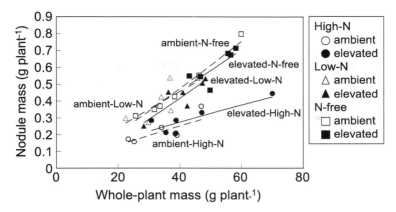

Fig. 6.1 Relationships between whole-plant mass and nodule mass per plant in *Alnus hirsuta*. *Lines* represent statistically significant ($P < 0.05$) power functional regression at each treatment [(nodule mass) = a × (whole-plant mass)^b]. Ambient CO_2, *dotted lines*; elevated CO_2, *solid lines*. Ambient CO_2, *open symbols*; elevated CO_2, *closed symbols*. High -N (*circles*), low-N (*triangles*), N-free (*squares*) (after Tobita et al. 2005; copyright Phyton)

legumes greater photoassimilate availability resulting from enhanced photosynthesis at elevated [CO_2] mediates N inhibition of nodulation and N_2-fixation. However, increased N availability has a negative effect on both nodule mass and nodule mass relative to the whole-plant mass ratio of the actinorhizal species *A. hirsuta*, regardless of [CO_2], demonstrating that the inhibitory effect of soil N availability on nodulation is retained even under elevated [CO_2] (Tobita et al. 2005) (Fig. 6.1). Other *Alnus* species have similarly inhibited nodulation under elevated [CO_2] (Temperton et al. 2003).

N_2-fixing plants have relatively high P requirements in relation to those that do not fix N, likely because of the high energy demands by N_2-fixation (Vitousek et al. 2002), growth, and root nodulation (Valverde et al. 2002; Gentili and Huss-Danell 2003); N_2-fixers thus respond strongly to P fertilization (Brown and Courtin 2003). Increased atmospheric N deposition might limit soil P for plant growth by disturbing soil nutrient balance (Gress et al. 2007; Hyvönen et al. 2007). Under P deficiency, N_2-fixation rates per plant in two *Alnus* species do not increase under elevated [CO_2] because plant growth is strongly suppressed and nodule biomass does not increase (Tobita et al. 2010) (Figs. 6.2 and 6.3). A_{sat} is not stimulated in these two alder species under elevated [CO_2] and insufficient P supply (Fig. 6.4), indicating obvious photosynthetic downregulation in response to elevated [CO_2] (Ainsworth and Rogers 2007). Rogers et al. (2009) suggest that in legumes growth under low P conditions limits the demand for N, resulting in fewer nodules and lower N_2-fixation rates.

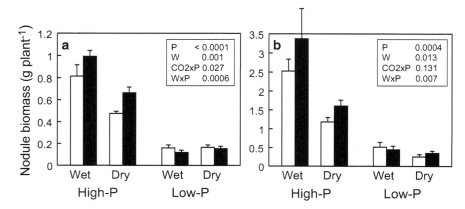

Fig. 6.2 Nodule biomass per plant in *Alnus hirsuta* (**a**) and *Alnus maximowiczii* (**b**). Values shown are means + SE ($n = 6$). Ambient CO_2, *open columns*; elevated CO_2, *closed columns*. High-P, high-P treatment; low-P, low-P treatment; wet, well-watered treatment; dry, drought treatment. Results of three-way ANOVA are shown when $P < 0.15$. W water treatment (after Tobita et al. 2010; copyright Symbiosis)

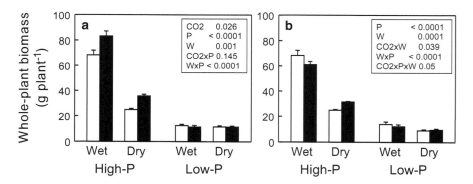

Fig. 6.3 Whole-plant biomass in *Alnus hirsuta* (**a**) and *Alnus maximowiczii* (**b**). Values shown are means + SE ($n = 6$). Ambient CO_2, *open columns*; elevated CO_2, *closed columns*. High-P, high-P treatment; low-P, low-P treatment; wet, well-watered treatment; dry, drought treatment. Results of three-way ANOVA are shown when $P < 0.15$. W water treatment (after Tobita et al. 2010; copyright Symbiosis)

6.3.2.2 Soil Moisture

NA is sensitive to drought in the short term (Brown and Courtin 2003). Higher air temperatures are predicted to reduce precipitation during the summer season (Calfapietra et al. 2010). Under sufficient P supply, A_{sat} and the growth of two *Alnus* species are enhanced by elevated [CO_2] even under dry soil conditions,

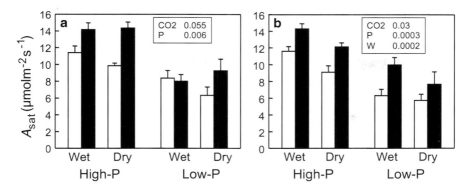

Fig. 6.4 Light-saturated photosynthetic rate (A_{sat}) under each ambient CO_2 levels in *Alnus hirsuta* (**a**) and *Alnus maximowiczii* (**b**). Values shown are means + SE ($n = 6$). Ambient CO_2 (36 Pa), *open columns*; elevated CO_2 (72 Pa), *closed columns*. High-P, high-P treatment; low-P, low-P treatment; wet, well-watered treatment; dry, drought treatment. Results of three-way ANOVA are also shown when $P < 0.15$. W water treatment (after Tobita et al. 2010; copyright Symbiosis)

though each value in dry soil is reduced compared to wet conditions (Tobita et al. 2010) (Figs. 6.3 and 6.4).

6.3.3 Ozone (O_3) Pollution

The concentration of surface background tropospheric ozone ([O_3]) over forested land in the Northern Hemisphere is also rising and now considered the most important air pollutant affecting vegetation (Kitao et al. 2012). *Alnus* species are considered to be relatively sensitive to O_3 exposure (Gunthardt-Goerg et al. 1997). *Alnus incana*, which is used as a bioindicator for [O_3] in Europe (Manning and Godzik 2004), decreases biomass and enhances leaf senescence under increased [O_3] (Mortensen and Skre 1990). In the absence of additional environmental modifications, such as elevated [CO_2], A_{sat} is reduced by 20 % under elevated [O_3] of 87 ppb (1 ppb = 1 nl l^{-1}) (Wittig et al. 2007).

6.3.4 Other Inorganic Chemicals

Calcium (Ca) has positive effects on *D. trinervis*, probably through promoting plant growth, rather than by direct effects on nodule growth and N_2-fixation (Valverde et al. 2009). Kitao et al. (1997) demonstrated that *Alnus hirsuta* accumulates manganese (Mn) in its leaves without a marked decline in photosynthesis, thereby suggesting that early successional species such as *Alnus* have a greater tolerance for excess Mn in leaves than mid- and late successional species. Bélanger et al. (2011b) showed that *A. glutinosa* and actinorhizal symbioses are not strongly affected by the presence of heavy metals, arsenic (As), Se, or vanadium (V), thus confirming potential applications for soil rehabilitation.

6.4 Biotic Factors

A variety of microbes are present in soil. The roots of terrestrial plants provide a specialized habitat for various soil microbes by supplying carbohydrates, organic acids, and amino acids or sloughed tissue. This zone of high soil microbe activity is called the rhizosphere; it is where fungi and bacteria obtain nutrients for growth. The simultaneous occurrence of mycorrhizal fungi and N_2-fixing bacteria has stimulated considerations of multipartite associations among plants, *Frankia*, and mycorrhizal fungi that promote both significant improvement of N_2-fixation by *Frankia* and the growth of actinorhizal plants.

6.4.1 Arbuscular Mycorrhizal Fungi

An arbuscular mycorrhiza (AM) is an endomycorrhiza formed by the members of Glomeromycota, which invade the cortical cells of plant roots without any fungal structure on the root surface; it is comparable to the fungal sheath of ectomycorrhizae (Smith and Read 2008). AM plants include angiosperms, gymnosperms, and the sporophytes of pteridophytes, as well as the gametophytes of some hepatics and ferns. AM may have originated between 462 and 353 million years ago, as estimated from SSU rRNA sequences (Simon et al. 1993), and this mycorrhizal association was likely important in plant colonization of land. The name "arbuscular" is derived from a characteristic structure (the "arbuscule") that forms within the cortical cells of infected roots. This type of mycorrhiza was also termed the vesicular arbuscular mycorrhiza (VAM) in the past, as another structure (the "vesicle") also develops within or between cortical cells. A large majority (about 80 %) of species are presently described to form both arbuscules and vesicles; the remainder do not form vesicles (Smith and Read 2008). Thus, the term "AM" is widely accepted to mean the type of mycorrhiza formed by members of Glomeromycota. AM fungi develop extrametrical hyphae in the soil for obtaining minerals and water and then transfer both to the associated plant via mycorrhizae.

All eight families of actinorhizal plants are AM plants (Gardner and Barrueco 1999). Rose (1980) demonstrated from observations of fungal structures in plant roots that most of the species of the actinorhizal plants he examined are AM. These AM fungi colonizing roots have yet to be successfully identified, as AM fungal taxa are described by the morphology of spores isolated from the soil or the morphology of extrametrical hyphae connected with roots and these are not always identical to those colonizing root tissue. Jansa et al. (2008) recently reported on the colonization of roots in pot culture using molecular techniques and various AM fungi.

AM fungi contribute to nutrient uptake of the associated plant by obtaining minerals and water from the soil via extrametrical hyphae, especially from root-distant soil not depleted of nutrients by the roots. Among the minerals essential for growth, P has strong effects on the growth and N status of actinorhizal plants (Prégent and Camiré 1985). Given the low solubility of P in soil, P uptake can be linked to the function of AM. Accordingly, the effects of AM on the growth of actinorhizal plants have been studied in inoculation experiments. Actinorhizal plants inoculated with both *Frankia* and AM fungi have better growth, enhanced P uptake, and better nodulation and N_2-fixation than those inoculated only with *Frankia* or only with AM fungi (Rose and Youngberg 1981; Jha et al. 1993; Tian et al. 2002; Yamanaka et al. 2005). Mycorrhizal infection rates in actinorhizal plants inoculated with both *Frankia* and AM fungi are higher than those inoculated only with AM fungi (Gardner and Barrueco 1999). Thus, AM fungi utilize fixed N, presumably via transfer through the actinorhizal plants.

6.4.2 Ectomycorrhizal Fungi

Most ectomycorrhizal (EM) fungi are members of Basidiomycotina and Ascomycotina. EM fungi form a fungal mantle (fungal sheath) that encloses the rootlets (Smith and Read 2008). From the mantle, extrametrical hyphae (often forming a rhizomorph or mycelial fan) spread into the soil. These hyphae penetrate inward between the root cells to form an intercellular structure (the "Hartig net"). There is little or no intracellular penetration by EM hyphae.

Many EM fungi form fruit bodies (by which species are identified), making it possible to obtain pure cultures of EM fungi. Field data on the occurrence of fruit bodies and data from inoculation studies using pure cultures of EM fungi indicate that plant species forming EM are restricted to those that are woody (Smith and Read 2008). Among actinorhizal species, the Betulaceae and Casuarinaceae are well known as EM. In addition, species of *Shepherdia* in the Elaeagnaceae, *Chaembatia* and *Dryas* in the Rosaceae, and the Myricaceae are also EM (Smith and Read 2008). The conventional method of identifying EM species by morphology is limited, but molecular techniques may be used to directly to identify EM in the absence of visible fruit bodies (Smith and Read 2008).

Similar to the AM fungi, EM fungi improve the growth of actinorhizal host plants. Among EM fungi associated with alder, *Alpova diplophloeus* Trappe is specific to alder (Molina 1981). In pot culture, this fungus improves the growth of *Frankia*-inoculated *Alnus* spp. (Yamanaka et al. 2003; Becerra et al. 2009; Yamanaka and Akama 2010), indicating a supplemental effect of EM fungi in the solubilization and gathering of minerals for N_2-fixation (Yamanaka et al. 2003). The effect of EM fungi on the growth of *Frankia*-inoculated alder differs by host species.

Fig. 6.5 Seedlings of *Alnus sieboldiana* inoculated with *Alpova* sp. (an ectomycorrhizal fungus, *left*) or without inoculation (control, *right*)

Alnus sieboldiana (common in dry soil in Japan) grows better after inoculation with *Alpova* (Fig. 6.5); however, *Alnus hirsuta* from wet soil in Japan is not affected by *Alpova* inoculation (Yamanaka and Akama 2010).

6.4.3 Soil Bacteria

In addition to mycorrhizal fungi, some rhizospheric bacteria also affect the actinorhizal association. *Pseudomonas cepacia* induces the deformation of root hairs in *Alnus rubra*, thus allowing intimate contact between *Frankia* hyphae and the hair wall (Knowlton and Dawson 1983). Some *Pseudomonas* species are able to solubilize rock minerals containing Fe and P, making nutrients available for uptake by plants (Bar-Ness et al. 1992; Li and Strzelczyk 2000). However, the co-inoculation of *Pseudomonas* with *Frankia* does not affect growth of actinorhizal plants (Rojas et al. 2002; Yamanaka et al. 2005). Rojas et al. (1992) examined the effects of actinomycetes isolated from the nodule surface, root surface, and soil in the root zone on the growth of nodulated *A. rubra* in pot culture. Thirteen of 30 isolates of these actinomycetes reduced the growth of *A. rubra*.

6.5 Conclusions

Physiological and genomic analyses have provided a better understanding of the effects of inorganic compounds on *Frankia* strains. Many recent studies that focused on the effects of elevated $[CO_2]$ on actinorhizal symbiosis suggest the need to take into consideration the physiological implications of interactive effects among several factors. Crosscutting research ranging from genomics to ecosystems will provide an opportunity to find optimal *Frankia*–actinorhizal plants symbioses (Anderson et al. 2009) and promote a deeper understanding of the mechanisms of succession (Seed and Bishop 2009).

References

Ainsworth EA, Rogers A (2007) The response of photosynthesis and stomatal conductance to rising [CO_2]: mechanisms and environmental interactions. Plant Cell Environ 30:258–270

Alloisio N, Felix S, Marechal J, Pujic P, Rouy Z, Vallenet D, Medigue C, Normand P (2007) *Frankia alni* proteome under nitrogen-fixing and nitrogen-replete conditions. Physiol Plant 130:440–453

Alloisio N, Queiroux C, Fournier P, Pujic P, Normand P, Vallenet D, Medigue C, Yamaura M, Kakoi K, Kucho K (2010) The *Frankia alni* symbiotic transcriptome. Mol Plant Microbe Interact 23:593–607

Anderson MD, Ruess RW, Myrold DD, Taylor DL (2009) Host species and habitat affect nodulation by specific *Frankia* genotypes in two species of *Alnus* in interior Alaska. Oecologia 160:619–630

Arnone JA III, Gordon JC (1990) Effect of nodulation, nitrogen fixation and CO_2 enrichment on the physiology, growth and dry mass allocation of seedlings of *Alnus rubra* Bong. New Phytol 116:55–66

Arnone JA, Kohls SJ, Baker DD (1994) Nitrate effects on nodulation and nitrogenase activity of actinorhizal *Casuarina* studied in split-root systems. Soil Biol Biochem 26:599–606

Baker DD, Schwintzer CR (1990) Introduction. In: Schwintzer CR, Tjepkema JD (eds) The biology of *Frankia* and actinorhizal plants. Academic, San Diego, CA

Bar-Ness E, Hadar Y, Chen Y, Römheld V, Marschner H (1992) Short-term effects of rhizosphere microorganisms on Fe uptake from microbial siderophores by maize and oat. Plant Physiol 100:451–456

Becerra AG, Menoyo E, Lett I, Li CY (2009) *Alnus acuminata* in dual symbiosis with *Frankia* and two different ectomycorrhizal fungi (*Alpova austronicola* and *Alpova diplophloeus*) growing in soilless growth medium. Symbiosis 47:85–92

Bélanger PA, Beaudin J, Roy S (2011a) High-throughput screening of microbial adaptation to environmental stress. J Microbiol Methods 85:92–97

Bélanger PA, Bissonnette C, Bernéche-D'Amours A, Bellenger JP, Roy S (2011b) Assessing the adaptability of the actinorhizal symbiosis in the face of environmental change. Environ Exp Bot 74:98–105

Berry AM, Harriott OT, Moreau RA, Osman SF, Benson DR, Jones AD (1993) Hopanoid lipids compose the *Frankia* vesicle envelope, presumptive barrier of oxygen diffusion to nitrogenase. Proc Natl Acad Sci USA 90:6091

Bolaños L, Redondo-Nieto M, Bonilla I, Wall LG (2002) Boron requirement in the *Discaria trinervis* (Rhamnaceae) and *Frankia* symbiotic relationship. Its essentiality for *Frankia* BCU110501 growth and nitrogen fixation. Physiol Plant 115:563–570

Brown KR, Courtin PJ (2003) Effects of phosphorus fertilization and liming on growth, mineral nutrition, and gas exchange of *Alnus rubra* seedlings grown in soils from mature alluvial *Alnus* stands. Can J For Res 33:2089–2096

Burleigh SH, Dawson JO (1991) *In vitro* sporulation of *Frankia* strain HFPCcI3 from *Casuarina cunninghamiana*. Can J Microbiol 37:897–901

Calfapietra C, Ainsworth EA, Beier C, Angelis PD, Ellsworth DS, Godbold DL, Hendrey GR, Hickler T, Hoosbeek MR, Karnosky DF, King J, Körner C, Leakey ADB, Lewin KF,

Liberloo M, Long SP, Lukac M, Matyssek R, Miglietta F, Nagy J, Norby RJ, Oren R, Percy KE, Rogers A, Mugnozza GS, Stitt M, Taylor G, Ceulemans R (2010) Challenges in elevated CO_2 experiments on forests. Trends Plant Sci 15:5–10

Chaia EE, Vobis G (2000) Seasonal change of the actinorhizal nodules and the movement of N in *Discaria trinervis*. In: Pedrosa FO et al (eds) Nitrogen fixation: from molecules to crop productivity. Kluwer Academic, Dordrecht

Claessens H, Oosterbaan A, Savill P, Rondeux J (2010) A review of the characteristics of black alder (*Alnus glutinosa* (L.) Gaertn.) and their implications for silvicultural practices. Forestry 83:163–175

Cusato MS, Tortosa RD, Valiente L, Barneix AJ, Puelles MM (2007) Effects of Zn^{2+} on nodulation and growth of a South American actinorhizal plant, *Discaria americana* (Rhamnaceae). World J Microbiol Biotechnol 23:771–777

Dawson JO (2008) Ecology of actinorhizal plants. In: Pawlowski K, Newton WE (eds) Nitrogen-fixing actinorhizal symbiosis. Springer, Dordrecht

Doyle JJ (1998) Phylogenetic perspectives on nodulation: evolving views of plants and symbiotic bacteria. Trends Plant Sci 3:473–478

Eguchi N, Karatsu K, Ueda T, Funada R, Takagi K, Hiura T, Sasa K, Koike T (2008) Photosynthetic responses of birch and alder saplings grown in a free air CO_2 enrichment system in northern Japan. Trees 22:437–447

Ekblad A, Lundquist P-O, Sjöström M, Huss-Danell K (1994) Day-to-day variation in nitrogenase activity of *Alnus incana* explained by weather variables: a multivariate time series analysis. Plant Cell Environ 17:319–325

Gardner IC, Barrueco CR (1999) Mycorrhizal and actinorhizal biotechnology–problems and prospects. In: Varma A, Hock B (eds) Mycorrhiza, 2nd edn. Springer, Berlin

Gentili F (2006) Phosphorus, nitrogen and their interactions affect N_2 fixation, N isotope fractionation and N partitioning in *Hippophae rhamnoides*. Symbiosis 41:39–45

Gentili F, Huss-Danell K (2002) Phosphorus modifies the effects of nitrogen on nodulation in split-root systems of *Hippophae rhamnoides*. New Phytol 153:53–61

Gentili F, Huss-Danell K (2003) Local and systemic effects of phosphorus and nitrogen on nodulation and nodule function in *Alnus incana*. J Exp Bot 54:2757–2767

Gentili F, Wall LG, Huss-Danell K (2006) Effects of phosphorus and nitrogen on nodulation are seen already at the stage of early cortical cell divisions in *Alnus incana*. Ann Bot 98:309–315

Gress SE, Nichols TD, Northcraft CC, Peterjohn WT (2007) Nutrient limitation in soil exhibiting differing nitrogen availabilities: what lies beyond nitrogen saturation? Ecology 88:119–130

Gunthardt-Goerg MS, McQuattie CJ, Scheidegger C, Rhiner C, Matyssek R (1997) Ozone-induced cytochemical and ultrastructural changes in leaf mesophyll cell walls. Can J For Res 27:453–463

Hafeez FY, Hameed S, Malik KA (1999) *Frankia* and *Rhizobium* strains as inoculum for growing trees in a saline environment. Pak J Bot 31:173–182

Hibbs DE, Cromack C Jr (1990) Actinorhizal plants in pacific northwest forests. In: Schwintzer CR, Tjepkema JD (eds) The biology of *Frankia* and actinorhizal plants. Academic, San Diego, CA

Hoosbeek MR, Lukac M, Velthorst E, Smith AR, Godbold DL (2011) Free atmospheric CO_2 enrichment increased above ground biomass but did not affect symbiotic N_2-fixation and soil carbon dynamics in a mixed deciduous stand in Wales. Biogeoscience 8:353–364

Hurd TM, Raynal DJ, Schwintzer CR (2001) Symbiotic N_2 fixation of *Alnus incana* ssp. *rugosa* in shrub wetlands of the Adirondack Mountains, New York, USA. Oecologia 126:94–103

Huss-Danell K (1990) The physiology of actinorhizal nodules. In: Schwintzer CR, Tjepkema JD (eds) The biology of *Frankia* and actinorhizal plants. Academic, San Diego, CA

Huss-Danell K (1997) Actinorhizal symbioses and their N_2 fixation. New Phytol 136:375–405

Hyvönen R, Ågren GI, Linder S, Persson T, Cotrufo F, Ekblad A, Freeman M, Grelle A, Janssens IA, Jarvis PG, Kellomäki S, Lindroth A, Loustau D, Lundmark T, Norby RJ, Oren R, Pilegaard K, Ryan MG, Sigurdsson BD, Stromgren M, van Oijen M, Wallin G (2007) The likely impact of elevated [CO_2], nitrogen deposition, increased temperature and management on carbon sequestration in temperate and boreal forest ecosystems: a literature review. New Phytol 173:463–480

Igual JM, Dawson JO (1999) Stimulatory effects of aluminum on *in vitro* growth of *Frankia*. Can J Bot 77:1321–1326

Igual JM, Rodriguez-Barrueco C, Cervantes E (1997) The effects of aluminium on nodulation and symbiotic nitrogen fixation in *Casuarina cunninghamiana* Miq. Plant Soil 190:41–46

IPCC (2007) Climate change 2007: the physical science basis. In: Solomon S, Qin D, Manning M, Chen Z, Marquis M, Averyt KB, Tignor M, Miller HL (eds) Contribution of working group I to the fourth assessment report of the intergovernmental panel on climate change. Cambridge University Press, Cambridge

Ishii G (2003) Sea buckthorn (*Hippophae rhamnoides*. L.) production manual. Misc Pub Natl Agric Res Cent Hokkaido Reg 62:1–32 (In Japanese)

Jansa J, Smith FA, Smith SE (2008) Are there benefits of simultaneous root colonization by different arbuscular mycorrhizal fungi? New Phytol 177:779–789

Jeong SC, Ritchie NJ, Myrold DD (1999) Molecular phylogenies of plants and *Frankia* support multiple origins of actinorhizal symbioses. Mol Phylogenet Evol 13:493–503

Jha DK, Sharma GD, Mishra RR (1993) Mineral nutrition in the tripartite interaction between *Frankia*, *Glomus* and *Alnus* at different soil phosphorus regimes. New Phytol 123:307–311

Kaelke CM, Dawson JO (2003) Seasonal flooding regimes influence survival, nitrogen fixation, and the partitioning of nitrogen and biomass in *Alnus incana* ssp. *rugosa*. Plant Soil 254:167–177

Kitao M, Lei TT, Koike T (1997) Comparison of photosynthetic responses to manganese toxicity of deciduous broad-leaved trees in northern Japan. Environ Pollut 97:113–118

Kitao M, Winkler JB, Löw M, Nunn AJ, Kuptz D, Haberle KH, Reiter IM, Matyssek R (2012) How closely does stem growth of adult beech (*Fagus sylvatica*) relate to net carbon gain under experimentally enhanced ozone stress. Environ Pollut 166:108–115

Knowlton S, Dawson JO (1983) Effects of *Pseudomonas cepacia* and cultural factors on the nodulation of *Alnus rubra* roots by *Frankia*. Can J Bot 61:2877–2882

Koike T, Kitao M, Maruyama Y, Mori S, Lei TT (2001) Leaf morphology and photosynthetic adjustments among deciduous broad-leaved trees within the vertical canopy profile. Tree Physiol 21:951–958

Körner C (2006) Plant CO_2 responses: an issue of definition, time and resource supply. New Phytol 172:393–411

Li CY, Strzelczyk E (2000) Belowground microbial processes underpin forest productivity. Phyton 40:129–134

Luo Y, Su B, Currie WS, Dukes JS, Finzi A, Hartwig U, Hungate B, McMurtrie RE, Oren R, Parton WJ, Pataki DE, Shaw MR, Zak DR, Field CB (2004) Progressive nitrogen limitation of ecosystem responses to rising atmospheric carbon dioxide. Bioscience 54:731–739

Manning WJ, Godzik B (2004) Bioindicator plants for ambient ozone in central and eastern Europe. Environ Pollut 130:33–39

Mattsson U, Sellstedt A (2002) Nickel affects activity more than expression of hydrogenase protein in *Frankia*. Current Microbiol 44:88–93

Molina R (1981) Ectomycorrhizal specificity in the genus *Alnus*. Can J Bot 59:325–334

Mortensen LM, Skre O (1990) Effects of low ozone concentrations on growth of *Betula pubescens* Ehrh., *Betula verrucosa* Ehrh. and *Alnus incana* (L.) Moench. New Phytol 115:165–170

Niemann J, Tisa LS (2008) Nitric oxide and oxygen regulate truncated hemoglobin gene expression in *Frankia* strain CcI3. J Bacteriol 190:7864–7867

Normand P, Lapierre P, Tisa LS, Gogarten JP, Alloisio N, Bagnarol E, Bassi CA, Berry AM, Bickhart DM, Choisne N, Couloux A, Cournoyer B, Cruveiller S, Daubin V, Demange N, Francino MP, Goltsman E, Huang Y, Kopp OR, Labarre L, Lapidus A, Lavire C, Marechal J, Martinez M, Mastronunzio JE, Mullin BC, Niemann J, Pujic P, Rawnsley T, Rouy Z, Schenowitz C, Sellstedt A, Tavares F, Tomkins JP, Vallenet D, Valverde C, Wall LG,

Wang Y, Medigue C, Benson DR (2007) Genome characteristics of facultatively symbiotic *Frankia* sp. strains reflect host range and host plant biogeography. Genome Res 17:7–15

Okamoto S, Ohnishi E, Sato S, Takahashi H, Nakazono M, Tabata S, Kawaguchi M (2009) Nod factor/nitrate-induced CLE genes that drive HAR1-mediated systemic regulation of nodulation. Plant Cell Physiol 50:67–77

Prégent G, Camiré C (1985) Mineral nutrition, dinitrogen fixation, and growth *Alnus crispa* and *Alnus glutinosa*. Can J For Res 15:855–861

Richards JW, Krumholz GD, Chval MS, Tisa LS (2002) Heavy metal resistance patterns of *Frankia* strains. Appl Environ Microbiol 68:923–927

Rogers A, Ainsworth EA, Leakey DB (2009) Will elevated carbon dioxide concentration amplify the benefits of nitrogen fixation in legumes? Plant Physiol 151:1009–1016

Rojas NS, Perry DA, Li CY, Friedman J (1992) Influence of actinomycetes on *Frankia* infection, nitrogenase activity and seedling growth of red alder. Soil Biol Biochem 24:1043–1049

Rojas NS, Perry DA, Li CY, Ganio LM (2002) Interactions among soil biology, nutrition, and performance of actinorhizal plant species in the H. J. Andrews Experimental Forest of Oregon. Appl Soil Ecol 19:13–26

Rose SL (1980) Mycorrhizal associations of some actinomycete nodulated nitrogen-fixing plants. Can J Bot 58:1449–1454

Rose SL, Youngberg CT (1981) Tripartite associations in snowbrush (*Ceanothus velutinus*): effect of vesicular-arbuscular mycorrhizae on growth, nodulation, and nitrogen fixation. Can J Bot 59:34–39

Santos CL, Vieira J, Sellstedt A, Normand P, Moradas-Ferreira P, Tavares F (2007) Modulation of *Frankia alni* ACN14a oxidative stress response: activity, expression and phylogeny of catalases. Physiol Plant 130:454–463

Sasakura F, Uchiumi T, Shimoda Y, Suzuki A, Takenouchi K, Higashi S, Abe M (2006) A class 1 hemoglobin gene from *Alnus firma* functions in symbiotic and nonsymbiotic tissues to detoxify nitric oxide. Mol Plant Microbe Interact 19:441–450

Seed JD, Bishop JG (2009) Low *Frankia* inoculation potentials in primary successional sites at Mount St. Helens, Washington, USA. Plant Soil 323:225–233

Sharma G, Sharma R, Sharma E (2010) Impact of altitudinal gradients on energetics and efficiencies of N_2-fixation in alder-cardamom agroforestry systems of the eastern Himalayas. Ecol Res 25:1–12

Simon L, Bousquet J, Lévesque RC, Lalonde M (1993) Origin and diversification of endomycorrhizal fungi and coincidence with vascular land plants. Nature 363:67–69

Singh A, Mishra AK, Singh SS, Sarma HK, Shukla E (2008) Influence of iron and chelator on siderophore production in *Frankia* strains nodulating *Hippophae salicifolia* D. Don. J Basic Microbiol 48:104–111

Smith SE, Read DJ (2008) Mycorrhizal symbiosis, 3rd edn. Elsevier, New York, NY

Son Y, Lee YY, Lee CY, Yi MJ (2007) Nitrogen fixation, soil nitrogen availability, and biomass in pure and mixed plantations of alder and pine in central Korea. J Plant Nutr 30:1841–1853

Swensen SM (1996) The evolution of actinorhizal symbioses: evidence for multiple origins of the symbiotic association. Am J Bot 83:1503–1512

Swensen SM, Benson DR (2008) Evolution of actinorhizal host plants and *Frankia* endosymbionts. In: Pawlowski K, Newton WE (eds) Nitrogen-fixing actinorhizal symbioses. Springer, Dordrecht

Tani C, Sasakawa H (2000) Salt tolerance of *Elaeagnus macrophylla* and *Frankia* Ema1 strain isolated from the root nodules of *E. macrophylla*. Soil Sci Plant Nutr 46:927–937

Tani C, Sasakawa H (2003) Salt tolerance of *Casuarina equisetifolia* and *Frankia* Ceq1 strain isolated from the root nodules of *C. equisetifolia*. Soil Sci Plant Nutr 49:215–222

Tateno M (2003) Benefit to N_2-fixing alder of extending growth period at the cost of leaf nitrogen loss without resorption. Oecologia 137:338–343

Tavares F, Santos CL, Sellstedt A (2007) Reactive oxygen species in legume and actinorhizal nitrogen-fixing symbioses: the microsymbiont's responses to an unfriendly reception. Physiol Plant 130:344–356

Temperton VM, Grayston SJ, Jackson G, Barton CVM, Millard P, Jarvis PG (2003) Effects of elevated carbon dioxide concentration on growth and nitrogen fixation in *Alnus glutinosa* in a long-term field experiment. Tree Physiol 23:1051–1059

Thomas RB, Bashkin MA, Richter DD (2000) Nitrogen inhibition of nodulation and N_2 fixation of a tropical N_2-fixing tree (*Gliricidia sepium*) grown in elevated atmospheric CO_2. New Phytol 145:233–243

Tian C, He X, Zhong Y, Chen J (2002) Effects of VA mycorrhizae and *Frankia* dual inoculation on growth and nitrogen fixation of *Hippophae tibetana*. For Ecol Manage 170:307–312

Tobita H, Kitao M, Koike T, Maruyama Y (2005) Effects of elevated CO_2 and nitrogen availability on nodulation of *Alnus hirsuta* Turcz. Phyton 45:125–131

Tobita H, Uemura A, Kitao M, Kitaoka S, Utsugi H (2010) Interactive effects of elevated CO_2, phosphorus deficiency, and soil drought on nodulation and nitrogenase activity in *Alnus hirsuta* and *Alnus maximowiczii*. Symbiosis 50:59–69

Tobita H, Uemura A, Kitao M, Kitaoka S, Maruyama Y, Utsugi H (2011) Effects of elevated [CO_2] and soil nutrients and water conditions on photosynthetic and growth responses of *Alnus hirsuta*. Funct Plant Biol 38:702–710

Uliassi DD, Ruess RW (2002) Limitation to symbiotic nitrogen fixation in primary succession on the Tanana river floodplain. Ecology 83:88–103

Valdes M (2008) *Frankia* ecology. In: Pawlowski K, Newton WE (eds) Nitrogen-fixing actinorhizal symbiosis. Springer, Dordrecht

Valverde C, Wall LG (2003) The regulation of nodulation, nitrogen fixation and assimilation under a carbohydrate shortage stress in the *Discaria trinervis-Frankia* symbiosis. Plant Soil 254:155–165

Valverde C, Ferrari A, Wall LG (2002) Phosphorus and the regulation of nodulation in the actinorhizal symbiosis, between *Discaria trinervis* (Rhamnaceae) and *Frankia* BCU110501. New Phytol 153:43–51

Valverde C, Ferrari A, Wall LG (2009) Effects of calcium in the nitrogen-fixing symbiosis between actinorhizal *Discaria trinervis* (Rhamnaceae) and *Frankia*. Symbiosis 49:151–155

Vitousek PM, Cassman K, Cleveland C, Crews T, Field CB, Grimm JI (2002) Towards an ecological understanding of biological nitrogen fixation. Biogeochemistry 57:1–45

Vogel CS, Curtis PS, Thomas RB (1997) Growth and nitrogen accretion of dinitrogen-fixing *Alnus glutinosa* (L.) Gaertn. under elevated carbon dioxide. Plant Ecol 130:63–70

Wall LG, Hellsten A, Huss-Danell K (2000) Nitrogen, phosphorus, and the ratio between them affect nodulation in *Alnus incana* and *Trifolium pratense*. Symbiosis 29:91–105

Wall LG, Valverde C, Huss-Danell K (2003) Regulation of nodulation in the absence of N_2 is different in actinorhizal plants with different infection pathways. J Exp Bot 54:1253–1258

Watanabe Y, Tobita H, Kitao M, Maruyama Y, Choi D, Sasa K, Funada R, Koike T (2008) Effects of elevated CO_2 and nitrogen on wood structure related to water transport in seedlings of two deciduous broad-leaved tree species. Trees 22:403–411

Wheeler CT, Hughes LT, Oldroyd J, Pulford ID (2001) Effects of nickel on *Frankia* and its symbiosis with *Alnus glutinosa* (L.) Gaertn. Plant Soil 231:81–90

Wittig VG, Ainthworth EA, Long SP (2007) To what extent do current and projected increases in surface ozone affect photosynthesis and stomatal conductance of trees? A meta-analytic review of the last 3 decades of experiments. Plant Cell Environ 30:1150–1162

Yamanaka T, Akama K (2010) Effects of *Alpova* inoculation on the growth of *Alnus sieboldiana* and *A. hirsuta*. Kanto J For Res 61:149–150 (in Japanese)

Yamanaka T, Okabe H (2008) Actinorhizal plants and *Frankia* in Japan. Bull FFPRI 7:67–80 (In Japanese with English summary)

Yamanaka T, Li CY, Bormann BT, Okabe H (2003) Tripartite associations in an alder: effects of *Frankia* and *Alpova diplophloeus* on the growth, nitrogen fixation and mineral acquisition of *Alnus tenuifolia*. Plant Soil 254:179–186

Yamanaka T, Akama A, Li CY, Okabe H (2005) Growth, nitrogen fixation and mineral acquisition of *Alnus sieboldiana* after inoculation of *Frankia* together with *Gigaspora margarita* and *Pseudomonas putida*. J For Res 10:21–26

Chapter 7
Diversity of *Frankia* Strains, Actinobacterial Symbionts of Actinorhizal Plants

Maher Gtari, Louis S. Tisa, and Philippe Normand

7.1 Introduction

In the last edition of the Bergey's Manual of Systematic Bacteriology, *Frankia* was the sole genus found in the family Frankiaceae. This single-genus family, together with Geodermatophilaceae, Nakamurellaceae, Sporichthyaceae, Acidothermaceae, and Cryptosporangiaceae, makes up the order Frankiales, an artificial taxon within the phylum Actinobacteria supported only marginally by 16S rDNA similarities (Normand and Benson 2012). Genus *Frankia* represents a monophyletic assemblage of soil actinobacteria distinguishable by multilocular sporangia-forming branched hyphae and their unique ability to grow in nitrogen-limited substrates due to their ability to fix nitrogen within specialized thick-walled structures, termed vesicles, produced in vitro and in planta. Diversity of *Frankia* has been noted in several separate reports on the presence of comparable root nodule structures induced in numerous trees and shrubs growing as aggressive pioneers in nitrogen-limited ecosystems such as glacial moraines, gravel slopes, volcanic ashes, mine spoils, or burned forests with a contributed input of 15 % of total biologically fixed nitrogen on earth (Quispel 1990). Despite the absence of apparent close kinship among these plants, they have been termed actinorhizal plants due to the nature of the root nodule-causing agents detected in plant tissues. Based on in planta morphology and the limited cross-infectivity of crushed-nodule inocula, Becking

M. Gtari (✉)
Laboratoire Microorganismes et Biomolécules Actives, Université de Tunis El-Manar (FST) et Université de Carthage (INSAT), Tunis 2092, Tunisia
e-mail: maher.gtari@fst.rnu.tn

L.S. Tisa
Department of Molecular, Cellular, and Biomedical Sciences, University of New Hampshire, 289 Rudman Hall, 46 College Road, Durham, NH 03824-2617, USA

P. Normand
Ecologie Microbienne, Centre National de la Recherche Scientifique UMR 5557, Université Lyon I, Villeurbanne 69622 cedex, France

(1970) grouped these microorganisms into the same genus, *Frankia*, and described ten species of uncultured endosymbionts that overlapped exactly the ten then known host plant genera. Since the first successfully validated *Frankia* isolation (Callaham et al. 1978), hundreds of *Frankia* isolates from five of the eight described actinorhizal families have been reported (Benson and Silvester 1993). However, grouping of these isolated *Frankia* strains based on host plant infectivity (Baker 1987) revealed less variation, providing a limited view on diversity. Despite that, 19 genospecies have been proposed based on DNA–DNA hybridization with only one species, *Frankia alni*, validated (Fernandez et al. 1989). The development of molecular tools, especially those based on polymerase chain reaction, has deepened knowledge on *Frankia* genetic structure and diversity in culture, root nodules, and in soils. Currently, the genus *Frankia* is described as a heterologous assemblage of saprophytic, facultative, and obligately symbiotic, sporangia-forming actinobacteria, and we consider that host plant specialization has selected strains leading to both erosion of the genetic diversity and reduction in lifestyle. This hypothesis is also supported by analysis of *Frankia* genomes that vary in size showing either contraction or expansions that are likely being driven by symbiotic or saprophytic lifestyles, respectively.

7.1.1 Frankia *the Microsymbiont of Actinorhizal Plants*

With regard to host plant interactions, several different parameters have been examined for *Frankia* strains including nodulation speed (Nesme et al. 1985) and efficiency (Normand and Lalonde 1982). Their influence on *Frankia* diversity will be addressed below.

7.1.1.1 Diversity Based on Host Plant Ranges

There are more than 200 species of actinorhizal plants, grouped into 22 genera and 8 families, that are able to share the symbiotic partnership with the actinobacterium *Frankia* (Table 7.1). With the exception of *Datisca*, all these plants are woody dicotyledonous trees and shrubs. Families of actinorhizal plants, together with legumes, the Ulmaceae *Parasponia*, and others, belong to a plant clade, called Rosids 1, that has been suggested to be linked to a 100 Ma-old common ancestor. This ancestral lineage eventually evolved to acquire a unique feature, that of being able to establish nitrogen-fixing nodulating symbioses, and has divergently evolved since 70 Ma, the age of the oldest putative actinorhizal genus, to the current ten plant families (Soltis et al. 1995; Oono et al. 2010; Doyle 2011).

Initial studies on host infectivity relied on the use of root nodules as sources of inoculum, but these cross-inoculation experiments are fastidious and unreliable (Torrey 1990). Prior to the availability of isolated *Frankia* strains, Becking (1970) proposed species status for the microsymbionts of each of the ten

Table 7.1 List of actinorhizal plants

Subclades	Families	Genera	Species[a]	Geographical distribution[b]	Biotope
Fagales	Betulaceae	*Alnus*	31	Circumpolar, mountains in South America	River banks, sand, mine spoils, burned forest soils
	Casuarinaceae	*Allocasuarina*	57	Oceania	Sandy beaches
		Casuarina	14	Oceania, tropical worldwide	Sandy beaches
		Ceuthostoma	2	Oceania	Sandy beaches
		Gymnostoma	13	Oceania	Sandy beaches
	Myricaceae	*Comptonia*	1	Eastern North America	Sandy open forest soils
		Morella	28	Global	Sandy soils
		Myrica	3	North America, Europe	Acidic peat bogs, lake shores
Rosales	Elaeagnaceae	*Elaeagnus*	13	Circumpolar	Sandy soils
		Hippophae	7	Asia, Europe	Sand beaches, mountain slopes
		Shepherdia	2	North America	Sandy open areas
	Rhamnaceae[c]	*Ceanothus*	59	Western North America	Dry chaparral
		Colletia	5	South America	Dry chaparral
		Discaria	8	South America	Dry chaparral
		Kentrothamnus	1	South America	Dry chaparral
		Trevoa (incl. *Talguena*)	1	South America	Dry chaparral
	Rosaceae	*Cercocarpus*	3	Western North America	Rocky substrate
		Chamaebatia	1	Western North America	Understory in open ponderosa pine (*Pinus ponderosa*) forests
		Dryas[d]	6	North America	Glacial moraines, nunataq gravel islands
		Purshia	2	Western North America	Burned forest soils
Cucurbitales	Coriariaceae	*Coriaria*	12	Disjunct (NZ, Japan, Mexico, Europe, North Africa)	Dry chaparral
	Datiscaceae	*Datisca*	2	California, Turkey to Pakistan	Mountain slopes

[a]Number of species according to the NCBI web site (http://www.ncbi.nlm.nih.gov/taxonomy)
[b]Dawson (2007) and Gtari and Dawson (2011)

(continued)

Table 7.1 (continued)

Subclades	Families	Genera	Species[a]	Geographical distribution[b]	Biotope

[c]Genus *Adolphia* described by Cruz-Cisneros and Valdés (1991) remains unconfirmed according to http://web2.uconn.edu/mcbstaff/benson/Frankia/Rhamnaceae.htm (Cruz-Cisneros, R. y M. Valdés. 1991. Actinorhizal root nodules on *Adolphia infesta* (H.B.K.) Meissner (Rhamnaceae) (Nitrogen Fixing Trees Research Report 9: 87–89)
[d]Only those North American species appear to have nodules; the European species, *D. octopetala*, does not have nodules

actinorhizal plant genera described at that time (Table 7.2). However, after *Frankia* isolates became available, inocula from crushed nodules were shown to be more restricted in their host plant ranges than isolated strains from nodules of similar plants (Lalonde 1979), as though host plant phenotypically "fixed" the microbial cells.

7.1.1.2 The Sporulation Phenotype: Spore+ and Spore−

Two nodule phenotypes have been described based on the extent of sporulation within host plant cells: spore-positive (Sp+) and spore-negative (Sp−) (Käppel and Wartenberg 1958; Van Dijk and Merkus 1976; Van Dijk 1978; Torrey 1987; Schwintzer 1990) (Table 7.2). Since sporangia formation is hypothesized to be driven mainly by *Frankia* strains and not the host plant (Huss-Danell 1997), the two nodule phenotypes have been assumed to represent two microsymbiont phenotypes with no intermediaries. Another study on Sp+ and Sp− *Frankia* phenotypes provided an alternative view. Sp− strains are saprophytes that can be quite readily isolated in pure cultures, while Sp+ strains are difficult to isolate and maintained in pure culture and have been considered as more or less obligate symbionts. The hypersporulation phenotype of Sp+ strains would make the nodules less efficient in nitrogen fixation than nodules containing the Sp− strains (Normand and Lalonde 1982). Moreover, the respiratory cost (moles CO_2 respired per mole C_2H_2 reduced) is considerably higher for Sp+ nodules than for Sp− nodules of *Myrica gale* (VandenBosch and Torrey 1984). One obvious explanation is that there is a trade-off between the number of sporangia and production of vesicles, with a reduction in the number of vesicles within Sp+ nodules of *Comptonia peregrina* (VandenBosch and Torrey 1985). The Sp+ phenotype is usually predominant in old actinorhizal stands at disturbed sites with mineral acidic soils, while Sp− phenotype strains are found widespread in soils, even those devoid of host plants (Schwintzer 1990). One working model suggests that Sp− *Frankia* strains maintain themselves at sites through saprophytic growth in soil and would nodulate newly established hosts, whereas Sp+ strains are maintained primarily as spores within nodules and thus would be dependent on the extended presence of the host. Up to date all cultivable strains have been assumed to belong to the Sp− phenotype, while

Table 7.2 Phenotypic diversity of *Frankia* in root nodules

		Vesicle			Sporangia		Isolate
	Mode of infection	Orientation	Shape	Septation	Spore+	Spore−	
Betulaceae	Root hairs	Periphery	Spherical	Septate	+	−	Typical and atypical
Casuarinaceae	Root hairs	Not applicable	NA	NA		−	Typical
Myricaceae	Root hairs	Central	Club-shaped	Septate	+	−	Typical and atypical
Elaeagnaceae	Intercellular penetration	Throughout	Spherical	Septate	+	−	Typical and atypical
Rhamnaceae	Intercellular penetration	Throughout periphery	Spherical	+/−		−	Typical and atypical
Coriariaceae		Perpendicular to central	Filamentous	Non septate		−	Atypical
Datiscaceae	Intercellular penetration	Perpendicular to central	Filamentous	Non septate		−	Atypical
Rosaceae		Throughout	Elliptical	Non septate		−	Atypical

Atypical isolate: noninfective and/or nonnitrogen-fixing in symbiosis

no Sp+ phenotype has so far been isolated in pure culture (Torrey 1987) although such a yes/no phenotype may be hard to defend in the absence of comparable phenomena in other bacteria.

As source of inocula, Sp+ phenotypes have consistently proven to be highly infective. In fact, Sp+ *Frankia* strains are more infective than Sp− strains (Torrey 1990; Akkermans and Van Dijk 1981; Van Dijk 1984). This observation raises the point of whether this characteristic has emerged once or several times through evolution. To answer that question will require a thorough genotyping and phylogenetic study of these two phenotypes as well as an understanding of the physiological bases underlying the phenomenon.

7.1.1.3 Diversity of the Microsymbiont Based on Vesicles

During the infection process that aims to establish a root nodule structure within the host plant, the bacteria must enter the plant and move to a specific location to develop this symbiosis. The hyphal form of *Frankia* moves through the plant tissues and differentiates into specialized nitrogen-fixing cells, termed vesicles, within the developing nodule structure. Vesicles contain thickened cell walls composed of hopanoid lipids that provide a barrier to oxygen diffusion (Berry et al. 1993) that would otherwise inactivate nitrogenase. In general, vesicles are found in actinorhizal nodules (Table 7.2). However, several notable exceptions are *Casuarina* and *Allocasuarina* and to a lower extent *Myrica*, where the host plants provide a microenvironment with low oxygen pressure through the synthesis of hemoglobin to deliver high amounts of oxygen (Bogusz et al. 1990) and the oxygen-impermeable suberin (Berg 1983). Before the first isolates became available (Callaham et al. 1978), the first attempts at species definition were made based on the size, shape, and orientation of vesicles in the infected cells; *Frankia* microsymbionts were classified into eight species that corresponded to eight actinorhizal plant families (Becking 1970). There are lines of evidence indicating that the host plant family determines the morphology and orientation of symbiotic vesicles. This has been shown for strains isolated from *Myrica gale* and inoculated on *Alnus*, *Elaeagnus*, *Hippophae*, and other *Myrica* species and for strain CcI3 inoculated on *Casuarina* and *Allocasuarina* or on *Myrica gale* (St-Laurent and Lalonde 1986; Torrey and Racette 1988). Infective and effective strains from *Elaeagnus* produce round-shaped vesicles within *Elaeagnus* nodules but produce ineffective *Alnus* nodules that contain no or only very few vesicles (Bosco et al. 1992). Some strains isolated from *Casuarina* nodules form vesicles *in culture* and within *Myrica* nodules, but not in symbiosis with Casuarinaceae, whereas others can fulfill Koch's postulates.

7.1.2 Diversity of Cultivable Frankia Strains

For several reasons including slow growth rates and potential unknown growth requirements, the isolation of the microsymbionts present in actinorhizal root nodules has been a limiting factor in their further investigations. For long time, either early efforts in the identification of these bacteria failed to fulfill Koch's postulates or the isolates were not maintained (Wheeler et al. 2008). The first successful isolation in 1978 (Callaham et al. 1978) opened up the field and has been followed by a series of successful isolations from other actinorhizal plants representing different plant genera (Table 7.2). Bacterial isolates obtained from root nodules of eight families of actinorhizal plants fall into two groups: (1) typical *Frankia* isolates that fulfill Koch's postulates and are able to re-infect their original host plant or other actinorhizal plants and (2) atypical *Frankia* isolates that are unable to re-infect their original host plant or other actinorhizal plants. *Frankia* isolates have been obtained from actinorhizal plants in the Betulaceae, Casuarinaceae, Elaeagnaceae, Rhamnaceae, and Myricaceae families. Besides this inability to re-infect, these isolates may lack one or more morphological and/or symbiotic features. Subsequent molecular studies based on 16S phylogeny indicate that these isolates are *Frankia* but represent a distinct cluster (Normand et al. 1996). Most of these isolates were primarily obtained from nodules of the Coriariaceae, Datiscaceae, Rosaceae, and Rhamnaceae (*Ceanothus*) families but also from other actinorhizal species. Furthermore, many other non-*Frankia* actinobacteria have been isolated from these actinorhizal root nodules (Gtari et al. 2012). The status of these actinobacteria, as nitrogen-fixing, exo- or endosymbiont, or as pathogene, is open to discussion (Ghodhbane-Gtari and Gtari, unpublished).

7.1.2.1 Phenotypic Diversity of *Frankia* Isolates

Morphologically, *Frankia* strains are developmentally complex and form three basic cell types: (1) filamentous and branched hyphae, (2) multilocular sporangia-containing nonmotile spores, and (3) vesicles, specialized thick-walled cells, which are the site of nitrogen fixation under aerobic conditions of low nitrogen availability (Lechevalier 1994). While the filamentous hyphae are the primary vegetative state of a growing culture in most actinobacterial genera, the other cell forms represent unique developmental structures that show variability among isolates and growth conditions. For instance, the extent of sporangia formation varies from isolate to isolate, ranging from being totally suppressed for some strains under certain growth conditions to being abundant in others (Hahn 2008). For example, EAN1pec will only sporulate in media containing carbon sources like glucose or sucrose, while EuI1c sporulation is temperature dependent (Tisa et al. 1983; Krumholz et al. 2003). In this respect, strain ARgP5 (Normand and Lalonde 1982) isolated from nodules containing many sporangia and thus labeled Sp+ was found to produce in vitro numerous sporangia of very large size, up to 50 μm in diameter. In general, vesicle development is induced under nitrogen-limited growth conditions that enable nitrogenase to function under atmospheric partial pressures of oxygen. However, several strains produce vesicles in the presence of fixed nitrogen source, but their numbers are increased dramatically under nitrogen-limited conditions. It has also been claimed that some phenolic compounds induced production of vesicles (Perradin et al. 1983) but this could not be confirmed. Despite the presence of sporangia, vesicles are totally absent in some atypical non-nitrogen-fixing strains such as EuI1c isolated from *Elaeagnus umbellata* that are in some instance able to induce nodules on actinorhizal plants (Baker et al. 1980). A similar lack of vesicles and concomitant absence of nitrogen fixation had been previously noted by Mian et al. (1976) when using *Myrica gale* crushed nodules on *M. faya*, showing that the ability to produce vesicle may be an intrinsic property of the bacterium or the result of defective signaling in the case of heterologous symbionts.

A chemotaxonomic approach has permitted to characterize *Frankia* as having a cell wall composition of type III (meso-diaminopimelic acid, glutamic acid, alanine, glucosamine, and muramic acid), the simplest and most common among actinobacteria. *Frankia* phospholipid pattern contains phosphatidylinositol, phosphatidylinositol mannosides, and diphosphatidylglycerol and is referred to as PI. The major menaquinones include MK-9(H4), MK-9(H6), and MK-9(H8) (Lechevalier 1994). The G + C content ranges from 66 to 73 mol% (as determined from genome sequences). However, Lechevalier (1994) considers that whole-cell sugar patterns of *Frankia* are diverse and differ from those of other actinobacteria. While most isolates have been reported of type D pattern, characterized by xylose as diagnostic constituent and the presence of minor amounts of arabinose, four additional types are also reported with fucose (type E), madurose (type B), galactose (type A), or glucose (type C). 2-O-methyl-D-mannose, detected in all strains tested, has been considered as a diagnostic sugar (Mort et al. 1983).

Another potential aspect of diversity centers on the host plant infection process, which occurs via an intracellular or an intercellular process (Berry and Sunell 1990). However, the infection mechanism appears to be a host plant-dependent process via intercellular penetration or via penetration of deformed root hairs (Table 7.2). Based on cross-inoculation experiments using typical and atypical isolated strains, four *Frankia* phenotypes or groups were recognized as (1) strains infective on *Alnus* and *Myrica*, (2) strains infective on *Casuarina* and *Myrica*, (3) strains infective on Elaeagnaceae and *Myrica*, and (4) strains infective only on *Elaeagnus* (Baker 1987; Torrey 1990). Besides these groups, the existence of some *Frankia* strains that are able to cross normal boundaries has been identified including some *Elaeagnus*-infective *Frankia* strains that are also able to nodulate *Alnus* (Margheri et al. 1985; Bosco et al. 1992). Furthermore, some *Casuarina*-infective strains will also induce nodules on *Alnus* (Nagashima et al. 2008). In these two instances, the nodules produced are ineffective on the heterologous host, and there are no vesicles produced while vesicles are produced in the homologous host. Lastly, some *Casuarina* isolates are unable to re-infect their original actinorhizal host plants while being able to do so on *Elaeagnus* (Diem and Dommergues 1983).

Besides symbiotic traits, efforts to group *Frankia* strains based on their morphology, physiology, cell chemistry, and other physiological characteristics (for a review, see Lechevalier 1994) have failed to reveal potentially useful traits for assessing diversity. Two physiological groups, A and B, have been proposed to differentiate among *Frankia* isolates (Lechevalier 1994; Lechevalier et al. 1983; Lechevalier and Lechevalier 1979). Group A strains are metabolically diverse, noninfective strains that are comparable to other slowly growing saprophytic actinobacteria. Compared to group B strains, they are aerobic, relatively fast growing, and physiologically active. Many group A strains have an array of hydrolytic enzyme activities and utilize a variety of monosaccharides and disaccharides with or without acid production. Group B strains are characterized as strictly microaerophilic slow-growing (doubling times, 2–7 days) bacteria that are physiologically more restricted being generally unable to use carbohydrates, proteins, or starch while preferentially growing on organic acids or Tween compounds as carbon sources. All of the group B strains are able to re-infect their host plants. Although Lechevalier (1994) did not exclude the presence of other types that differ from these two groups (Lechevalier and Ruan 1984), this potential scheme of strain categorization needs to be reexamined and incorporated with the new physiological and genomic information. Besides physiological traits, Dobritsa (1998) clustered 39 selected *Frankia* strains based on their susceptibility to antibiotics, their pigment production, and their host specificity. The majority of studied strains fell into three groups, A, E, and C, corresponding to the three host specificity groups, *Alnus, Elaeagnus,* and *Casuarina*, grouping that has of course no relation to the A/B categories described above. These studies provide an important amount of background information that may help develop a clearer picture of *Frankia* diversity (Table 7.3).

7.1.2.2 Frankia *Genetic* Diversity

Based on their DNA–DNA relatedness, more than 19 genomic species have been delineated among the *Frankia* isolates (Fernandez et al. 1989; Akimov and Dobritsa 1992; Lumini et al. 1996). Although a unified proposal on species is not possible from these different nonoverlapping sets of studies on isolates, the following genomic species were proposed: seven genospecies among *Alnus*-compatible strains, 11 genospecies among Elaeagnaceae-compatible strains, and one genospecies among Casuarinaceae-compatible strains. Some concordance occurs between these described genospecies and the subsequent phenotypic clustering proposed by Dobritsa (1998) and may reflect the taxonomic structure of the genus *Frankia*. Further data or analysis is required to support this model; for instance, amplified fragment length polymorphism (AFLP) not only provides a resolution equivalent to DNA–DNA hybridization but has two added advantages: (1) it permits inferring phylogenetic relationships between genospecies (Coenye et al. 1999; Rademaker et al. 2000; Portier et al. 2006) and (2) is useful in assigning uncultured *Frankia* strains present in nodules containing a mixture of the microbe

Table 7.3 Characteristics of *Frankia* based on symbiotic relatedness, phenotypic, phylogenetic, genospecies, and genomic analysis

Frankia	Host infection phenotype	Biogeography[a]	Number of phenotypic clusters[b]	Phylogenetic clusters[c]	Number of genospecies[d]/AFLP[e]	Genome[f]	Size (Mb)	GC (%)	rRNA
Cluster 1	Betulaceae/Myricaceae	Cosmopolitan	4	1	7	ACN14a	7.49	72.83	6
	Casuarina/Allocasuarina/Myricaceae	Restricted to host plant	1		1	QA3	In progress	70.08	6
						CcI3	5.43		
Cluster 3	Elaeagnaceae/Rhamnaceae/*Gymnostoma*/Myricaceae	Cosmopolitan	8	3	11	EAN1pec	8.98	71.15	9
						EUN1f	9.32	70.82	9
						BMG5.12	In progress		
						BCU110501	In progress		
Cluster 2	Coriariaceae/Datiscaceae/Rosaceae/*Ceanothus*	Restricted to host plant	Not determined	2	Not determined	Dg	5.32	70.04	6
Cluster 4	Infective and nonnitrogen-fixing	Cosmopolitan	5	4	Not determined	Eu1lc	8.81	72.31	9
	Noninfective					CN3	9.97	71.81	9
						DC12	In progress		
	Nitrogen-fixing and noninfective					NRRLB 16386	In progress		

[a]Benson et al. (2004)
[b]Dobritsa (1998)
[c]Normand et al. (1996), Ghodhbane-Gtari et al. (2010), and Nouioui et al. (2011)
[d]Fernandez et al. (1989), Akimov and Dobritsa (1992), and Lumini et al. (1996)
[e]Bautista et al. (2011)
[f]http://www.ncbi.nlm.nih.gov/genome/genomes/13514

Table 7.4 Genetic diversity of Frankia

	Geographic origin	Sample	Typing method	Number of haplotype/sequence	References
Betulaceae					
Alnus	France, Canada	Isolate	16S rDNA sequencing	5	Nazaret et al. (1991)
	New Zealand	Nodule	16S rDNA sequencing	3	Clawson et al. (1997)
	Netherlands	Soil	16S rDNA sequencing	1	Wolters et al. (1997)
	USA	Isolate	16S rDNA sequencing	6	Clawson et al. (1998)
	Finland	Soil	23S rDNA sequencing, rep-PCR	3	Maunuksela et al. (1999)
	Italy, France, Canada, USA	Isolate	PCR-RFLP ITS 16S–23S rDNA and IGS nifD-K	6/7	Lumini and Bosco (1999)
	USA, UK	Nodule	PCR-RFLP ITS 16S–23S rDNA and 16S rDNA sequencing	3	Huguet et al. (2001)
	China	Nodule	PCR-RFLP ITS 16S–23S rDNA and IGS nifD-K PCR-RFLP and	20	Dai et al. (2004)
	UK	Nodule	PCR-RFLP ITS 16S–23S rDNA	10	Ridgway et al. (2004)
	China	Nodule	PCR-RFLP nifD-K	21	Dai et al. (2005)
	Spain	Nodule	PCR-RFLP and sequencing 16S rDNA	2	Gtari et al. (2007c)
	Tunisia	Nodule	PCR-RFLP ITS 16S–23S rDNA, GlnII sequencing	2	Igual et al. (2006)
	India	Nodule	PCR-RFLP and sequencing ITS 16S–23S rDNA	17	Khan et al. (2007)
	Argentina	Isolate	Rep-PCR	?	Valdés La Hens (2007)
	USA	Nodule	NifH sequencing	2	Kennedy et al. (2010)
	USA (Alaska)	Nodule	PCR-RFLP and sequencing nifD-K	3	Anderson et al. (2009)
	USA	Nodule	NifH sequencing	45	Welsh et al. (2009b)
	USA	Nodule	NifH sequencing, rep-PCR	3	Pokharel et al. (2010)
Casuarinaceae					
Allocasuarina/Casuarina	Australia, Senegal, Thailand, USA, Madagascar, Brazil	Isolate	Nif hybridization probe/16S rDNA sequencing	1	Nazaret et al. (1989, 1991)

(continued)

Table 7.4 (continued)

	Geographic origin	Sample	Typing method	Number of haplotype/sequence	References
	Australia, Senegal, Thailand, USA, Madagascar, Brazil	Isolate	PCR-RFLP ITS 16S–23S rDNA and IGS nifD-K	1	Rouvier et al. (1996)
	Australia, Madagascar	Nodule	PCR-RFLP ITS 16S–23S rDNA and IGS nifD-K	5	Rouvier et al. (1996)
	Australia, Madagascar	Nodule	PCR-RFLP ITS 16S–23S rDNA and IGS nifD-K	7	Simonet et al. (1999)
	Pakistan, Uruguay, Sri Lanka, Zimbabwe, Argentina, Tunisia, USA, Kenya, Mexico, Uganda			1	
	Mexico	Isolate	NifH and 16S rDNA probes, PCR-RFLP ITS 16S–23S rDNA	1	Pérez et al. (1999)
	Tunisia	Nodule	PCR-RFLP and sequencing 16S rDNA	1	Gtari et al. (2007c)
	Tunisia	Nodule	PCR-RFLP ITS 16S–23S rDNA and IGS nifD-K	1	Nasr et al. (2007)
	India, China, Brazil, Taiwan, Congo, South Africa	Nodule	PCR-RFLP ITS 16S–23S rDNA and IGS nifD-K		
Gymnostoma	New Caledonia	Soil	PCR-RFLP ITS 16S–23S rDNA	17	Navarro et al. (1999)
Myricaceae	USA	Isolate	RFLP on total DNA	9	Bloom et al. (1989)
	USA, UK	Nodule	16S rDNA sequencing	36	Clawson and Benson (1999)
	USA, Sweden	Nodule	PCR-RFLP ITS 16S–23S rDNA	9	Huguet et al. (2001)
	China	Nodule	NifD-K sequencing	10	He et al. (2004)
	Belgium, France, Spain	Nodule	PCR-RFLP ITS 16S–23S rDNA	9	Huguet et al. (2004)
	Canary Islands, Hawaii, Rwanda, Hungary,	Nodule	PCR-RFLP ITS 16S–23S rDNA	11	Huguet et al. (2005b)
	Japan, Alaska, Peru	Soil	NifH sequencing	73	Welsh et al. (2009a, b)

Family	Location	Source	Method	Number	Reference
Elaeagnaceae Rhamnaceae	France, USA, Argentina, China	Isolate	16S rDNA sequencing	5	Nazaret et al. (1991)
	France	Isolate			Bosco et al. (1992)
		Isolate	PCR-RFLP IGS nifD-K	7	Jamann et al. (1992, 1993)
		Isolate	16S rDNA sequencing		Clawson et al. (1997)
	France	Isolate	PCR-RFLP and sequencing IGS nifD-K	7	Nalin et al. (1997)
	Chile, France, USA, NewZealand, Canada	Isolate/Nodule	16S rDNA sequencing	9	Clawson et al. (1998)
					Lumini and Bosco (1999)
	Chile, France, Italy, USA, Russia, Argentina	Isolate/Nodule	PCR-RFLP ITS 16S–23S rDNA and IGS nifD-K	16/22	Huguet et al. (2001)
	Tunisia	Isolate	16S RDNA and glnII sequencing, rep-PCR	8	Gtari et al. (2004)
	Tunisia	Isolate/Nodule	PCR-RFLP and sequencing nifH	8	Gtari et al. (2007a, b)
	Chili	Nodule	PCR-RFLP ITS 16S–23S rDNA and IGS nifD-K	15	Chávez and Carú (2007)
	Tunisia	Nodule	PCR-RFLP and sequencing 16S rDNA	3	Gtari et al. (2007b)
	USA	Nodule	NifH sequencing	7	Mirza et al. (2009)
	India	Nodule	PCR-RFLP and sequencing ITS 16S–23S	3	Khan et al. (2009)
Ceanothus	USA	Nodule	NifDH probe	?	Baker and Mullin (1994)
	USA	Nodule	rep-PCR	3	Murry et al. (1997)
	New Zealand	Nodule	16S rDNA sequencing	1	Benson et al. (1996)
	USA, New Zealand	Nodule	16S rDNA sequencing	1	Clawson et al. (1998)
	USA	Nodule	16S rDNA sequencing	1	Ritchie and Myrold (1999a)
	USA	Nodule	PCR-RFLP ITS 16S–23S rDNA	4	Ritchie and Myrold (1999b)
	USA	Nodule	Rep-PCR, ITS 16S–23S rDNA sequencing	3/3	Jeong and Myrold (1999)
	USA	Nodule	ITS 16S–23S rDNA sequencing	2	Oakley et al. (2004)
	USA	Nodule	16S rDNA, glnA and ITS 16S–23S rDNA sequencing	1	Vanden Heuvel et al. (2004)

(continued)

Table 7.4 (continued)

	Geographic origin	Sample	Typing method	Number of haplotype/sequence	References
Coriariaceae	France, Italy, New Zealand, Mexico	Nodule	16S rDNA sequencing	1	Nick et al. (1992)
Datiscaceae	Italy	Nodule	16S rDNA sequencing	1	Bosco et al. (1994)
Rosaceae	Pakistan	Nodule/atypical isolate	16S rDNA and nifH sequencing	1	Mirza et al. (1994a)
	Pakistan	Nodule	16s rDNA probe	1	Mirza et al. (1994b)
	New Zealand	Nodule	PCR-RFLP and sequencing 16S rDNA	1	Benson et al. (1996)
	USA, New Zealand	Nodule	16S rDNA sequencing	1	Clawson et al. (1998)
	USA	Nodule	16S rDNA, glnA and ITS 16S–23S rDNA sequencing	1	Vanden Heuvel et al. (2004)

and the plant DNA (Bautista et al. 2011). The nine genospecies as defined by Fernandez et al. (1989) have been validated and supplementary strains have been assigned to genospecies (Table 7.3) (Bautista et al. 2011).

PCR-based approaches have been extensively used to investigate *Frankia* diversity (Jamann et al. 1993; Hahn et al. 1999). Throughout the years, several primers combinations targeting different gene sets with different levels of resolution have been developed for PCR-based analyses of *Frankia* in culture and directly in root nodules. Table 7.4 briefly summarizes data on genetic diversity for *Frankia* in culture, in field-collected nodules, or in soil through baiting-capturing assays. Despite the incomplete information on several actinorhizal species and from several geographic provenances, *Frankia* associated with actinorhizal plants from Fagales (except *Casuarina* and *Allocasuarina*) and Rhamnales (except *Ceanothus*) are highly diverse genetically compared to those associated with plant from *Datisca, Coriaria, Rosaceae, Ceanothus,* and to *Casuarina* and *Allocasuarina*. One primary explanation for this observation is that the distinctive native range of actinorhizal plant species varies from global to very limited. Secondarily, compatible *Frankia* strains develop an apparently diverse degree of lifestyle and host plant infeodation. Through the use of the highly variable ITS 16S–23S rDNA sequence for 67 *Frankia* strains, Ghodhbane-Gtari et al. (2010) showed a different degree of variability within and between *Frankia* clusters. These ITS sequences were widely diverse among cultivable *Frankia* strains belonging to clusters 1 (except casuarinas strains), 3, and 4 but showed a limited diversity in cluster 2 and among "casuarina strains" of cluster 1. From a geographic point of view, *Frankia* cluster 1 (except "casuarina strains") and 3 are globally distributed in soils regardless of the presence of a suitable host plant suggesting that these strains have a relative high saprophytic potential (Benson et al. 2004). In contrast, "casuarina strains" and the uncultivable *Frankia* cluster 2 have a narrow geographic distribution and are highly dependent on their respective host plants for their distribution (Benson et al. 2004). Furthermore, "casuarina strains" represent the most difficult group to isolate in pure culture among the genus and generated only one cultivable genotype among the seven *Casuarina* plant-nodulating strains (Rouvier et al. 1996; Simonet et al. 1999). Despite many attempts by several investigators, cluster 2 strains have not been isolated in pure culture and have been considered obligate symbionts. From molecular studies, these strains are considered a very homogenous group. Hence, the degree of diversity among *Frankia* genus is consistent with the idea of a gradual erosion of *Frankia* diversity concomitantly with a shift from saprophytic noninfective/noneffective to facultative and symbiotic lifestyle.

The actinorhizal symbiosis is based on an exchange with the plants feeding the microbe with photosynthates and the microbe feeding the plant with nitrogenous compounds. In the case of *Frankia*, it has also been shown that some *Alnus*-infective *Frankia* strains were more effective by 30 % than others, and it was hypothesized that this could be linked to the intensity of sporulation (Normand and Lalonde 1982). Another factor that appears to influence the N_2 efficiency of strains is their hydrogenase activity that recycles the hydrogen, a by-product of nitrogen fixation (Sellstedt and Winship 1990). There have been otherwise no

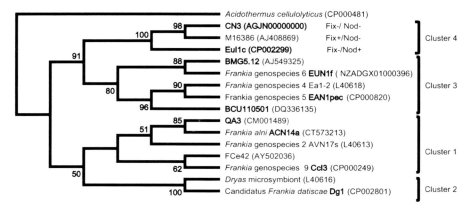

Fig. 7.1 16S rDNA-based phylogeny of *Frankia* genus showing the four clusters with representative strains. *Bold acronyms* indicate genomes completed or in progress

comprehensive studies to correlate symbiotic efficiency with specific metabolic activity, which would help understand the contrasted ecological and evolutionary pressures shaping the symbiosis.

7.1.2.3 Frankia *Phylogeny*

The first studies of *Frankia* using 16S rRNA gene sequences and oligonucleotide cataloguing permitted to show that *Frankia* was close to *Geodermatophilus* and *Blastococcus* and, despite all having multilocular sporangia, away from *Dermatophilus,* leading to a modified definition of the family Frankiaceae (Hahn et al. 1989).

Based on entire 16S rRNA gene sequences, four main subdivisions or clusters are defined for the genus *Frankia* (Normand et al. 1996) (Fig. 7.1): (1) cluster 1, *Frankia* infective on *Alnus* (including *Alnus rugosa* Sp+ microsymbionts that are seldom isolated in pure culture), *Casuarina*, and *Myrica*; (2) cluster 2, uncultured microsymbionts from Rosaceae (*Dryas, Cercocarpus, Chamaebatia, Purshia* etc.), *Coriaria*, and *Datisca* species; (3) cluster 3, *Elaeagnus*-infective strains; and (4) cluster 4, "atypical" strains which are unable to re-infect actinorhizal host plants or form noneffective root nodule structures that are unable to fix nitrogen. This grouping is in concordance with comparative sequence analysis of an actinobacteria-specific insertion in domain III of the 23S rRNA (Honerlage et al. 1994). The four established clusters have been further extended based on 16S rDNA sequence analysis to include additional *Frankia* strains from cluster 3 that are *Gymnostoma*-infective strains (Navarro et al. 1997) as well as Rhamnaceae except for *Ceanothus*-infective strains (Clawson et al. 1998) and cluster 2 strains that are Rosaceae microsymbionts (Clawson et al. 1998). Huguet et al. (2005a) reexamined the specificity and the phylogeny of Myricaceae-infective strains in their natural habitats and showed that the plant genus *Myrica*, including *M. gale* and *M. hartwegii*, and genus *Comptonia*, including *C. peregrina*, belong to a phylogenetic cluster distinct from the other *Myrica* species that were thus transferred into a new genus, *Morella*. Moreover, *Frankia* cluster 1 strains include *Alnus*-infective strains, *Comptonia*, and *Myrica*-infective strains, while *Frankia* cluster 3 strains include Elaeagnaceae- and Rhamnaceae-infective strains as well as *Morella*-infective strains under natural conditions. This *Frankia* clustering scheme, first proposed by Normand et al. (1996), has been confirmed through an analysis of the combined partial 16S rDNA and *glnA* sequences (Clawson et al. 2004), *glnII* sequences (Cournoyer and Lavire 1999; Gtari et al. 2004), *nifH* sequences (Jeong and Myrold 1999; Gtari et al. 2007a; Mirza et al. 2009), and the ITS 16S–23S rDNA sequences (Vanden Heuvel et al. 2004; Ghodhbane-Gtari et al. 2010). While the overall clustering remains conserved between each locus analyzed showing different levels of resolution, a clear evolutionary history of the four clusters still remains unclear and inconsistent. A promising evolutionary hypothesis was generated from an analysis of concatenated *gyrB*, *nifH*, and *glnII* sequences (Nouioui et al. 2011). Cluster 4 (noninfective and/or nonnitrogen-fixing *Frankia*) was positioned as the deepest branch close to the common ancestor followed by cluster 3 (Rhamnaceae- and Elaeagnaceae-infective *Frankia*) branching, while cluster 2 represents uncultured *Frankia* microsymbionts of the Coriariaceae, Datiscaceae, Rosaceae, and Ceanothus sp. (Rhamnaceae) and cluster 1 (Betulaceae-, Myricaceae-, and Casuarinaceae-infective *Frankia*) appears to have diverged more recently. This topology should soon be testable using multigene approaches, made possible by the numerous genomes presently being acquired.

The question of the origin of the genus remains debatable. As explained above, the Frankiales is not a coherent taxon when using several protein-based phylogenies (Wu et al. 2009). Beyond the 16S rRNA gene phylogeny, *Acidothermus*, which lives in hydrothermal springs, and Geodermatophilaceae, present in stone surfaces and desert soil, are not convincing sister groups to *Frankia* and require further analysis at a genome-wide basis. Culture-independent molecular analysis of soils may yield other more likely sister groups as suggested from the amplification of soil from the rhizosphere of *Alnus* (Normand and Chapelon 1997) or bioweathered feldspar mineral (HQ674869, unpublished). Metagenomic approaches may also provide further baseline information on *Frankia* diversity and its sister organisms. For example, by analyzing different foliose green algal lichen species (*Parmelia sulcata*, *Rhizoplaca chrysoleuca*, *Umbilicaria americana*, and *Umbilicaria phaea*) that are often in symbiosis with bacteria, Bates et al. (2011) detected many sequences closely related to *Frankia*. The metagenome of *Trewia nudiflora* stem bark (EU289513) shows also the existence of a highly diverse array of uncultivated and nonsymbiotic soil bacteria that is thus gaining recognition. Such soil saprophytic actinobacteria would thus be the likely ancestor of the *Frankia* genus.

7.1.2.4 Frankia *Genomes*

The development of novel next-generation sequencing methods has increased our ability to sequence whole genomes. The elucidation of *Frankia* genomes has provided a wealth of information and a baseline to understand the bacterium. Since the first published report on genomes of three *Frankia* strains (Normand et al 2007), there have been several other genomes representing the four major lineages that are either sequenced or in progress (Table 7.3). At present this group includes three genomes from cluster 1 (ACN14a, CcI3, and QA3), one from cluster 2 (Dg1) (Persson et al. 2011), four from cluster 3 (EAN1pec, EUN1f, BCU110501, and BMG5.12), and three from cluster 4 (EuI1c, CN3, and DC12). An initial analysis of these draft genomes has provided insight on these bacteria and their diversity. First, there is a correlation between host-specific range and genome size. Those strains that have a wide host plant range had very large genomes (8.8–9.1 Mb range), while those strains with a narrow host plant range that are potentially obligate symbiont have the smallest genomes (5.4 Mb). The medium host-range strains had a genome size that was in between the previous two (7.5 Mb). Cluster 4 strains had the largest genome sizes (9.1–9.5 Mb). Furthermore, the metabolically diverse *Frankia* groups (clusters 3 and 4) had larger genomes than the other two clusters. Among these genomes, a core of *Frankia* genes was identified that may provide tools to further access *Frankia* diversity including *Frankia*-unique genes that may serve as monochronometers and molecular markers. One gene that we are currently investigating is a *Frankia*-specific *whi*B-like gene that codes for an Fe-S-caged transcriptional regulator found in all of the *Frankia* genomes. The WhiB family of transcriptional regulator is uniquely found in members of the Actinobacteria (den Hengst and Buttner 2008), and at least six different clusters of the gene are found among the *Frankia* with five having homologues (Wishart and Tisa, unpublished) in the well-studied *Mycobacteria* (Alam et al. 2009).

Concordantly with the very low DNA–DNA homology displayed by some *Frankia* genospecies (Fernandez et al. 1989), *Frankia* genomes revealed a remarkably large divergence in genome size, the largest reported so far for a bacterial genus. Due to the continuous gene gains, gene losses, and lateral gene transfers, bacterial genomes are considered as dynamic and plastic entities (Snel et al. 2002; Daubin et al. 2003; Lawrence and Hendrickson 2005; Ochman 2005). Depending to the bacterial lifestyle, the resulting genome expansion or reduction is mainly driven, respectively, by adaptation to extensive environmental changes of free-living bacteria or to stable conditions within associated host, as parasitic or symbiotic, bacteria (Gao and Gupta 2012). One of the most explicit example is represented by *Frankia* where extended symbiotic or saprophytic lifestyle favors, respectively, genome contraction or expansion. One inconsistency of this hypothesis might be the relatively small genome sizes (2.4–5.3 Mb) observed for other nonsymbiotic Frankiales such as *Acidothermus* (Barabote et al. 2009), *Geodermatophilus* (Ivanova et al. 2010), *Blastococcus* (Chouaia et al. 2012), and *Modestobacter* (Normand et al. 2012). However, this can be minimized by the fact that excepting *Frankia* genus, species of Frankiales do not form a coherent phylogenetic lineage and the taxonomic regrouping of the order is artificial and only supported by 16S rDNA (Gao and Gupta 2012; Normand and Benson 2012; Zhi et al. 2009). Furthermore, *Frankia* is the sole genus that inhabits for most of its life cycle in the soil, a biotope shown by Konstantinidis and Tiedje (2004) to be the most permissive one for slow growth and large genome microbes.

7.2 Conclusion

Despite the lack of data at local scale for several host plant species and areas, *Frankia* diversity studies indicate that host plant evolution and speciation have resulted in a sustained erosion of diversity and a reduction in genome size. *Frankia* strains in symbiosis with host plants with a narrow geographic distribution have the smallest population diversity and genome size, while those in symbiosis with cosmopolitan host plants have an expanded breath of population diversity and maintain larger genome sizes. Likewise, atypical *Frankia* (noninfective and/or non-nitrogen-fixing strains) present the highest population diversity and largest genome sizes.

Early attempts to discriminate among *Frankia* genus based on general taxonomic methods are now considered to be in need of reassessment, based on the data published by Dobritsa (1998); for example, we can anticipate that old interpretations were speculative. Our recently increased knowledge on *Frankia* diversity will not only help species definition according to conventional nomenclature. One obstacle is the continuity of speculative conclusions and misinterpreted results present in the published data. For example, host specificity groups are not as clear as reported for some groups. Strains that are strictly infective on Elaeagnaceae/Rhamnaceae are not distinguishable among currently available culture collections from those infective on Elaeagnaceae/Rhamnaceae and Fagales, especially on Myricaceae. Moreover, are all Myricaceae plant host members promiscuous with respect to *Frankia* clusters? The chemotaxonomic positioning of *Frankia* strains representative of the four phylogenetic clusters is missing. Thus, the issue of *Frankia* speciation has been neglected and needs to be addressed in the light of the recent increase of genomic and diversity baseline data, made possible by the advent of NGS approaches.

Acknowledgments MG was supported in part by a grant from the Ministère de l'Enseignement Supérieur et de la Recherche Scientifique–Tunisia (LabMBA-206) and a Visiting Scientist Program administered by the NH Agricultural Experimental Station at the University of New Hampshire. LST was supported in part by the Agriculture and Food Research Initiative Grant 2010-65108-20581 from the USDA National Institute of Food and Agriculture, the Hatch grant NH530, and the College of Life Sciences and Agriculture at the University of New Hampshire, Durham, NH. PN acknowledges receiving grant SESAM (2011–2013) from the French Agence Nationale de la Recherche (ANR).

References

Akimov VN, Dobritsa SV (1992) Grouping of *Frankia* strains on the basis of DNA relatedness. Syst Appl Microbiol 15:372–379

Akkermans ADL, Van Dijk C (1981) Non-leguminous root-nodule symbioses with actinomycetes and *Rhizobium*. In: Broughton JW (ed) Nitrogen fixation, vol 1. Academic, London, pp 57–103

Alam MS, Garg SK, Agrawal P (2009) Studies on structural and functional divergence among seven WhiB proteins of *Mycobacterium tuberculosis* H37Rv. FEBS J 276:76–93

Anderson MD, Ruess RW, Myrold DD, Taylor DL (2009) Host species and habitat affect nodulation by specific *Frankia* genotypes in two species of *Alnus* in interior Alaska. Oecologia 160:619–630

Baker DD (1987) Relationships among pure cultured strains of *Frankia* based on host specificity. Physiol Plant 70:245–248

Baker DD, Mullin BC (1994) Diversity of *Frankia* nodule endophytes of the actinorhizal shrub *Ceanothus* as assessed by RFLP patterns from single nodule lobes. Soil Biol Biochem 26:547–552

Baker DD, Newcomb W, Torrey JG (1980) Characterization of an ineffective actinorhizal microsymbiont, *Frankia* sp. EuI1. Can J Microbiol 26:1072–1089

Barabote RD, Xie G, Leu DH, Normand P, Necsulea A, Daubin V et al (2009) Complete genome of the cellulolytic thermophile *Acidothermus cellulolyticus* 11B provides insights into its ecophysiological and evolutionary adaptations. Genome Res 19:1033–1043

Bates ST, Cropsey GW, Caporaso JG, Knight R, Fierer N (2011) Bacterial communities associated with the lichen symbiosis. Appl Environ Microbiol 77:1309–1314

Bautista GH, Cruz HA, Nesme X, Valdés M, Mendoza HA, Fernandez MP (2011) Genomospecies identification and phylogenomic relevance of AFLP analysis of isolated and non-isolated strains of *Frankia* spp. Syst Appl Microbiol 34:200–206

Becking JH (1970) Frankiaceae fam. nov. (Actinomycetales) with one new combination and six new species of the genus *Frankia* Brunchorst 1886, 174. Int J Syst Bacteriol 20:201–220

Benson DR, Silvester WB (1993) Biology of *Frankia* strains, actinomycete symbionts of actinorhizal plants. Microbiol Rev 57:293–319

Benson DR, Stephens DW, Clawson ML, Silvester WB (1996) Amplification of 16S rRNA genes from *Frankia* strains in root nodules of *Ceanothus griseus, Coriaria arborea, Coriaria plumosa, Discaria toumatou,* and *Purshia tridentata*. Appl Environ Microbiol 62:2904–2909

Benson DR, Vanden Heuvel BD, Potter D (2004) Actinorhizal symbioses: diversity and biogeography. In: Gillings M (ed) Plant microbiology. BIOS Scientific, Oxford

Berg RH (1983) Preliminary evidence for the involvement of suberization in infection of *Casuarina*. Can J Bot 61:2910–2918

Berry A, Sunell L (1990) The infection process and nodule development. In: Schwintzer C, Tjepkema JD (eds) The biology of *Frankia* and actinorhizal plants. Academic, San Diego, CA, pp 61–81

Berry AM, Harriott OT, Moreau RA, Osman SF, Benson DR, Jones AD (1993) Hopanoid lipids compose the *Frankia* vesicle envelope, presumptive barrier of oxygen diffusion to nitrogenase. Proc Natl Acad Sci USA 90:6091–6094

Bloom RA, Mullin BC, Tate RL III (1989) DNA restriction patterns and DNA-DNA solution hybridization studies of *Frankia* isolates from *Myrica pensylvanica* (bayberry). Appl Environ Microbiol 55:2155–2160

Bogusz D, Llewellyn DJ, Craig S, Dennis ES, Appleby CA, Peacock WJ (1990) Nonlegume hemoglobin genes retain organ-specific expression in heterologous transgenic plants. Plant Cell 2:633–641

Bosco M, Fernandez MP, Simonet P, Materassi R, Normand P (1992) Evidence that some *Frankia* sp. strains are able to cross boundaries between *Alnus* and *Elaeagnus* host specificity groups. Appl Environ Microbiol 58:1569–1576

Bosco M, Jamann S, Chapelon C, Simonet P, Normand P (1994) *Frankia* microsymbiont in *Dryas drummondii* nodules is closely related to the microsymbiont of *Coriaria* and genetically distinct from other characterized *Frankia* strains. In: Hegazi HA, Fayez M, Monib M (eds) Nitrogen fixation with non-legumes. The American University in Cairo Press, Cairo, pp 173–183

Callaham D, Del Tredici P, Torrey JG (1978) Isolation and cultivation in vitro of the actinomycete causing root nodulation in *Comptonia*. Science 199:899–902

Chávez M, Carú M (2007) Genetic diversity of Frankia microsymbionts in root nodules from Colletia hystrix (Clos.) plants by sampling at a small-scale. World J Microbiol Biotechnol 22:813–820

Chouaia B, Crotti E, Brusetti L, Daffonchio D, Essoussi I, Nouioui I, Sbissi I, Ghodhbane-Gtari F, Gtari M, Vacherie B, Barbe V, Médigue C, Gury J, Pujic P, Normand P (2012) Genome sequence of Blastococcus saxobsidens DD2, a stone-inhabiting bacterium. J Bacteriol 194:2752–2753

Clawson ML, Benson DR (1999) Natural diversity of *Frankia* strains in actinorhizal root nodules from promiscuous hosts in the family Myricaceae. Appl Environ Microbiol 65:4521–4527

Clawson ML, Benson DR, Stephens DW, Resch SC, Silvester WB (1997) Typical *Frankia* infect actinorhizal plants exotic to New Zealand. New Zealand J Bot 35:361–367

Clawson ML, Carú M, Benson DR (1998) Diversity of *Frankia* strains in root nodules of plants from the families Elaeagnaceae and Rhamnaceae. Appl Environ Microbiol 64:3539–3543

Clawson ML, Bourret A, Benson DR (2004) Assessing the phylogeny of *Frankia*-actinorhizal plant nitrogen-fixing root nodule symbioses with *Frankia* 16S rRNA and glutamine synthetase gene sequences. Mol Phylogenet Evol 31:131–138

Coenye T, Schouls LM, Govan JR, Kersters K, Vandamme P (1999) Identification of Burkholderia species and genomovars from cystic fibrosis patients by AFLP fingerprinting. Int J Syst Bacteriol 4:1657–1666

Cournoyer B, Lavire C (1999) Analysis of *Frankia* evolutionary radiation using *gln*II sequences. FEMS Microbiol Letts 177:29–34

Dai Y, Cao J, Tang X, Zhang C (2004) Diversity of *Frankia* in nodules of *Alnus nepalensis* at Gaoligong mountains revealed by IGS, PCR-RFLP analysis. Chin J Appl Ecol 15:186–190

Dai Y, Zhang C, Xiong Z, Zhang Z (2005) Correlations between the ages of Alnus host species and the genetic diversity of associated endosymbiotic Frankia strains from nodules. Sci China C Life Sci 48:76–81

Daubin V, Lerat E, Perrière G (2003) The source of laterally transferred genes in bacterial genomes. Genome Biol 4:R57

Dawson JO (2007) The ecology of actinorhizal plants. In: Pawlowski K, Newton WE (eds) Nitrogen-fixing actinorhizal symbioses nitrogen fixation: applications and research progress, vol 6. Springer, Dordrecht, pp 199–234

den Hengst CD, Buttner MJ (2008) Redox control in actinobacteria. Biochim Biophys Acta 1780:1201–1216

Diem H, Dommergues Y (1983) The isolation of *Frankia* from nodules of *Casuarina*. Can J Bot 61:2822–2825

Dobritsa SV (1998) Grouping of *Frankia* strains on the basis of susceptibility to antibiotics, pigment production and host specificity. Int J Syst Evol Microbiol 48:1265–1275

Doyle JJ (2011) Phylogenetic perspectives on the origins of nodulation. Mol Plant Microbe Interact 24:1289–1295

Fernandez MP, Meugnier H, Grimont PAD, Bardin R (1989) Deoxyribonucleic acid relatedness among members of the genus *Frankia*. Int J Syst Bacteriol 39:424–429

Gao B, Gupta RS (2012) Phylogenetic framework and molecular signatures for the main clades of the phylum Actinobacteria. Microbiol Mol Biol Rev 76:66–112

Ghodhbane-Gtari F, Nouioui I, Chair M, Boudabous A, Gtari M (2010) 16S-23S rRNA intergenic spacer region variability in the genus *Frankia*. Microb Ecol 60:487–495

Gtari M, Dawson JO (2011) An overview of actinorhizal plants in Africa. Funct Plant Biol 38:653–661

Gtari M, Brusetti L, Skander G, Mora D, Boudabous A, Daffonchio D (2004) Isolation of Elaeagnus-compatible Frankia from soils collected in Tunisia. FEMS Microbiol Lett 234:349–355

Gtari M, Brusetti L, Hassen A, Mora D, Daffonchio D, Boudabous A (2007a) Genetic diversity among *Elaeagnus* compatible *Frankia* strains and sympatric-related nitrogen-fixing actinobacteria revealed bynifH sequence analysis. Soil Biol Biochem 39:372–377

Gtari M, Daffonchio D, Boudabous A (2007b) Assessment of the genetic diversity of Frankia microsymbionts of *Elaeagnus angustifolia* L. plants growing in a Tunisian date-palm oasis by analysis of PCR amplified nifD-K intergenic spacer. Can J Microbiol 53:440–445

Gtari M, Daffonchio D, Boudabous A (2007c) Occurrence and diversity of Frankia in Tunisian soil. Physiol Plant 130:372–379

Gtari M, Ghodhbane-Gtari F, Nouioui I, Beauchemin N, Tisa LS (2012) Phylogenetic perspectives of nitrogen-fixing actinobacteria. Arch Microbiol 194:3–11

Hahn D (2008) Polyphasic taxonomy of the genus *Frankia*. In: Pawlowski K, Newton WE (eds) Nitrogen-fixing actinorhizal symbioses. Springer, Dordrecht, pp 25–47

Hahn D, Lechevalier M, Fischer A, Stackebrandt E (1989) Evidence for a close phylogenetic relationship between members of the genera *Frankia, Geodermatophilus*, and "Blastococcus" and emendation of the family Frankiaceae. Syst Appl Microbiol 11:236–242

Hahn D, Nickel A, Dawson JO (1999) Assessing *Frankia* populations in plants and soil using molecular methods. FEMS Microbiol Ecol 29:215–227

He XH, Chen LG, Hu XQ, Asghar S (2004) Natural diversity of nodular microsymbionts of Myrica rubra. Plant Soil 262:229–239

Honerlage W, Hahn D, Zepp K, Zyer J, Normand P (1994) A hypervariable region provides a discriminative target for specific characterization of uncultured and cultured *Frankia*. Syst Appl Microbiol 17:433–443

Huguet V, McCray Batzli J, Zimpfer JF, Normand P, Dawson JO, Fernandez MP (2001) Diversity and specificity of *Frankia* strains in nodules of sympatric *Myrica gale, Alnus incana,* and *Shepherdia canadensis* determined by *rrs* gene polymorphism. Appl Environ Microbiol 67:2116–2122

Huguet V, Mergeay M, Cervantes E, Fernandez MP (2004) Diversity of *Frankia* strains associated to *Myrica gale* in Western Europe: impact of host plant (*Myrica* vs. *Alnus*) and of edaphic factors. Environ Microbiol 6:1032–1041

Huguet V, Gouy M, Normand P, Zimpfer JF, Fernandez MP (2005a) Molecular phylogeny of Myricaceae: a reexamination of host-symbiont specificity. Mol Phylogenet Evol 34:557–568

Huguet V, Land EO, Casanova JG, Zimpfer JF, Fernandez MP (2005b) Genetic diversity of Frankia microsymbionts from the relict species *Myrica faya* (Ait.) and *Myrica rivas-martinezii* (S.) in Canary Islands and Hawaii. Microb Ecol 49:617–625

Huss-Danell K (1997) Actinorhizal symbioses and their N2 fixation. New Phytol 136:375–405

Igual JM, Valverde A, Velázquez E, Regina IS, Rodríguez-Barrueco C (2006) Natural diversity of nodular microsymbionts of *Alnus glutinosa* in the Tormes River Basin. Plant Soil 280:373–383

Ivanova N, Sikorski J, Jando M, Munk C, Lapidus A, Glavina Del Rio T, Copeland A, Tice H, Cheng JF, Lucas S, Chen F, Nolan M, Bruce D, Goodwin L, Pitluck S, Mavromatis K, Mikhailova N, Pati A, Chen A, Palaniappan K, Land M, Hauser L, Chang YJ, Jeffries CD, Meincke L, Brettin T, Detter JC, Rohde M, Göker M, Bristow J, Eisen JA, Markowitz V, Hugenholtz P, Kyrpides NC, Klenk HP (2010) Complete genome sequence of *Geodermatophilus obscurus* type strain (G-20). Stand Genomic Sci 2:158–167

Jamann S, Fernandez MP, Moiroud A (1992) Genetic diversity of Elaeagnaceae-infective *Frankia* strains isolated from various soils. Acta Oecol 13:395–405

Jamann S, Fernandez MP, Normand P (1993) Typing method for N2-fixing bacteria based on PCR-RFLP–application to the characterization of Frankia strains. Mol Ecol 2:17–26

Jeong SC, Myrold DD (1999) Genomic fingerprinting of *Frankia* microsymbionts from *Ceanothus* copopulations using repetitive sequences and polymerase chain reactions. Can J Bot 77:1220–1230

Käppel M, Wartenberg H (1958) Der Formenwechsel des Actinomyces alni Peklo in den M, Wurzeln von Alnus glutinosa Gaertner. Arch Mikrobiol 30:46–63

Kennedy PG, Schouboe JL, Rogers RH, Weber MG, Nadkarni NM (2010) *Frankia* and *Alnus rubra* canopy roots: an assessment of genetic diversity, propagule availability, and effects on soil nitrogen. Microb Ecol 59:214–220

Khan A, Myrold DD, Misra AK (2007) Distribution of *Frankia* genotypes occupying *Alnus nepalensis* nodules with respect to altitude and soil characteristics in the Sikkim Himalayas. Physiol Plant 130:364–371

Khan A, Myrold DD, Misra AK (2009) Molecular diversity of Frankia from root nodules of Hippophae salicifolia D.Don found in Sikkim. Indian J Microbiol 49:196–200

Konstantinidis KT, Tiedje JM (2004) Trends between gene content and genome size in prokaryotic species with larger genomes. Proc Natl Acad Sci USA 101:3160–3165

Krumholz GD, Chval MS, McBride MJ, Tisa LS (2003) Germination and physiological properties of *Frankia* spores. Plant Soil 254:57–67

Lalonde M (1979) Immunological and ultrastructural demonstration of nodulation of the European *Alnus glutinosa* (L.) Gaertn. host plant by an actinomycetal isolate from the North American *Comptonia peregrina* (L.) Coult. root nodule. Bot Gaz 140(suppl):S35–S43

Lawrence JG, Hendrickson H (2005) Genome evolution in bacteria: order beneath chaos. Curr Opin Microbiol 8:572–578

Lechevalier MP (1994) Taxonomy of the genus *Frankia (Actinomycetales)*. Int J Syst Bacteriol 44:1–8

Lechevalier MP, Lechevalier HA (1979) The taxonomic position of the actinomycetic endophytes. In: Wheeler CT, Perry DA, Gordon JC (eds) Symbiotic nitrogen fixation in the management of temperate forests. Forest Research Laboratory, Oregon State University, Corvallis, OR, pp 111–122

Lechevalier MP, Ruan JS (1984) Physiology and chemical diversity of *Frankia* spp. isolated from nodules of *Comptonia peregrina* (L.) Coult. and *Ceanothus americanus* L. Plant Soil 78:15–22

Lechevalier MP, Baker D, Horriere F (1983) Physiology, chemistry, serology, and infectivity of two *Frankia* isolates from *Alnus incana* subsp. *rugosa*. Can J Bot 61:2826–2833

Lumini E, Bosco M (1999) Polymerase chain reaction-restriction fragment length polymorphisms for assessing and increasing biodiversity of *Frankia* culture collections. Can J Bot 77:1261–1269

Lumini E, Bosco M, Fernandez MP (1996) PCR-RFLP and total DNA homology revealed three related genomic species among broad-host-range *Frankia* strains. FEMS Microbiol Ecol 21:303–311

Margheri MC, Vagnoli L, Favilli F, Sili C (1985) Proprieta morfofisiologiche di *Frankia* ceppo EanII57 da *Elaeagnus angustifolia*, infettivo su *Alnus glutinosa*. Ann Microbiol 35:143–153

Maunuksela L, Zepp K, Koivula T, Zeyer J, Haahtela K, Hahn D (1999) Analysis of *Frankia* populations in three soils devoid of actinorhizal plants. FEMS Microbiol Ecol 28:11–21

Mian S, Bond G, Rodriguez-Barrueco C (1976) Effective and ineffective root nodules in *Myrica faya*. Proc R Soc Lond B 194:285–293

Mirza MS, Hameed S, Akkermans ADL (1994a) Genetic diversity of *Datisca cannabina*-compatible *Frankia* strains as determined by sequence analysis of the PCR-amplified 16S rRNA gene. Appl Environ Microbiol 60:2371–2376

Mirza MS, Akkermans WM, Akkermans ADL (1994b) PCR-amplified 16S rRNA sequence analysis to confirm nodulation of *Datisca cannabina* L. by the endophyte of *Coriaria nepalensis* Wall. Plant Soil 160:147–152

Mirza BS, Welsh A, Rasul G, Rieder JP, Paschke MW, Hahn D (2009) Variation in Frankia populations of the Elaeagnus host infection group in nodules of six host plant species after inoculation with soil. Microb Ecol 58:384–393

Mort A, Normand P, Lalonde M (1983) 2-O-Methyl-D-Mannose, a key sugar in the taxonomy of *Frankia*. Can J Microbiol 29:993–1002

Murry MA, Konopka AS, Pratt SD, Vandergon TL (1997) The use of PCR-based typing methods to assess the diversity of *Frankia* nodule endophytes of the actinorhizal shrub *Ceanothus*. Physiol Plant 99:714–721

Nagashima Y, Tani C, Yamamoto M, Sasakawa H (2008) Host range of Frankia strains isolated from actinorhizal plants growing in Japan and their relatedness based on 16S rDNA. Soil Sci Plant Nutr 54:379–386

Nalin R, Normand P, Domenach AM (1997) Distribution and N2-fixing activity of *Frankia* strains in relation with soil depth. Physiol Plant 99:732–738

Nasr H, Domenach AM, Ghorbel MH, Benson DR (2007) Divergence in symbiotic interactions between same genotypic PCR–RFLP *Frankia* strains and different Casuarinaceae species under natural conditions. Physiol Plant 130:400–408

Navarro E, Nalin R, Gauthier D, Normand P (1997) The nodular microsymbionts of *Gymnostoma* spp. are *Elaeagnus*-infective *Frankia* strains. Appl Environ Microbiol 63:1610–1616

Navarro E, Jaffre T, Gauthier D, Gourbiere F, Rinaudo G, Simonet P, Normand P (1999) Distribution of *Gymnostoma* spp. Microsymbiotic Frankia strains in New Caledonia is related to soil type and to host-plant species. Mol Ecol 8:1781–1788

Nazaret S, Simonet P, Normand P, Bardin R (1989) Genetic diversity among *Frankia* strains isolated from *Casuarina* nodules. Plant Soil 118:241–247

Nazaret S, Cournoyer B, Normand P, Simonet P (1991) Phylogenetic relationships among *Frankia* genomic species determined by use of amplified 16S rDNA sequences. J Bacteriol 173:4072–4078

Nesme X, Normand P, Tremblay FM, Lalonde M (1985) Nodulation speed of *Frankia* sp. on *Alnus glutinosa*, *Alnus crispa*, and *Myrica gale*. Can J Bot 63:1292–1295

Nick G, Paget E, Simonet P, Moiroud A, Normand P (1992) The nodular endophytes of *Coriaria* sp. form a distinct lineage within the genus *Frankia*. Mol Ecol 1:175–181

Normand P, Benson DR (2012) Order XVI Frankiales. In: Goodfellow M et al (eds) Bergey's manual of systematic bacteriology, vol 5, 2nd edn, The actinobacteria. Springer, New York, NY, pp 508–510

Normand P, Chapelon C (1997) Direct characterization of *Frankia* and of close phylogenetic neighbors from an *Alnus viridis* rhizosphere. Physiol Plant 99:722–731

Normand P, Lalonde M (1982) Evaluation of *Frankia* strains isolated from provenances of two *Alnus* species. Can J Microbiol 28:1133–1142

Normand P, Orso S, Cournoyer B, Jeannin P, Chapelon C, Dawson J, Evtushenko L, Misra AK (1996) Molecular phylogeny of the genus *Frankia* and related genera and emendation of family Frankiaceae. Int J Syst Bacteriol 46:1–9

Normand P, Lapierre P, Tisa LS, Gogarten JP, Alloisio N, Bagnarol E, Bassi CA, Berry AM, Bickhart DM, Choisne N, Couloux A, Cournoyer B, Cruveiller S, Daubin V, Demange N, Francino MP, Goltsman E, Huang Y, Kopp OR, Labarre L, Lapidus A, Lavire C, Marechal J, Martinez M, Mastronunzio JE, Mullin BC, Niemann J, Pujic P, Rawnsley T, Rouy Z, Schenowitz C, Sellstedt A, Tavares F, Tomkins JP, Vallenet D, Valverde C, Wall LG, Wang Y, Medigue C, Benson DR (2007) Genome characteristics of facultatively symbiotic *Frankia* sp. strains reflect host range and host plant biogeography. Genome Res 17:7–15

Normand P, Gury J, Pujic P, Chouaia B, Crotti E, Brusetti L, Daffonchio D, Vacherie B, Barbe V, Médigue C, Calteau A, Ghodhbane-Gtari F, Essoussi I, Nouioui I, Abbassi-Ghozzi I, Gtari M (2012) Genome sequence of radio-resistant *Modestobacter marinus* strain BC501, a representative actinobacterium thriving on calcareous stone surfaces. J Bacteriol 194(17):4773–4774. doi:10.1128/JB.01029-12

Nouioui I, Ghodhbane-Gtari F, Beauchemin NJ, Tisa LS, Gtari M (2011) Phylogeny of members of the Frankia genus based on gyrB, nifH and glnII sequences. Antonie Van Leeuwenhoek 100:579–587

Oakley B, North M, Franklin JF, Hedlund BP, Staley JT (2004) Diversity and distribution of *Frankia* strains symbiotic with *Ceanothus* in California. Appl Environ Microbiol 70:6444–6452

Ochman H (2005) Genomes on the shrink. Proc Natl Acad Sci USA 102:11959–11960

Oono R, Schmitt I, Sprent JI, Denison RF (2010) Multiple evolutionary origins of legume traits leading to extreme rhizobial differentiation. New Phytol 187:508–520

Pérez NO, Olivera H, Vásquez L, Valdés M (1999) Genetic characterization of Mexican *Frankia* strains nodulating *Casuarina equisetifolia*. Can J Bot 77:1214–1219

Perradin Y, Mottet M, Lalonde M (1983) Influence of phenolics on in vitro growth of *Frankia* strains. Can J Bot 61:2807–2814

Persson T, Benson DR, Normand P, Vanden Heuvel B, Pujic P, Chertkov O, Teshima H, Bruce DC, Detter C, Tapia R, Han S, Han J, Woyke T, Pitluck S, Pennacchio L, Nolan M, Ivanova N, Pati A, Land ML, Pawlowski K, Berry AM (2011) Genome sequence of "Candidatus Frankia datiscae" Dg1, the uncultured microsymbiont from nitrogen-fixing root nodules of the dicot Datisca glomerata. J Bacteriol 193:7017–7018

Pokharel A, Mirza BS, Dawson JO, Hahn D (2010) *Frankia* populations in soil and root nodules of sympatrically grown *Alnus* taxa. Microb Ecol 61:92–100

Portier P, Fischer-LeSaux M, Mougel C, Lerondelle C, Chapulliot D, Thioulouse J, Nesme X (2006) Identification of genomic species in *Agrobacterium* Biovar 1 by AFLP genomic markers. Appl Environ Microbiol 72:7123–7131

Quispel A (1990) Discoveries, discussions, and trends in research on actinorhizal root nodule symbioses before 1978. In: Schwintzer CR, Tjepkema JD (eds) The biology of *Frankia* and actinorhizal plants. Academic, New York, NY, pp 15–33

Rademaker JLW, Hoste B, Louws FJ, Kersters K, Swings J, Vauterin L, Vauterin P, DeBruijn FJ (2000) Comparison of AFLP and rep-PCR genomic fingerprinting with DNA-DNA homology studies: *Xanthomonas* as a model system. Int J Syst Evol Microbiol 50:665–677

Ridgway KP, Marland LA, Harrison AF, Wright J, Young JPW, Fitter AH (2004) Molecular diversity of Frankia in root nodules of Alnus incana grown with inoculum from polluted urban soils. FEMS Microbiol Ecol 50:255–263

Ritchie NJ, Myrold DD (1999a) Phylogenetic placement of uncultured *Ceanothus* microsymbionts using 16S rRNA gene sequences. Can J Bot 77:1208–1213

Ritchie NJ, Myrold DD (1999b) Geographic distribution and genetic diversity of *Ceanothus*-infective *Frankia* strains. Appl Environ Microbiol 65:1378–1383

Rouvier C, Prin Y, Reddell P, Normand P, Simonet P (1996) Genetic diversity among *Frankia* strains nodulating members of the family Casuarinaceae in Australia revealed by PCR and restriction fragment length polymorphism analysis with crushed root nodules. Appl Environ Microbiol 62:979–985

Schwintzer CR (1990) Spore-positive and spore-negative nodules. In: Schwintzer CR, Tjepkema JD (eds) The biology of *Frankia* and actinorhizal plants. Academic, New York, NY, pp 177–193

Sellstedt A, Winship LJ (1990) Acetylene, not ethylene, inactivates the uptake hydrogenase of actinorhizal nodules during acetylene reduction assays. Plant Physiol 94:91–94

Simonet P, Navarro E, Rouvier C, Reddell P, Zimpfer J, Dommergues Y, Bardin R, Combarro P, Hamelin J, Domenach AM, Gourbiere F, Prin Y, Dawson JO, Normand P (1999) Co-evolution between *Frankia* populations and host plants in the family Casuarinaceae and consequent patterns of global dispersal. Environ Microbiol 1:525–533

Snel B, Bork P, Huynen MA (2002) Genomes in flux: the evolution of archaeal and proteobacterial gene content. Genome Res 12:17–25

Soltis DE, Soltis PS, Morgan DR, Swensen SM, Mullin BC, Dowd JM, Martin PG (1995) Chloroplast gene sequence data suggest a single origin of the predisposition for symbiotic nitrogen fixation in angiosperms. Proc Natl Acad Sci USA 92:2647–2651

St-Laurent L, Lalonde M (1986) Isolation and characterization of *Frankia* strains isolated from *Myrica gale*. Can J Bot 65:1356–1363

Tisa L, McBride M, Ensign J (1983) Studies of growth and morphology of *Frankia* strains EAN1pec, EuI1c, CpI1 and ACN1AG. Can J Bot 61:2768–2773
Torrey JG (1987) Endophyte sporulation in root nodules of actinorhizal plants. Physiol Plant 70:279–288
Torrey JG (1990) Cross-inoculation groups within *Frankia*. In: Schwintzer CR, Tjepkema JD (eds) The biology of *Frankia* and actinorhizal plants. Academic, New York, NY, pp 83–106
Torrey JG, Racette S (1988) Specificity among the Casuarinaceae in root nodulation by *Frankia*. Plant Soil 118:157–164
Valdés La Hens (2007) Aislamiento y caracterización de actinomycetes simbióticos de nódulos de *Alnus acuminata*. PhD thesis. University of Quilmes, Argentina, p 312
Van Dijk C (1978) Spore formation and endophyte diversity in root nodules of *Alnus glutinosa* (L.) Vill. New Phytol 81:601–615
Van Dijk C (1984) Ecological aspects of spore formation in the Frankia-Alnus symbiosis. PhD thesis, State University, Leiden, The Netherlands
Van Dijk C, Merkus E (1976) A microscopic study of the development of a spore-like stage in the life cycle of the root nodule endophyte of *Alnus glutinosa* (L.) Vill. New Phytol 77:73–90
Vanden Heuvel BD, Benson DR, Bortiri E, Potter D (2004) Low genetic diversity among *Frankia* spp. strains nodulating sympatric populations of actinorhizal species of Rosaceae, *Ceanothus* (Rhamnaceae) and *Datisca glomerata* (Datiscaceae) west of the Sierra Nevada (California). Can J Microbiol 50:989–1000
VandenBosch KA, Torrey JG (1984) Consequences of sporangial development for nodule function in root nodules of *Comptonia peregrina* and *Myrica gale*. Plant Physiol 76:556–560
VandenBosch KA, Torrey JG (1985) Development of endophytic *Frankia* sporangia in field- and laboratory-grown nodules of *Comptonia peregrina* and *Myrica gale*. Am J Bot 72:99–108
Welsh A, Mirza BS, Rieder JP, Paschke M, Hahn D (2009a) Diversity of frankiae in root nodules of *Morella pensylvanica* grown in soils from five continents. Syst Appl Microbiol 32:201–210
Welsh AK, Dawson JO, Gottfried GJ, Hahn D (2009b) Diversity of Frankia populations in root nodules of geographically isolated Arizona alder trees in central Arizona (United States). Appl Environ Microbiol 75:6913–6918
Wheeler CT, Akkermans ADL, Berry AB (2008) *Frankia* and actinorhizal plants: a historical perspective. In: Pawlowski K, Newton WE (eds) Nitrogen-fixing actinorhizal symbioses. Springer, Berlin, pp 1–24
Wolters DJ, Van Dijk C, Zoetendal EG, Akkermans ADL (1997) Phylogenetic characterization of ineffective *Frankia* in *Alnus glutinosa* (L.) Gaertn. nodules from wetland soil inoculants. Mol Ecol 6:971–981
Wu D, Hugenholtz P, Mavromatis K, Pukall R, Dalin E, Ivanova NN, Kunin V, Goodwin L, Wu M, Tindall BJ, Hooper SD, Pati A, Lykidis A, Spring S, Anderson IJ, D'haeseleer P, Zemla A, Singer M, Lapidus A, Nolan M, Copeland A, Han C, Chen F, Cheng JF, Lucas S, Kerfeld C, Lang E, Gronow S, Chain P, Bruce D, Rubin EM, Kyrpides NC, Klenk HP, Eisen JA (2009) A phylogeny-driven genomic encyclopaedia of Bacteria and Archaea. Nature 462:1056–1060
Zhi XY, Li WJ, Stackebrandt E (2009) An update of the structure and 16S rRNA gene sequence-based definition of higher ranks of the class Actinobacteria, with the proposal of two new suborders and four new families and emended descriptions of the existing higher taxa. Int J Syst Evol Microbiol 59:589–608

Part III
Endophytic Plant Growth-Promoting Rhizobacteria (PGPR)

Chapter 8
Abiotic Stress Tolerance Induced by Endophytic PGPR

Patricia Piccoli and Rubén Bottini

8.1 Introduction

Plant roots release substantial amounts of compounds into the surrounding soil and as consequence a dense microflora colonize the roots and use the root exudates and lysates to propagate, survive, and disperse along the growing root (Ryan and Delhaize 2001). Among the root-zone microflora, plant growth-promoting rhizobacteria (PGPR) colonize the rhizosphere of many plant species and confer beneficial effects, such as increased plant growth and reduced susceptibility to diseases caused by plant pathogenic fungi, bacteria, viruses, and nematodes (Kloepper et al. 2004). Some PGPR also elicit physical or chemical changes related to plant defense, a process referred to as "induced systemic resistance" (ISR, Kloepper et al. 2004; Van Loon and Glick 2004). However, few reports have been published on PGPR as elicitors of tolerance to abiotic stresses, such as drought, salt, and nutrient deficiency or excess. Recent work by several groups shows that PGPR also elicit the so-called induced systemic tolerance to salt and drought (Yang et al. 2009 and references included therein).

Although the potential use of PGPR (Kloepper et al. 1991; Bashan and Holguin 1998) as plant growth and yield enhancers (either not under abiotic or biotic stresses) has been known for several decades (Döbereiner et al. 1976; Okon and Labandera-González 1994; Glick et al. 1999), there has been limited success in practical use. A considerable amount of literature has been produced in the meantime relating the understanding of the mechanisms involved in the purported beneficial effects (Dimkpa et al. 2009), but the yield increases obtained by inoculation with PGPR are rather modest. As an example, the effect of long-living spore

---

P. Piccoli • R. Bottini (✉)
Laboratorio de Bioquímica Vegetal, Instituto de Biología Agrícola de Mendoza (IBAM), Facultad de Ciencias Agrarias, Consejo Nacional de Investigaciones Científicas y Técnicas-Universidad Nacional de Cuyo, Almirante Brown 500, M5528AHB Chacras de Coria, Argentina
e-mail: rbottini@fca.uncu.edu.ar

formulations of the *Bacillus* strain FZB24 and FZB42 produced and commercialized in Germany were tested in extended and long-lasting field trials with potatoes under practical farming conditions. Forty-eight field trials performed with FZB24 under standard conditions (250 g/ha) during 1995–1998 resulted in an increase of tuber potatoes yield of 8.3 % as the mean value. Similar experiments conducted in 2002, 2003, and 2004 confirmed that FZB strains in fact enhance productivity of potatoes about 7.5–10 % (mean value of 87 independent trials). Best results were obtained if application of bacilli was combined with use of fungicides. In such cases an increase of tuber yield up to 40 % was obtained (Choudhary and Johri 2009), although the question rises how much of the effect was due to the inoculation.

PGPR might also increase nutrient uptake from soils, thus reducing the need for fertilizers and preventing the accumulation of nitrates and phosphates in agricultural soils. A reduction in fertilizer use would lessen the effects of water contamination from fertilizer runoff and lead to savings for farmers (Gyaneshwar et al. 2002; Mantelin and Touraine 2004). Some studies tested the hypothesis that PGPR might enable agricultural plants to maintain productivity with reduced rates of fertilizer application. In field-grown maize partial replacement of N fertilization with *Azospirillum* sp., bacterization has been obtained (Fulchieri and Frioni 1994), although under conditions of limited soil fertility. In another field study with *Triticum aestivum* L. (Shaharoona et al. 2008), the yield for plants that were given 75 % of the recommended amount of N–P–K fertilizer plus a PGPR strain was equivalent to the yield for plants that were given the full amount of fertilizer but without PGPR. In tomato, the dry weight of tomato transplants grown in the greenhouse was significantly greater with two PGPR strains and 75 % fertilizer than with the full amount of fertilizer and without PGPR; after transplanting to the field, the yields for some combinations of PGPR and mycorrhizal fungi with fertilizers at 50 % were greater than the yield of the 100 % fertilizer control without microbes.

8.2 Plants Are Mostly Subjected to Several Stresses That Act as Signals Preparing the Organism to Afford More Stress

Plants growing in natural environments are frequently exposed to different stresses, even in environments where there is no apparent reason. For instance, one of the most common stresses is water restriction during mid-afternoon hours even in humid environments (Granier and Tardieu 1999). Plants respond to such stressful situations by modifying their metabolism in a way that they become more tolerant to that situation but also to other stresses. As an example, the plant hormone abscisic acid (ABA) is produced by dehydrating roots and transported to leaves as a sensitive indicator of degree of water deficits in the soils (Zhang and Davies

1989). Thus leaves from plants growing under water-stress conditions usually show an abrupt rise in ABA level which produce stomata closure and therefore decreasing water losses that prepare the plant to cope with other stresses (Huang and Zhu 2004). Also, grape plants perceiving relatively high levels of UV-B radiation (potentially harmful for the plant's tissues) responded by enhancing ABA levels, which in turn promote increase of polyphenols that filter UV-B at epidermal level, trigger antioxidant enzymatic mechanisms and the sterol-structural defense of membranes (Berli et al. 2010). By consequence, acclimation to relatively high UV-B conditions prepares plants to better afford drought, and, conversely, water-stress situations may promote UV-B tolerance.

8.3 Beneficial Bacteria Are Present in Plants and They Often Have the Capability to Stimulate the Host's Growth and Yield

In nature, beneficial endophytic bacteria play a fundamental role in plant adaptation to the environment (Hallman et al. 1997). They can pre-sensitize plant cell metabolism, so that upon exposure to stress primed plants are able to respond more quickly and efficiently than non-primed individuals (Compant et al. 2005). Moreover, microorganisms from the rhizosphere or tissues of a specific plant may be better adapted to that plant and environment conditions, and they may therefore provide better control of diseases than organisms originally belonging from other rhizosphere (Cook 1993). Since PGPR were characterized as beneficial for plants, genera like *Azospirillum, Herbaspirillum, Bacillus, Burkholderia, Pseudomonas, Gluconacetobacter*, and others have been tested to improve growth and yield in different crops (Egamberdiyeva and Höflich 2004). Oliveira et al. (2002) and Muthukumarasamy et al. (2006) found that *Gluconacetobacter* and *Herbaspirillum* improved N uptake and increased biomass in sugar cane. *Burkholderia phytofirmans* is able to colonize several parts of grapevines cv. Chardonnay (Compant et al. 2005), to increase root and shoot dry weight, and to induce growth of secondary roots (Ait Barka et al. 2000; Compant et al. 2005). Among the PGPR *Azospirillum* sp. may be considered the most important genus for improving plant growth or crop yield worldwide, under a variety of environmental and soil conditions (Bashan and de-Bashan 2010). The main visible effects of inoculations with *Azospirillum* sp. and other PGPR are in the plant root system. *Azospirillum* sp. can promote root elongation (Levanony and Bashan 1989; Dobbelaere et al. 1999), formation of lateral and adventitious roots (Creus et al. 2005; Molina-Favero et al. 2008), root hairs (Hadas and Okon 1987; Fulchieri et al. 1993), and branching of root hairs (Jain and Patriquin 1985), which increase the root area active in water and nutrient uptake. It has also been proposed that some PGPR increase the plant tolerance against abiotic stresses such as drought, salinity, and metal toxicity (Creus et al. 1997; Mayak et al. 2004; Cohen et al. 2009). In addition to their usefulness in agriculture, they possess potential in solving environmental problems by improving growth of desert plants and by reducing pollution through phytoremediation that decontaminates soils and waters (de-Bashan et al. 2011).

Among the mechanisms that explain the beneficial effects on plant growth and yield promotion by *Azospirillum* sp. is the production of phytohormones (Costacurta and Vanderleyden 1995; Bastián et al. 1998; Bloemberg and Lugtenberg 2001; Bottini et al. 2004), mainly the auxin indol-3-acetic acid (IAA; Crozier et al. 1988; Patten and Glick 2002), gibberellins (GAs; Bottini et al. 1989, 2004; Fulchieri et al. 1993), and cytokinins (Arkhipova et al. 2005). Results of several researches indicated that the most important mechanism in plant-growth promotion by PGPR is production of phytohormones and/or enhancement of phytohormones synthesis by the plant tissues [for reviews see Bottini et al. (2004), Spaepen et al. (2007), Bashan and de-Bashan (2010)]. Also production of the stress-related hormones, salicylic acid (De Meyer and Höfte 1997; Forchetti et al. 2010), ABA (Cohen et al. 2008, 2009) and jasmonic acid (Forchetti et al. 2010; Piccoli et al. 2011), and of the signaling molecule nitric oxide (Creus et al. 2005), have also been indicated as possible players in the game. Complementary, impairment in ethylene production via 1-aminocyclopropane-1-carboxylicacid (ACC) deaminase activity may also be involved in plant growth promotion by PGPR (Belimov et al. 2009).

Considering the numerous interactions that exist among the different hormonal signaling pathways in plants, it is difficult to assess which of these pathways is the primary target of PGPR. It is known that many signals affect root architecture and branching, conspicuously plant hormones that regulate initiation and growth of lateral roots (Nibau et al. 2008). As an example, the formation of lateral roots which is an important postembryonic event that is vital for the growth of plants is primarily regulated by auxins (Casimiro et al. 2003), and PGPR are well-known auxin producers (Patten and Glick 2002). More likely, the rhizobacteria alters not just a single but several hormonal pathways, even in a pleiotropic manner, which could account for the different morphological changes observed, for example lateral root elongation and root hair development (Fulchieri et al. 1993; Dobbelaere and Okon 2007). Also, lateral root primordial emergence is repressed by limiting water supply (Deak and Malamy 2005); therefore PGPR-produced ABA may account for plant's growth under water restriction (Cohen et al. 2008, 2009). However, De Smet et al. (2006) observed that Arabidopsis seedlings grown on medium containing exogenous ABA did not form clearly visible lateral roots, and roots of aba2-1 plants (deficient in ABA) had a larger number of lateral roots that grew longer than those of the wild type (Deak and Malamy 2005). Notwithstanding, *Azospirillum* sp. increased the number of lateral roots and root's fresh weight in Arabidopsis aba2-1 plants suggesting the effects may be in turn mediated by IAA and GAs (Cohen, Bottini, Pontin, Berli, Moreno, Travaglia, Boccalandro, Piccoli, unpublished results), which is sustained by the fact that *Azospirillum* reversed the dwarf phenotype dx and dy in rice mutants (Cassán et al. 2001a, b). Indeed, it has been proven that *Azospirillum* sp. produce both IAA (Crozier et al. 1988) and GAs

(Bottini et al. 1989). For instance, elicitation of *Arabidopsis thaliana* growth promotion by PGPR involved signaling of cytokinins, brassinosteroids, auxin, salicylic acid, and GAs (Ryu et al. 2003, 2005, 2007). López-Bucio et al. (2007) found that the bacteria *Bacillus megaterium* is able to increase lateral root number and root hair length in *A. thaliana* plants through auxin- and ethylene-independent mechanisms. It has also been found that *Bacillus* sp. modulates the root-system architecture in *A. thaliana* through the emission of volatiles (Gutiérrez-Luna et al. 2010).

A. piechaudii ARV8 confer tolerance to water stress in tomatoes and peppers and therefore can promote growth under such situation (Mayak et al. 2004). It has also been reported that *Bacillus pumilus* and *Bacillus licheniformis* increased leaf area in dwarfed alder seedlings (Gutiérrez-Mañero et al. 2001) and in Arabidopsis (Ryu et al. 2007), and *Azospirillum lipoferum* USA 5b augmented leaf area in both well-watered and water-stressed individuals (Cohen et al. 2009). Zhang et al. (2008) reported that *Bacillus subtilis* strain GB03 can stimulate growth of *Arabidopsis* by the emission of organic compounds and increase photosynthesis through modulation of ABA signaling (Xie et al. 2009). Increased chlorophyll, and consequently, enhanced photosynthesis, is a known response of plant to inoculation with several PGPR (Deka and Dileep 2002), including *Azospirillum* sp. (Bashan et al. 2006).

Yield increases ranging from 10 to 20 % with PGPR applications have also been documented for several agricultural crops (Kloepper et al. 1991). Increases in seed yield had also been observed in lettuce inoculated with *Bacillus* sp. (Arkhipova et al. 2005). Wheat grain yield was increased by up to 30 % (Okon and Labandera-González 1994) by inoculation with *Azospirillum brasilense*, and partial replacement of N fertilization with *Azospirillum* sp. bacterization has been obtained in maize (Fulchieri and Frioni 1994).

8.4 PGPR Enhance Stress Tolerance in Plants Subjected to Abiotic Stresses

Drought is one of the main stressful environmental conditions that reduce crop yield worldwide. It has been shown that diurnal water stress is a condition normally found in most species growing in temperate climates during the noon and afternoon hours, even though the soil water status may be at field capacity. This temporary stress might then affect the growth rate (Granier and Tardieu 1999). In fact, mild water deficits that cause reduction of the plant tissues turgidity (equivalent to a reduction of 10–15 % in the plant water content) result in large changes in growth and metabolism. The plant's tolerance to water stress results from both morphological adaptation and responses at biochemical and genetic levels. The central response to water deficits however is the increase in ABA biosynthesis and/or a decrease in ABA breakdown (Bray 2002). In plants experiencing drought, it is assumed that ABA acts as the signal that prepares the plant to resist the water deficit, mainly by controlling stomata closure and water loss (Zhang and Outlaw 2001). Also, there is evidence suggesting that ABA plays a role in root branching, improving the plant water uptake capacity (De Smet et al. 2006), and it has been demonstrated that ABA sprayed onto leaves promotes vegetative growth in *Ilex paraguariensis* plants by alleviating diurnal water stress (Sansberro et al. 2004). In wheat and soybean, applications of ABA increase leaf carotenoid content and favor the allocation of carbohydrates into grains (Travaglia et al. 2007, 2009). ABA treatments also augment yield in wheat cultivated under moderate water restriction (Travaglia et al. 2010) and enhance fruit set (and so yield) in grapevines (Quiroga et al. 2009). Otherwise, gibberellin A_3 (GA_3) and ABA treatments promote carbon allocation in roots and berries of grapevines (Moreno et al. 2011).

ABA has been characterized by full scan mass spectrometry as a by-product of chemically defined growth cultures of *A. brasilense* Sp 245. ABA production by the bacteria increased when NaCl was added to the culture medium, and ABA levels were enhanced in *A. thaliana* seedlings inoculated with *A. brasilense* Sp 245 (Cohen et al. 2008). As well, *A. lipoferum* inoculated to 45-day-old maize plants increased ABA levels and reversed the effects of applied inhibitors of ABA and GAs synthesis, fluridone, and prohexadione-Ca, respectively, on the hormone levels and the plant's drought tolerance. That is, ABA and GA_3 contributed to water-stress alleviation of maize plants by *A. lipoferum* (Cohen et al. 2009). Although reports on the effects of plant hormones produced by PGPR on plants are abundant, information regarding the mechanism involved in ABA effects (including *A. thaliana* as a model) after bacterization with PGPR under drought conditions is scarce. Cho et al. (2012) proposed an ABA-independent stomata closure in Arabidopsis bacterized with *Pseudomonas chlororaphis* but without further information regarding the mechanism involved. In a recent work (Cohen, Bottini, Pontin, Berli, Moreno, Travaglia, Boccalandro, Piccoli, unpublished results), Arabidopsis was used as a model system to further analyze the physiological basis by which *A. brasilense* Sp 245 affects the plant's response to water restriction. In an agar-grown system inoculation with *A. brasilense* of Arabidopsis wild-type Col-0, the mutant aba2-1 and the transgenic pGL2::GUS genotypes during early growth increased both the aerial part and roots through modifications in root architecture, that is, increase in lateral root number and length (both main and lateral roots). Inoculation with *A. brasilense* also increased photosynthetic and photoprotective pigments and retarded water loss in *A. thaliana* wild-type plants and augmented ABA levels in both the wild-type and aba2-1 mutant plants; that is, *A. brasilense* has the ability to produce ABA in vivo and restore the wild-type phenotype. Also, inoculation with *A. brasilense* and application of GA_3 increased leaf trichomes, and *Azospirillum*, IAA, and GA_3 increased the number of lateral roots in transgenic pGL2::GUS plants. Furthermore, *Azospirillum* increased growth, survival, seed yield, and ABA and proline levels and decreased stomatal conductance in Arabidopsis plants subjected or not to drought. In a recent study (Salomon, Cohen, Bottini, Gil, Moreno, Piccoli, unpublished results), several bacteria strains were isolated and characterized from roots and rhizosphere of *Vitis vinifera* cv. Malbec. Two of them, *P. fluorescens* and *B. licheniformis*, elicitate production of defense-related terpenes purportedly involved in protection of dehydrated membranes (Beckett et al. 2012).

Salinity is an important stress that hinders crop yield in many parts of the world, mainly via enhanced biosynthesis of ethylene that inhibits root growth (Feng and Barker 1992). To overcome the ethylene-induced root inhibition is a requirement for successful production, and studies have shown that ethylene level in plants is regulated by ACC deaminase, which is present in some PGPR (Belimov et al. 2009; Nadeem et al. 2010). However, the effects of salinity are far more complicated than the simple augmentation of ethylene synthesis, and a main component in salt effects is drought, so phytohormones provided (or their synthesis being induced) by PGPR may be essential in the stress alleviation (Creus et al. 1997; Mayak et al. 2004; Cohen et al. 2008, 2009; Piccoli et al. 2011). As well, salinity increases oxidative stress so any mechanism that collaborates in avoiding such oxidative situation may help the plant defense (Grassmann et al. 2002). Apart, PGPR are identified that stimulate plant roots to excrete organic acids that chelate Na excess in the soil solution as a mechanism that protects plants against salinity (Li et al. 2007). Production of siderophores by PGPR, that is, substances that increase nutrient uptake under mineral shortage (Bagg and Neilands 1987), may in turn favor heavy (or excess of) metals sequestration, in a sort of homeostatic balance for the rhizospheric environment (Khan et al. 2009).

Also, other studies have provided new insights into the phytoremediation of metal-contaminated soils by PGPR (Zhuang et al. 2007). A metal-tolerant PGPR, *Enterobacter* sp., was able to enhance extraction of Ni, Zn, and Cr in *Brassica juncea*, along with plant growth increase by IAA production and with ACC deaminase activity (Kumar et al. 2008). A similar capacity was found for two strains of a metal resistant strain of *Pseudomonas* on the plant growth and the uptake of Ni, Cu, and Zn by *Ricinus communis* (Rajkumar and Freitas 2008) and for several PGPR in *Brassica* spp. (Ma et al. 2009). In the hyperaccumulating plant, *Sedum alfredii*, inoculation with *Burkholderia cepacia* enhanced plant growth submitted to excess of Zn, Pb, and Cd, in correlation with stimulation of organic acid production by the plant roots (Li et al. 2007). Co-inoculation of lupines with a mix of metal-resistant PGPR (including *Bradyrhizobium* sp., *Pseudomonas* sp., and *Ochrobactrum cytisi*) improved plant biomass and a decrease in metal accumulation due to a protective effect on the rhizosphere (Dary et al. 2010). The secretion of organic acids appears to be a functional metal resistance mechanism that chelates the metal ions extracellularly, reducing their uptake and subsequent impacts on root physiological processes (Li et al. 2007). In fact, nutrient-solubilizing activity of PGPR has been associated with the release of organic acids and a drop in the pH that may have the additional effect of metal quelation (Dastager et al. 2010). In other words, the mechanisms involved in metal detoxification appear to be the same as per salinity, that is, metal quelation, extrusion pumps, and ACC deaminase activity, with other side effects like nutrient solubilization.

Regarding extreme temperatures, detrimental situations are mostly related with water restriction, and few reports deal specifically with the subject. In this respect,

B. phytofirmans has been reported as enhancing plant resistance to low temperatures (Ait Barka et al. 2002, 2006).

Solar ultraviolet-B radiation (UV-B, wavelength range 280–315 nm), even in relatively small amounts, is potentially harmful for plants (Frohnmeyer and Staiger 2003). The syntheses promoted by UV-B of terpenic compounds (Gil et al. 2012) that have the capacity of repelling the attack and propagation of pathogens (Vögeli and Chappell 1988; Escoriaza et al. 2013) are in turn enhanced by PGPR (Salomon, Cohen, Bottini, Gil, Moreno, Piccoli, unpublished results). Therefore PGPR prepare the plant tissues to afford UV-B damage. As well, considering that biosynthesis of membrane-related sterols is enhanced by ABA (Berli et al. 2010), an additional benefit of bacterial-produced ABA may be the increase in sterols that are associated with stability and integrity of membranes, and attenuation of oxidative damage.

8.5 Prospective

Most of the studies analyzed the plant/PGPR interaction based on one plant species interacting with one bacterial strain (or a few strains of the same species) under controlled conditions looking for some specific mechanism, which is not the case in nature where a plant population is interrelating with a whole bunch of microorganism of both deleterious or beneficial characteristics along with many environmental signals (some stressful like water restriction, some trophic like photosynthetic light, some just signaling other factors like light quality that indicates neighbors, etc.). In other words, the interaction of so many factors produces a "holistic" response by the plant in which some of these factors may be predominant (like water availability), whereas some others are not (like the relative presence of PGPR). Therefore, no general mechanisms can account for improvement in growth and yield of crops. That is, one may expect a noticeable beneficial effect of PGPR in precise situations in which bacterization becomes effective in alleviating specific environmental treats. In this regard, future research should be directed in addressing specific questions posed for definite conditions in order to develop technologies effective in abiotic stress tolerance induced by endophytic PGPR.

References

Ait Barka E, Belarbi A, Hachet C, Nowak J, Audran JC (2000) Enhancement of *in vitro* growth and resistance to gray mold of *Vitis vinifera* co-cultured with plant growth-promoting rhizobacteria. FEMS Microbiol Lett 186:91–95

Ait Barka E, Gognies S, Nowak J, Audran JC, Belarbi A (2002) Inhibitory effect of endophyte bacteria on *Botrytis cinerea* and its influence to promote the grapevine growth. Biol Control 24:135–142

Ait Barka E, Nowak J, Clément C (2006) Enhancement of chilling resistance of inoculated grapevine plantlets with a plant growth-promoting rhizobacterium, *Burkholderia phytofirmans* strain PsJN. Appl Environ Microbiol 72:7246–7252

Arkhipova TN, Veselov SU, Melentiev AI, Martynenko EV, Kudoyarova GR (2005) Ability of bacterium *Bacillus subtilis* to produce cytokinins and to influence the growth and endogenous hormone content of lettuce plants. Plant Soil 272:201–209

Bagg A, Neilands JB (1987) Molecular mechanism of regulation of siderophore-mediated iron assimilation. Microbiol Rev 51:509–518

Bashan Y, de-Bashan L (2010) How the plant growth-promoting bacterium *Azospirillum* promotes plant growth—a critical assessment. Adv Agron 108:77–136

Bashan Y, Holguin G (1998) Proposal for the division of plant growth promoting rhizobacteria into two classifications: biocontrol-PGPB (plant growth-promoting bacteria) and PGPB. Soil Biol Biochem 30:1225–1228

Bashan Y, Bustillos JJ, Leyva LA, Hernandez J-P, Bacilio M (2006) Increase in auxiliary photoprotective photosynthetic pigments in wheat seedlings induced by *Azospirillum brasilense*. Biol Fertil Soils 42:279–285

Bastián F, Cohen AC, Piccoli P, Luna V, Baraldi R, Bottini R (1998) Production of indole-3-acetic acid and gibberellins A_1 and A_3 by *Acetobacter diazotrophicus* and *Herbaspirillum seropedicae* in chemically-defined culture media. Plant Growth Regul 24:7–11

Beckett M, Loreto F, Velikova V, Brunetti C, Di Ferdinando M, Tattini M, Calfapietra C, Farrant JM (2012) Photosynthetic limitations and volatile and non-volatile isoprenoids in the poikilochlorophyllous resurrection plant *Xerophyta humilis* during dehydration and rehydration. Plant Cell Environ 35:2061–2074. doi:10.1111/j.1365-3040.2012.02536.x

Belimov AA, Dodd IC, Hontzeas N, Theobald JC, Safronova VI, Davies WJ (2009) Rhizosphere bacteria containing ACC deaminase increase yield of plants grown in drying soil via both local and systemic hormone signaling. New Phytol 181:413–423

Berli FJ, Moreno D, Piccoli P, Hespanhol-Viana L, Silva MF, Bressan-Smith R, Cavagnaro JB, Bottini R (2010) Abscisic acid is involved in the response of grape (*Vitis vinifera* L.) cv. Malbec leaf tissues to filtration of UV-B radiation by enhancing UV-absorbing compounds, antioxidant enzymes and membrane sterols. Plant Cell Environ 33:1–10

Bloemberg GV, Lugtenberg BJJ (2001) Molecular basis of plant growth promotion and biocontrol by rhizobia. Curr Opin Plant Biol 4:343–350

Bottini R, Fulchieri M, Pearce D, Pharis RP (1989) Identification of gibberellins A_1, A_3 and iso-A_3 in cultures of *Azospirillum lipoferum*. Plant Physiol 90:45–47

Bottini R, Cassán F, Piccoli P (2004) Gibberellin production by bacteria and its involvement in plant growth promotion and yield increase. Appl Microbiol Biotechnol 65:497–503

Bray EA (2002) Abscisic acid regulation of gene expression during water-deficit stress in the era of the *Arabidopsis* genome. Plant Cell Environ 25:153–161

Casimiro I, Beeckman T, Graham N, Bhalerao R, Zhang H, Casero P, Sandberg G, Bennett MJ (2003) Dissecting Arabidopsis lateral root development. Trends Plant Sci 8:165–171

Cassán F, Bottini R, Schneider G, Piccoli P (2001a) *Azospirillum brasilense* and *Azospirillum lipoferum* hydrolyze conjugates of GA_{20} and metabolize the resultant aglycones to GA_1 in seedlings of rice dwarf mutants. Plant Physiol 125:2053–2058

Cassán F, Lucangeli C, Bottini R, Piccoli P (2001b) *Azospirillum* spp. metabolize [17,17-$^2H_2$] gibberellin A_{20} to [17,17-$^2H_2$]gibberellin A_1 *in vivo* in *dy* rice mutant seedlings. Plant Cell Physiol 42:763–767

Cho SM, Kang BR, Kim JJ, Kim YC (2012) Induced systemic drought and salt tolerance by *Pseudomonas chlororaphis* O6 root colonization is mediated by ABA-independent stomatal closure. Plant Pathol J 28:202–206

Choudhary DV, Johri BD (2009) Interactions of *Bacillus* spp. and plants with special reference to induced systemic resistance (ISR). Microbiol Res 164:493–513

Cohen AC, Bottini R, Piccoli P (2008) *Azospirillum brasilense* Sp 245 produces ABA in chemically-defined culture medium and increases ABA content in *Arabidopsis* plants. Plant Growth Regul 54:97–103

Cohen AC, Travaglia C, Bottini R, Piccoli P (2009) Participation of abscisic acid and gibberellins produced by entophytic *Azospirillum* in the alleviation of drought effects in maize. Botany 87:455–462

Compant S, Reiter B, Sessitsch A, Nowak J, Clément C, Barka EA (2005) Endophytic colonization of *Vitis vinifera* L. by plant growth-promoting bacterium *Burkholderia* sp. strain PsJN. Appl Environ Microbiol 71:1685–1693

Cook RJ (1993) Making greater use of introduced microorganisms for biological control of plant pathogens. Annu Rev Phytopathol 31:53–80

Costacurta A, Vanderleyden J (1995) Synthesis of phytohormones by plant-associated bacteria. Crit Rev Microbiol 21:1–18

Creus C, Sueldo R, Barassi C (1997) Shoot growth and water status in *Azospirillum*-inoculated wheat seedlings grown under osmotic and salt stresses. Plant Physiol Biochem 35:939–944

Creus CM, Graziano M, Casanovas EM, Pereyra MA, Simontacchi M, Puntarulo S, Barassi CA, Lamattina L (2005) Nitric oxide is involved in the *Azospirillum brasilense*-induced lateral root formation in tomato. Planta 221:297–303

Crozier A, Arruda P, Jasmim JM, Monteiro AM, Sandberg G (1988) Analysis of indole-3-acetic acid and related indoles in culture medium from *Azospirillum lipoferum* and *Azospirillum brasilense*. Appl Environ Microbiol 54:2833–2837

Dary M, Chamber-Pérez MA, Palomares AJ, Pajuelo E (2010) "In situ" phytostabilisation of heavy metal polluted soils using *Lupinus luteus* inoculated with metal resistant plant-growth promoting rhizobacteria. J Hazard Mater 177:323–330

Dastager SG, Deepa CK, Pandey A (2010) Isolation and characterization of novel plant growth promoting *Micrococcus* sp. NII-0909 and its interaction with cowpea. Plant Physiol Biochem 48:987–992

De Meyer G, Höfte M (1997) Salicylic acid produced by the rhizobacterium *Pseudomonas aeruginosa* 7NSK2 induces resistance to leaf infection by *Botrytis cinerea* on bean. Phytopathology 87:588–593

De Smet I, Zhang H, Inzé D, Beeckman T (2006) A novel role for abscisic acid emerges from underground. Trends Plant Sci 11:434–439

Deak KI, Malamy J (2005) Osmotic regulation of root system architecture. Plant J 43:17–28

de-Bashan L, Hernandez JP, Bashan Y (2011) The potential contribution of plant growth-promoting bacteria to reduce environmental degradation – a comprehensive evaluation. Appl Soil Ecol 61:171–189

Deka BHP, Dileep KBS (2002) Plant disease suppression and growth promotion by a fluorescent *Pseudomonas strain*. Folia Microbiol 47:137–143

Dimkpa C, Weinand T, Asch F (2009) Plant-rhizobacteria interactions alleviate abiotic stress conditions. Plant Cell Environ 32:1682–1694

Dobbelaere S, Okon Y (2007) The plant growth promoting effects and plant responses. In: Elmerich C, Newton WE (eds) Associative and endophytic nitrogen-fixing bacteria and cyanobacterial associations. Springer, Heidelberg

Dobbelaere SA, Croonenborghs A, Thys A, Vande Broek A, Vanderleyden J (1999) Phytostimulatory effect of *Azospirillum brasilense* wild type and mutant strain altered in IAA production on wheat. Plant Soil 212:155–164

Döbereiner J, Mariel IE, Nery M (1976) Ecological distribution of *Spirillum lipoferum* Beinjerick. Can J Microbiol 22:1464–1473

Egamberdiyeva D, Höflich G (2004) Effect of plant growth-promoting bacteria on growth and nutrient uptake of cotton and pea in a semi-arid region of Uzbekistan. J Arid Environ 56:293–301

Escoriaza G, Sansberro P, García Lampasona S, Gatica M, Bottini R, Piccoli P (2013) In vitro cultures of *Vitis vinifera* L. cv. Chardonnay synthesize the phytoalexin nerolidol upon infection by *Phaeoacremonium parasiticum*. Phytopatol Mediterr (in press)

Feng J, Barker AV (1992) Ethylene evolution and ammonium accumulation by tomato plants under water and salinity stresses. Part II. J Plant Nutr 15:2471–2490

Forchetti G, Masciarelli O, Izaguirre MJ, Alemano S, Álvarez D, Abdala G (2010) Endophytic bacteria improve seedling growth of sunflower under water stress, produce salicylic acid, and inhibit growth of pathogenic fungi. Curr Microbiol 61:485–493

Frohnmeyer H, Staiger D (2003) Ultraviolet-B radiation-mediated responses in plants. Balancing damage and protection. Plant Physiol 133:1420–1428

Fulchieri M, Frioni L (1994) *Azospirillum* inoculation on maize (*Zea mays*): effect on yield in a field experiment in central argentina. Soil Biol Biochem 26:921–923

Fulchieri M, Lucangeli C, Bottini R (1993) Inoculation with *Azospirillum lipoferum* affects growth and gibberellins content of corn seedling roots. Plant Cell Physiol 34:1305–1309

Gil M, Pontin M, Berli F, Bottini R, Piccoli P (2012) Metabolism of terpenes in the response of grape (*Vitis vinifera* L.) leaf tissues to UV-B radiation. Phytochemistry 77:89–98

Glick BR, Patten CL, Holguin G, Penrose DM (1999) Biochemical and genetic mechanisms used by plant growth promoting bacteria. Imperial College Press, London

Granier C, Tardieu F (1999) Water deficit and spatial pattern of leaf development. Variability in responses can be stimulated using a simple model of leaf development. Plant Physiol 119:609–619

Grassmann J, Hippeli S, Elstner EF (2002) Plant's defence and its benefits for animals and medicine: role of phenolics and terpenoids in avoiding oxygen stress. Plant Physiol Biochem 40:471–478

Gutiérrez-Luna FM, López-Bucio J, Altamirano-Hernández J, Valencia-Cantero E, Reyes de la Cruz H, Macías-Rodríguez L (2010) Plant growth-promoting rhizobacteria modulate root-system architecture in *Arabidopsis thaliana* through volatile organic compound emission. Symbiosis 51:75–83

Gutiérrez-Mañero FJ, Ramos-Solano B, Probanza A, Mehouachi J, Tadeo FR, Talon M (2001) The plant-growth-promoting rhizobacteria *Bacillus pumilus* and *Bacillus licheniformis* produce high amounts of physiologically active gibberellins. Physiol Plant 111:206–211

Gyaneshwar P, Naresh Kumar G, Parekh LJ, Poole PS (2002) Role of soil microorganisms in improving P nutrition of plants. Plant Soil 245:83–93

Hadas R, Okon Y (1987) Effect of *Azospirillum brasilense* inoculation on root morphology and respiration in tomato seedlings. Biol Fertil Soils 5:241–247

Hallman J, Quadt-Hallman A, Mahafee WF, Kloepper JW (1997) Bacterial endophytes in agricultural crops. Can J Microbiol 43:895–914

Huang J, Zhu J-K (2004) Plant responses to stress: abscisic acid and water stress. In: Goodman RM (ed) Encyclopedia of plant and crop science. Dekker, New York, NY

Jain DK, Patriquin DG (1985) Characterization of a substance produced by *Azospirillum* which causes branching of wheat root hairs. Can J Microbiol 31:206–210

Khan MS, Zaidi A, Wani PA, Oves M (2009) Role of plant growth promoting rhizobacteria in the remediation of metal contaminated soils. Environ Chem Lett 7:1–19

Kloepper JW, Zablotowiez RM, Tipping EM, Lifshitz R (1991) Inorganic plant growth promotion mediated by bacterial rhizosphere colonizer. In: Keister KL, Gregan PB (eds) The rhizosphere and plant growth. Kluwer, Dordrecht

Kloepper JW, Ryu CM, Zhang S (2004) Induced systemic resistance and promotion of plant growth by *Bacillus* species. Phytopathology 94:1259–1266

Kumar KV, Singh N, Behl HM, Srivastava S (2008) Influence of plant growth promoting bacteria and its mutant on heavy metal toxicity in *Brassica juncea* grown in fly ash amended soil. Chemosphere 72:678–683

Levanony H, Bashan Y (1989) Enhancement of cell division in wheat root tips and growth of root elongation zone induced by *Azospirillum brasilense* Cd. Can J Bot 67:2213–2216

Li WC, Ye ZH, Wong MH (2007) Effects of bacteria on enhanced metal uptake of the Cd/Zn-hyperaccumulating plant, Sedum alfredii. J Exp Bot 58:4173–4182

López-Bucio J, Campos-Cuevas JC, Hernández-Calderón E, VelásquezBecerra C, Farías-Rodríguez R, Macías-Rodríguez LI, Valencia-Cantero E (2007) Bacillus megaterium Rhizobacteria promote growth and alter root-system architecture through an auxin- and ethylene-independent signaling mechanism in Arabidopsis thaliana. Mol Plant Microbe Interact 20:207–217

Ma Y, Rajkumar M, Freitas H (2009) Isolation and characterization of Ni mobilizing PGPB from serpentine soils and their potential in promoting plant growth and Ni accumulation by Brassica spp. Chemosphere 75:719–725

Mantelin S, Touraine B (2004) Plant growth-promoting bacteria and nitrate availability impacts on root development and nitrate uptake. J Exp Bot 55:27–34

Mayak S, Tirosh T, Glick B (2004) Plant growth promoting bacteria confer resistance in tomato plants to salt stress. Plant Physiol Biochem 42:565–572

Molina-Favero C, Creus CM, Simontacchi M, Puntarulo S, Lamattina L (2008) Aerobic nitric oxide production by Azospirillum brasilense Sp 245 and its influence on root architecture in tomato. Mol Plant Microbe Interact 21:1001–1009

Moreno D, Berli F, Piccoli PN, Bottini R (2011) Gibberellins and abscisic acid promote carbon allocation in roots and berry of grapevines. J Plant Growth Regul 30:220–228

Muthukumarasamy R, Govindarajan M, Vadivelu M, Revathi G (2006) N-fertilizer saving by the inoculation of Gluconacetobacter diazotrophicus and Herbaspirillum sp. in micropropagated sugarcane plants. Microbiol Res 161:238–245

Nadeem SM, Zahir ZA, Naveed M, Ashraf M (2010) Microbial ACC-deaminase: prospects and applications for inducing salt tolerance in plants. Crit Rev Plant Sci 29:360–393

Nibau C, Gibbs DJ, Coates JC (2008) Branching out in new directions: the control of root architecture by lateral root formation. New Phytol 179:595–614

Okon Y, Labandera-González C (1994) Agronomic applications of Azospirillum: an evaluation of 20 years worldwide field inoculation. Soil Biol Biochem 26:1591–1601

Oliveira ALM, Urquiaga S, Döbereiner J, Baldani JI (2002) The effect of inoculating endophytic N_2-fixing bacteria on micropropagated sugarcane plants. Plant Soil 242:205–215

Patten CL, Glick BR (2002) Role of Pseudomonas putida indoleacetic acid in development of the host plant root system. Appl Environ Microbiol 68:3795–3801

Piccoli P, Travaglia C, Cohen A, Sosa L, Cornejo P, Masuelli R, Bottini R (2011) An endophytic bacterium isolated from roots of the halophyte Prosopis strombulifera produces ABA, IAA, gibberellins A_1 and A_3 and jasmonic acid in chemically-defined culture medium. Plant Growth Regul 64:207–210

Quiroga AM, Berli F, Moreno D, Cavagnaro JB, Bottini R (2009) Abscisic acid sprays significantly increase yield per plant in vineyard-grown wine grape (Vitis vinifera L.) cv. Cabernet Sauvignon through increased berry set with no negative effects on anthocyanin content and total polyphenol index of both juice and wine. J Plant Growth Regul 28:28–35

Rajkumar M, Freitas H (2008) Influence of metal resistant-plant growth-promoting bacteria on the growth of Ricinus communis in soil contaminated with heavy metals. Chemosphere 71:834–842

Ryan PR, Delhaize E (2001) Function and mechanism of organic anion exudation from plant roots. Annu Rev Plant Physiol Plant Mol Biol 52:527–560

Ryu CM, Farag MA, Hu CH, Reddy MS, Wei H-X, Paré PW, Kloepper JW (2003) Bacterial volatiles promote growth in Arabidopsis. Proc Natl Acad Sci USA 100:4927–4932

Ryu CM, Hu CH, Locy R, Kloepper JW (2005) Study of mechanisms for plant growth promotion elicited by rhizobacteria in Arabidopsis thaliana. Plant Soil 268:285–292

Ryu CM, Murphy JF, Reddy MS, Kloepper JW (2007) A two-strain mixture of rhizobacteria elicits induction of systemic resistance against Pseudomonas syringae and Cucumber mosaic virus coupled to promotion of plant growth on Arabidopsis thaliana. J Microbiol Biotechnol 17:280–286

Sansberro P, Mroginski L, Bottini R (2004) Abscisic acid promotes growth of *Ilex paraguariensis* plants by alleviating diurnal water stress. Plant Growth Regul 42:105–111

Shaharoona B, Naveed M, Arshad M, Zahir ZA (2008) Fertilizer-dependent efficiency of pseudomonads for improving growth, yield and nutrient use efficiency of wheat (*Triticum aestivum* L.). Appl Microbiol Biotechnol 79:147–155

Spaepen S, Vanderleyden J, Remans R (2007) Indole-3-acetic acid in microbial and microorganism-plant signaling. FEMS Microbiol Rev 31:425–448

Travaglia C, Cohen AC, Reinoso H, Castillo C, Bottini R (2007) Exogenous abscisic acid increases carbohydrate accumulation and redistribution to the grains in wheat grown under field conditions of soil water restriction. J Plant Growth Regul 26:285–289

Travaglia C, Reinoso H, Bottini R (2009) Application of abscisic acid promotes yield in field-cultured soybean by enhancing production of carbohydrates and their allocation in seed. Crop Pasture Sci 60:1131–1136

Travaglia C, Reinoso H, Cohen AC, Luna C, Castillo C, Bottini R (2010) Exogenous ABA increases yield in field-grown wheat with a moderate water restriction. J Plant Growth Regul 29:366–374

Van Loon LC, Glick DR (2004) Increased plant fitness by rhizobacteria. In: Sandermann H (ed) Molecular ecotoxicology of plants. Springer, Berlin

Vögeli U, Chappell J (1988) Induction of sesquiterpene cyclase and suppression of squalene synthetase activities in plant cell cultures treated with fungal elicitor. Plant Physiol 88:1291–1296

Xie X, Zhang H, Paré PW (2009) Sustained growth promotion in Arabidopsis with long-term exposure to the beneficial soil bacterium *Bacillus subtilis* (GB03). Plant Signal Behav 4:948–953

Yang J, Kloepper JW, Ryu CM (2009) Rhizosphere bacteria help plants tolerate abiotic stress. Trends Plant Sci 14:1–4

Zhang J, Davies WJ (1989) Abscisic acid produced in dehydrating roots may enable the plant to measure de water status in the soil. Plant Cell Environ 12:73–81

Zhang SQ, Outlaw WH Jr (2001) Abscisic acid introduced into the transpiration stream accumulates in the guard-cell apoplast and causes stomatal closure. Plant Cell Environ 24:1045–1054

Zhang H, Xie X, Kim M-S, Kornyeyev DA, Holaday S, Pare PW (2008) Soil bacteria augment Arabidopsis photosynthesis by decreasing glucose sensing and abscisic acid levels in planta. Plant J 56:264–273

Zhuang X, Chen J, Shim H, Bai Z (2007) New advances in plant growth-promoting rhizobacteria for bioremediation. Environ Int 33:406–413

Chapter 9
Fighting Plant Diseases Through the Application of *Bacillus* and *Pseudomonas* Strains

Sonia Fischer, Analía Príncipe, Florencia Alvarez, Paula Cordero, Marina Castro, Agustina Godino, Edgardo Jofré, and Gladys Mori

9.1 Soilborne Fungal Pathogens Affect the Yield and Quality Crops: How to Control the Diseases They Produce?

The plant rhizosphere is an environment where both pathogenic and beneficial microorganisms influence plant growth and health. Pathogenic microorganisms adversely affecting plants are fungi, oomycetes, bacteria, and nematodes (Raaijmakers et al. 2009). Soilborne plant pathogens can cause serious damage to agriculture and significantly reduce the yield and quality of crops. This section will discuss selected examples of the pathogenic fungi that produce substantial economic losses in crops of consequential relevance to human consumption.

The *Fusarium* head blight (FHB) of grain cereals and the ear rot in maize are significant diseases all over the world, with more than 20 *Fusarium* species having been associated with those diseases. In addition, *Fusarium* species produce mycotoxins that are hazardous to both animal and human health (Chulze et al. 1996; Edwards 2004). If infected cereal grain is stored or transported with too high a moisture content, a postharvest growth of the fungus occurs along with an increase in the levels of mycotoxin (Magan et al. 2010). *Fusarium graminearum* and *Fusarium culmorum* are the most destructive FHB pathogens of wheat, as these species are more pathogenic than others and produce higher levels of mycotoxins (Edwards 2004). The infections of maize are predominantly caused by *Fusarium verticillioides* (the pink ear rot) and *F. graminearum* (the red ear rot). In Argentina, the most prevalent *Fusarium* in freshly harvested maize is *F. verticillioides*,

---

S. Fischer (✉) · A. Príncipe · F. Alvarez · P. Cordero · M. Castro · A. Godino · E. Jofré · G. Mori
Facultad de Ciencias Exactas, Físico-Químicas y Naturales, Universidad Nacional de Río Cuarto, Ruta 36-Km 601, 5800 Río Cuarto, Córdoba, Argentina
e-mail: sfischer@exa.unrc.edu.ar

resulting in contamination of corn with the mycotoxin fumonisin (Chulze et al. 1996).

Another example of economically significant soilborne fungi is *Sclerotinia sclerotiorum* (Lib.) de Bary, which is one of the most nonspecific and widely successful plant pathogens. This success results from the broad range of hosts infected by *S. sclerotiorum* infects (more than 400 plant species at all stages of growth, development, and harvested products) along with the persistence or survival of the sclerotia (resting bodies) in the soil for long periods: depending on environmental conditions, the sclerotia can subsequently germinate either carpogenically to produce airborne ascospores that travels long distances or myceliogenically to infect the roots of hosts (Bardin and Huang 2001). Moreover, this fungus is the causal agent of more than 60 diseases, including stem rot, stalk rot, head rot, pod rot, and wilt (Purdy 1979). This extensive broad range of pathologies occurs in almost every country of the world but more often in the temperate regions. For example, the stem rot of the soybean was ranked as one of the most severe diseases of that crop in Argentina and the second most devastating disease in the United States during the 1990s (Wrather et al. 2001).

Magnaporthe oryzae is another phytopathogenic fungus that produces significantly economical losses in the world. This filamentous ascomycete is the causal agent of the rice blast, the most destructive disease of rice worldwide. The significance of this pathogenesis lies in the dependence of half the world's population primarily on rice as a source of calories. All foliar tissues are susceptible to infection; infection of the panicle, however, can lead to a complete loss of the grain (Dean et al. 2012). A predominant place is also occupied by *Botrytis cinerea*, known as the gray mold. This fungus is the most destructive on mature or senescent tissues of dicotyledonous hosts and infects more than 200 plant species. The costs of *Botrytis* damage are most difficult to estimate because of the fungus's broad host range (Williamson et al. 2007; Dean et al. 2012).

A common fungal disease on wheat and grasses is the rust. The stem (black) rust is caused by *Puccinia graminis*, the stripe (yellow) rust is caused by *Puccinia striiformis* and the leaf (brown) rust by *Puccinia triticina*. *Puccinia* spp. are biotrophic basidiomycetes that spread by airborne spores over long distances. A high efficiency in the dissemination, in addition to a prolific sporulation, a pathogenic variability, and the conditions of cultivation of wheat all contribute to the destructive potential of these rusts (Hovmøller et al. 2011; Dean et al. 2012).

Two possibilities are currently open to farmers for fighting against pests that affect crops. Although modern agriculture uses chemical pesticides for controlling plant diseases and increasing crop productivity, the abuse of chemical pesticides has a major impact on the environment and often on human health; furthermore, repeated chemical treatments can result in the development of pest resistance. A desirable alternative strategy for pest management is the use of bacterial strains or their metabolites as biocontrol agents against soilborne phytopathogens. Microbial activities can be exploited as a low-cost biotechnology and an environmentally friendly practice for the protection of agriculture and the natural ecosystem (Barea et al. 2005).

This chapter will therefore focus on the use of bacteria, particularly strains of the *Bacillus* and *Pseudomonas* genera, as prime candidates for application as biologic control agents.

9.2 Plant Growth-Promoting Bacteria

9.2.1 Mechanisms Used by PGPB

Plant growth-promoting bacteria (PGPBs) benefit plants directly by stimulating plant growth and/or indirectly through a reduction in the incidence of plant disease. The best known group of PGPBs is the plant growth-promoting rhizobacteria (PGPRs) that colonize the root surfaces and the rhizosphere. Some of these PGPRs are able to enter the root interior and establish endophytic populations. Despite their different ecologic niches, the free-living rhizobacteria and the endophytic bacteria use some of the same mechanisms to promote plant growth and control phytopathogens (Compant et al. 2005).

Depending on the direct mechanism used, PGPBs can be categorized into the following different groups (1) biofertilizers, (2) phytostimulators, (3) stress controllers, and (4) rhizoremediators (Lugtenberg and Kamilova 2009). (1) Biofertilizers increase plant nutrition, for example, the diazotrophic PGPBs fix atmospheric N_2, thus efficiently transferring nitrogen to the plant. Endophytic diazotrophs—e.g., *Pseudomonas stutzeri* A15 (formerly *Alcaligenes faecalis*) (Vermeiren et al. 1999)—encounter more favorable conditions, since they colonize the interior of plants, for effective N_2 fixation and the delivery of fixed nitrogen to plant cells. Phosphorus is another essential plant macronutrient, but its compounds are in insoluble forms in the soil and therefore are not directly available to the plant. Some PGPBs of the genus *Bacillus*, *Pseudomonas*, or *Enterobacter* can solubilize phosphates from the soil through of the production of organic acids, acid phosphatases, and phytases (Drogue et al. 2012). (2) The second group, the phytostimulators, produces phytohormones like auxins, cytokinins, and gibberellins that stimulate the growth of plants. For example, the auxin indole-3-acetic acid (IAA) increases the biomass of the root system and therefore improves the plant's uptake of nutrients from the environment (Lugtenberg and Kamilova 2009; Drogue et al. 2012). Several endophyte strains of the genus *Pseudomonas* were reported to be able to promote root growth in *Solanum nigrum*, and this effect was associated with the production of IAA (Long et al. 2008).

(3) The stress controllers are able to regulate the level of the ethylene hormone in a plant through the action of 1-aminocyclopropane-1-carboxylate (ACC) deaminase (encoded by the *acdS* locus), which enzyme catalyzes the conversion of ACC (the precursor of ethylene) into NH_3 and α-ketobutyrate, thus reducing the level of ethylene in a plant. During a biotic and an abiotic stress, high amounts of ethylene are synthesized in plant tissues and this hormone produces senescence in plants.

Therefore, bacteria possessing ACC deaminase activity enhance a plant's stress tolerance (Saravanakumar and Samiyappan 2007; Drogue et al. 2012). The *acdS* gene is widely distributed in rhizobacteria and especially in bacterial endophytes, for which the expression of ACC deaminase is determinant for endophyte competence (Drogue et al. 2012). Several *Pseudomonas* endophytes possessing ACC deaminase activity have been isolated from field-grown *S. nigrum* (Long et al. 2008).

Finally, (4) the rhizoremediators are contaminant-degrading rhizobacteria that can be included among the PGPBs because the presence of contaminants reduces the growth of the plant so that the elimination of the pollutants becomes beneficial (Segura and Ramos 2013). For example, the inoculation of plants with the endophyte *Pseudomonas putida* W619-TCE promotes plant growth and reduces trichloroethylene (TCE) phytotoxicity by diminishing the levels of that compound present in the leaves (Weyens et al. 2010).

Moreover, some PGPB strains play a prominent role in the suppression of a broad spectrum of phytopathogenic fungi (Ryan et al. 2008). Mechanisms contributing to this biocontrol activity are (a) competition, (b) antibiosis against the fungi, and (c) the induction of a systemic resistance in the host.

(a) Competition takes place for space and nutrients at the root surface; competitive colonization of the rhizosphere and successful establishment in the root zone are a prerequisite for effective biocontrol. The treatment of plants or seeds with a nonpathogenic strain that can outcompete the pathogenic organism could thus prove effective. For example, endophytic bacteria belonging to the genera *Bacillus* and *Pseudomonas* (Melnick et al. 2008; Prieto and Mercado-Blanco 2008; Prieto et al. 2011) have been used for the biologic control of various plant diseases. These bacterial endophytes are suitable biocontrol agents because they have the advantage of being relatively protected from the competitive soil environment; moreover, they usually grow in the same plant tissue where pathogens are detected (Ryan et al. 2008; Bulgari et al. 2009). Motility, chemotactic responses, and the synthesis of siderophores (especially under iron-limiting conditions) are effective bacterial properties for occupying the space and limiting the growth of pathogens in the rhizosphere (Lugtenberg et al. 2001; Compant et al. 2005).

(b) The production of antifungal substances can also be highly effective in plant disease suppression. Many microbes produce and secrete one or more compounds with antifungal activity (Shahraki et al. 2009). Natural antibiotics are thought to function in defense, fitness, competitiveness, biocontrol activity, communication, and regulation of gene expression (Mavrodi et al. 2012). The best known antibiotics produced by *Pseudomonas* spp. or *Bacillus* spp. are 2,4-diacetyl phloroglucinol (2,4-DAPG), phenazines (PHZ), pyrrolnitrin (PRN), pyoluteorin (PLT), zwittermycin A, kanosamine, and lipopeptide biosurfactants (Lugtenberg and Kamilova 2009).

Many microorganisms are also able to produce other metabolites that can interfere with pathogen growth and their activities. Lytic enzymes are among these metabolites that can break down polymeric compounds, such as chitin, proteins, celluloses, or hemicellulose. Thus, the production of extracellular enzymes such as chitinases, cellulases, β-1,3-glucanases, proteases, or lipases is a desirable characteristic for a biocontrol agent since these hydrolases are able to lyse certain fungal cells and in doing so suppress pathogen growth (Babalola 2010). For example, *P. stutzeri* produces extracellular chitinase and laminarinase that lyse the mycelia of *Fusarium solani* (Mauch et al. 1988).

Volatile organic compounds (VOCs) are small organic molecules (of molecular mass usually <300 Da) that characteristically have a high vapor pressure and are therefore easily volatilized. Some VOCs produced by rhizobacteria have an antifungal potential and can act over long distances via diffusion in air and through soil pores (Wheatley 2002; Kai et al. 2007). A given VOC produced by a particular bacterial strain does not cause the same effects or the same degree of inhibition in all fungi; rather, the responses depend on the specific fungus–bacterium combination. These differences may occur because not all fungi respond to the same VOCs, the sites of action may be different, or the fungi might possess different abilities to detoxify a given volatile metabolite. In this regard, rhizobacteria nevertheless produce different VOCs, depending on the growth medium and conditions (Kai et al. 2009). Carbon and nitrogen sources increase VOC production, for example, tryptic soy agar—it is richer in carbon and nitrogen compounds than nutrient agar medium, induces more VOC production, and results in greater antifungal activity (Fernando et al. 2005). In addition to biocontrol characteristics, the VOCs produced by PGPBs can elicit the so-called induced systemic tolerance to abiotic stress, including salt, drought (Cho et al. 2008; Zhang et al. 2008), or acid-soil (Raudales et al. 2009) stress.

(c) Many rhizospheric microorganisms can induce a systemic response in plants, activating plant defense mechanisms against pathogens attack. Inoculation with nonpathogenic root zone bacteria, for example, can trigger signaling pathways that lead to a stronger mechanism of defense in the host against the pathogen—the so-called induced systemic resistance (ISR; Jetiyanon and Kloepper 2002; Krause et al. 2003; Horst et al. 2005; Ongena et al. 2005). Beneficial rhizobacteria trigger ISR by priming the plants for a potentiated activation of various cellular defense responses, which are subsequently induced upon pathogen arrival (Conrath et al. 2006). In plants exhibiting an ISR, a number of reactions have been observed such as deposition of callose, lignin, and phenolics beyond the infection sites (Duijff et al. 1997); increased activities of chitinase, peroxidase, polyphenol oxidase, and phenylalanine ammonia lyase (Chen et al. 2000; Magnin-Robert et al. 2007); an enhanced phytoalexin production (Van Peer et al. 1991); and an induced expression of stress-related genes (Verhagen et al. 2004). The enhanced defense capability of induced plants does not necessarily require de-novo reaction since the induction can result in faster and/or stronger expression of basal defense responses (Conrath et al. 2002). Many reports have described the elicitation of ISR by several Gram-positive strains such as *Bacillus* spp. as well as by Gram negatives such as *Pseudomonas* spp. (Kloepper et al. 2004; Van Loon and Bakker 2006). ISR is phenotypically similar to what is referred to as systemic acquired resistance (SAR), but the molecular events leading to the induction are different. For example, the transduction pathways of the ISR and the SAR differ, but both routes involve the regulatory protein known as the nonexpressor of pathogenesis-related proteins 1, or NPR1 (Pieterse et al. 1998).

SAR is a systemic defense network induced in plants in response to attack by certain pathogens, usually biotrophs (Heil and Bostock 2002; Bari and Jones 2009), and controlled by the salicylic acid (SA)-dependent signaling pathway. The onset of SAR involves local and systemic increases in endogenously synthesized SA leading to activation of the NPR1 regulatory protein and the subsequent NPR1-dependent expression of genes encoding pathogenesis-related (PR) proteins—e.g., PR1, PR2, and PR5 (Ward et al. 1991; Van Loon and Van Strien 1999). In contrast to SAR, ISR is often regulated by jasmonic acid or its derivative, methyl jasmonate, and the ethylene signaling pathway (Pieterse et al. 2007; Bari and Jones 2009) and is associated with the expression of the gene encoding plant defensin 1.2 (Van Oosten et al. 2008). New reports, however, have shown that ISR can also be mediated by SA. The PGPBs *Bacillus cereus* AR156 and *Bacillus subtilis* FB17 induced resistance in *Arabidopsis thaliana* plants to the pathogen *Pseudomonas syringae* pv. *tomato* DC3000. Furthermore, the strain AR156 triggered ISR via simultaneous activation of both the SA and the jasmonic-acid–ethylene signaling pathways, thus resulting in an enhanced level of induced protection in *Arabidopsis* (Rudrappa et al. 2010; Niu et al. 2011).

9.2.2 Quorum Sensing: A Cell–Cell Communication

Biocontrol agents or plant growth-promoting bacteria need to be successful in colonizing plant roots in order to exert their beneficial effects. Rhizobacteria form microcolonies or biofilms at preferred sites of root exudation and such microcolonies are sites where bacteria communicate with each other and act in a coordinated manner (Choudary and Johri 2009). This cell–cell communication is known as quorum sensing (QS), and it is mediated by small diffusible signaling molecules (autoinducers) that are produced by bacteria during their growth and then secreted into the environment to accumulate there in proportion to the cell-population density. When the extracellular concentration of autoinducers reaches a threshold level (at which point the population is considered to be—as it were—"quorate"), the bacteria detect and respond to this signal by altering their gene expression (Fuqua et al. 1994; Galloway et al. 2012).

In Gram-negative bacteria, the most common autoinducer is N-acyl homoserine lactones (AHLs) produced by the LuxI-type synthase enzymes. When a critical concentration of AHL is reached, AHL is internalized and interacts direct,y with the LuxR-type protein. The resulting LuxR–AHL complexes bind to DNA and activate the transcription of specific genes (Fuqua et al. 2001; Venturi 2006). Since the LuxR-type proteins usually recognize a specific AHL, this mechanism has normally been associated with intraspecies signaling.

Nevertheless, because some LuxR-type proteins recognize more than one AHL, those proteins are primarily involved in interspecies signaling (Reading and Sperandio 2006).

The Gram-positive bacteria use peptide-signaling molecules for QS that are exported from the bacterium by transporters. At or above the threshold level, those peptides are recognized by membrane-associated sensor kinases that initiate phosphotransfer to a regulatory protein to activate an intracellular response cascade. Similar to the mechanism employed by the Gram-negative bacteria, each Gram-positive bacterium uses a signal different from the ones used by other bacteria; and since cognate receptors are sensitive to the signals' structures, this mechanism constitutes a form of intraspecies communication (Reading and Sperandio 2006).

In addition, a family of molecules called autoinducer-2 is a class of non-species-specific autoinducers, capable of mediating intra- and interspecies communications among Gram-negative and Gram-positive bacteria (Galloway et al. 2012).

Plant–microbe interactions such as biofertilization, rhizoremediation, biocontrol, and phytostimulation may be QS dependent (Babalola 2010). For example, certain metabolites involved in biocontrol—e.g., surfactin in *Bacillus* or phenazine in *Pseudomonas*—are regulated by QS. These examples are described in more detail in later sections.

Plants can, however, perceive bacterial QS molecules at or below the threshold concentrations used by bacteria and then respond to these signals. Different AHLs have been shown to elicit several different responses in plants, thus indicating that the plant can distinguish among AHLs of different structures. In addition, plants can also produce and secrete compounds that mimic the bacterial QS signals and thus control quorum-regulated bacterial responses (Bauer and Mathesius 2004).

Finally, certain soil bacteria also have the ability to disrupt QS molecules by degrading autoinducers, a process known as quorum quenching. For example, *Bacillus thuringiensis* produces a lactonase enzyme that hydrolyzes the lactone ring of the AHLs. This lactonase probably interferes with the AHL signaling by other bacterial species with which *Bacillus* competes in nature (Dong et al. 2001).

9.3 The Role of the Genus *Bacillus* in Biologic Control

The members of the genus *Bacillus* possess several advantages that make them good candidates for use as biologic control agents. First, some of these bacteria produce several different types of antimicrobial compounds. Second, they induce growth and defense responses in the host plant. Third, since *Bacillus* species can produce spores that resist adverse environmental conditions and thus enable their facile formulation and storage in commercial products, the bacilli are among the beneficial bacteria most exploited as microbial pesticides, fungicides, or fertilizers and as such offer a promising alternative to the use of chemicals (Schallmey et al. 2004; Francis et al. 2010; Pérez-García et al. 2011).

A large number of reports have described the beneficial effects of several *Bacillus* spp. against diseases elicited by the oomycetes and other fungal pathogens that are major threats to crops and plant production. Examples that can be cited are the suppression of root disease (avocado root rot, tomato damping-off, and wheat take-all), foliar disease (cucurbit and the strawberry powdery mildews), and post-harvest diseases (the green, gray, and blue molds; Cazorla et al. 2007; Romero et al. 2007; Liu et al. 2009; Arrebola et al. 2010). Indeed, several commercial products based on various species of *Bacillus*—e.g., *B. amyloliquefaciens*, *B. licheniformis*, *B. pumilus* and *B. subtilis*—have been marketed as biofungicides (Fravel 2005).

Among the mechanisms studied on the genus *Bacillus*, the cell wall-degrading enzymes (chitinases, glucanases, and proteases), the peptide antibiotics, and even small molecules (such as the VOCs) have been shown to either suppress the pathogen directly (Shoda 2000) or else act indirectly through a stimulation of the inducible plant defense mechanisms (ISR). For instance, the VOCs 2,3-butanediol and acetoin from the endophytic *Bacillus* strain IN937a were shown to confer a protection against pathogen infection in *A. thaliana* through the ethylene-dependent signaling pathway (Ryu et al. 2003, 2004).

The cyclic lipopeptides (CLPs) are among the antibiotics most frequently produced by *Bacillus* species and most extensively studied. These compounds are amphiphilic and share a common structure consisting in a lipid tail linked to a short cyclic oligopeptide. CLPs are nonribosomally synthesized compounds and are classified into three families depending on their structure: the iturins, the fengycins, and the surfactins. The iturin family comprises iturins A, C, D, and E; bacillomycins D, F, and L; bacillopeptin; and mycosubtilin (Moyne et al. 2004).

The iturins are heptapeptides bonded to a β-amino-fatty-acid chain 14–17 carbons in length. This family has antifungal capabilities against a wide variety of yeast and other fungi but only limited antibacterial activity and no antiviral action (Ongena and Jacques 2008)—for example, the iturin produced by the endophytic *Bacillus* sp. CY22 has antifungal activity against *Rhizoctonia solani*, *Pythium ultimum*, and *Fusarium oxysporum* (Cho et al. 2003). The iturin operon of iturin A of *B. subtilis* strain RB14 consists in four open reading frames (ORFs), referred to as *ituDABC* (Tsuge et al. 2001), and the iturin A gene cluster is inserted at exactly the same position as the bacillomycin (*bmy*) operon in *B. amyloliquefaciens* strain FZB42 (Chen et al. 2009).

The fengycins show a strong fungitoxic activity, specifically against filamentous fungi (Ongena and Jacques 2008), and are likewise synthesized by nonribosomal peptide synthetases encoded by an operon of five ORFs, the *fenA–E* (or the *ppsA–E*) (Steller et al. 1999).

Surfactin is a heptapeptide linked to a β-hydroxy fatty acid consisting of 13–15 carbon atoms to form a cyclic lactone structure. The surfactin operon contains four ORFs encoding three multifunctional proteins: SrfA–C and an external thioesterase/acyltransferase enzyme SrfD (Chen et al. 2009). Surfactin behaves as a highly powerful biosurfactant and possesses several other relevant biologic activities, producing mainly an alteration in membrane integrity as a consequence of the establishment of strong interactions with the constituent of phospholipids (Ahimou et al. 2000; Carrillo et al. 2003).

Surfactin biosynthesis is QS dependent and is closely related to the competence-development pathway. In *B. subtilis*, two signaling molecules for QS have been identified, the pheromones ComX and PhrC. The main competence-stimulating gene product, ComX, is accumulated in the growth medium at the threshold level (i.e., that quantity determining a quorum). Then, the membrane protein kinase ComP sensing the accumulation of ComX in the medium responds by phosphorylating ComA. Phosphorylated ComA, in turn, activates the transcription of *srfA*. Interestingly, the gene *comS*, involved competence development, is located within the second open reading frame of the *srfA* operon. Therefore, *B. subtilis* uses a single QS pathway for two different adaptive processes (Hamoen et al. 2003). The other pheromone involved in the regulation of surfactin by QS, PhrC, also requires ComP and ComA for its activity. The stimulation of *srfA* expression by PhrC, however, also requires the permease Spo0K, suggesting that sensing of this pheromone occurs intracellularly. Why *B. subtilis* uses a double QS pathway for stimulating the expression *srfA*/*comS* still remains unclear (Lazazzera et al. 1997; Hamoen et al. 2003).

Besides antibiosis, the lipopeptides produced by certain strains of *Bacillus* can suppress the plant diseases caused by root and foliar pathogens through the mechanism of ISR (Choudary and Johri 2009). In this regard, surfactin and fengycin in particular have been identified as bacterial determinants responsible for the elicitation of host-plant ISR (Ongena et al. 2007).

Over the years, many bacterial isolates have been evaluated as potential biocontrol agents against soilborne fungal phytopathogens. For this reason, our lab established a collection of bacterial isolates from different environments and host plants, screened those bacteria for several biocontrol-associated traits, and in the end selected two strains that we designed as *B. amyloliquefaciens* strains $ARP_2 3$ and $MEP_2 18$ (Príncipe et al. 2007). Both of these showed strong antagonism in vitro against different species of the genera *Sclerotinia* and *Fusarium* on potato dextrose agar medium. Further analysis revealed that the biocontrol activity could be observed in cell-free supernatants, indicating that biocontrol metabolites were being secreted into the culture medium. Moreover, the antifungal activity was resistant to high temperature and protease activity (Príncipe et al. 2007). The antifungal compounds produced by those two strains were identified by reverse-phase high-performance liquid chromatography and matrix-assisted laser desorption/ionization time-of-flight mass spectrometry. The lipopeptides thus identified were surfactins (isoforms C_{14} and C_{15}) and fengycins A (C_{16} and C_{17}) and B (C_{16}) in the strain $ARP_2 3$ and only the lipopeptide iturin A (C_{15}) in the strain $MEP_2 18$. In addition, the presence of the nonribosomal peptide synthetase genes involved in CLP synthesis was confirmed in the genome of those strains by a PCR assay conducted with specific primers (Alvarez et al. 2012).

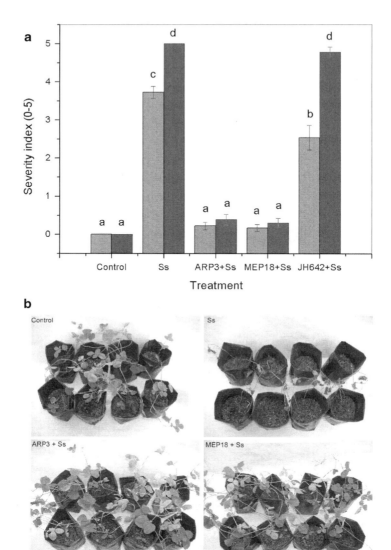

Fig. 9.1 Effects of the inoculation of *Bacillus amyloliquefaciens* strains ARP$_2$3 and MEP$_2$18 on stem rot symptoms in soybean caused by *Sclerotinia sclerotiorum*. (**a**) Severity index of *Sclerotinia* stem rot disease was evaluated after 7 (*light gray bars*) and 21 days (*dark gray bars*)

The ability of various *Bacillus* strains to control fungal soilborne, foliar, and postharvest diseases has been attributed mostly to the iturins and fengycins (Romero et al. 2007; Ongena and Jacques 2008; Arrebola et al. 2010). In view of this association, the effectiveness of the lipopeptide-producing *B. amyloliquefaciens* strains ARP_23 and MEP_218 in controlling sclerotinia stem rot was evaluated under growth-chamber conditions. Plants infected with *S. sclerotiorum* and inoculated with strains ARP_23 or MEP_218 showed a significant reduction in disease severity at 7 days postinoculation compared either to untreated plants or to those infected with *S. sclerotiorum* and inoculated with *B. subtilis* JH642—it being a strain unable to produce CLPs. These results showed that foliar application of *B. amyloliquefaciens* strains ARP_23 or MEP_218 conferred a protective effect on soybean plants against the sclerotinia stem rot disease (Fig. 9.1; Alvarez et al. 2012). Moreover, the CLPs produced by *B. amyloliquefaciens* strains ARP_23 and MEP_218 showed a significant inhibition in vitro of the germination of *Sclerotinia* spp. sclerotia. This ability to inhibit the germination of sclerotia is particularly advantageous because the soybean sclerotinia stem rot disease is caused by infection through ascospores produced from those same germinating sclerotia (Fig. 9.2). Therefore, a decrease in sclerotia production or a delayed germination of those sclerotia through the application of *B. amyloliquefaciens* strains in the field could constitute a strategy for reducing primary infection. Notable too is that, in addition to the inhibition of sclerotia germination, the CLPs produced by *B. amyloliquefaciens* strains affected *S. sclerotiorum* mycelial morphology: a granulated and vesicular cytoplasm became manifest compared to the hyaline and healthy cytoplasm of the untreated control hyphae (Fig. 9.3; Alvarez et al. 2012). Similar effects have been reported by other authors on additional phytopathogenic fungi, thus suggesting that the biosurfactant lipopeptides are able to traverse the fungal-cell wall and induce severe alterations in plasma-membrane permeability (Romero et al. 2007; Tendulkar et al. 2007).

All these results demonstrate the potential of the *B. amyloliquefaciens* strains ARP_23 and MEP_218 for controlling the various plant diseases caused by *S. sclerotiorum*.

Fig. 9.1 (continued) postinoculation. Bars represent the average $\pm$ SE per treatment ($n = 18$) of three independent experiments. Except for the uninfected control plants, leaves were first sprayed with 48-h-old cultures of *Bacillus* strains diluted 20-fold and supplement with 0.004 % (w/v) adherent humectant (NITRAP; Bioverde, Buenos Aires, Argentina) (2.5×10^7 CFU per plant). After air-drying for 1 h, the leaves were treated with *S. sclerotiorum* mycelia adjusted to an optical density of about 1.5 at 600 nm. Key to abbreviations: *S. sclerotiorum* (Ss), *B. amyloliquefaciens* ARP_23 (ARP3), *B. amyloliquefaciens* MEP_218 (MEP18), and *B. subtilis* JH642 (JH642). Treatments indicated with a common letter were not significantly different ($P = 0.05$) according to the Kruskal–Wallis test. (**b**) Representative soybean plants of each treatment at 21 days postinoculation (taken from Alvarez et al. 2012)

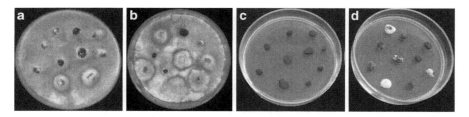

Fig. 9.2 Inhibition of the sclerotia germination of *S. sclerotiorum* on potato-dextrose-agar (PDA) medium supplemented with concentrated cell-free supernatants from either *B. amyloliquefaciens* ARP$_2$3 (**c**) or *B. amyloliquefaciens* MEP$_2$18 (**d**). PDA (**a**) or PDA-containing concentrated medium optimum for lipopeptide production (**b**) was used as a control. Final concentration of cell-free supernatants in the medium was 0.5× (taken from Alvarez et al. 2012)

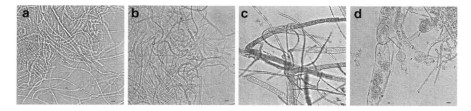

Fig. 9.3 Effect of antifungal compounds present in concentrated cell-free supernatants from *B. amyloliquefaciens* strains on *S. sclerotiorum* mycelia. Untreated (**a**) and mycelia treated with 100 µl of fivefold concentrated medium optimum for lipopeptide production (**b**) or 100 µl of fivefold concentrated cell-free supernatants from *B. amyloliquefaciens* strains ARP$_2$3 (**c**) and MEP$_2$18 (**d**) were evaluated by light microscopy after 5 days of incubation. Bars: 10 µm (taken from Alvarez et al. 2012)

9.4 The Role of the Genus *Pseudomonas* in Biologic Control

Pseudomonas spp. are aerobic, Gram-negative bacteria, and ubiquitous in nature, and many strains of this genus have been studied as biologic control agents against soilborne plant pathogens (Haas and Défago 2005). The traits that make *Pseudomonas* spp. effective potential biocontrol agents are the ability to grow rapidly in vitro, to quickly utilize root exudates, to colonize and multiply in the rhizosphere, to produce a wide spectrum of bioactive metabolites (e.g., antibiotics, siderophores, volatiles, and extracellular enzymes), to compete aggressively with other microorganisms, and to adapt to environmental stresses. The principal weakness of pseudomonads as biocontrol agents is their inability to produce spores (as do many *Bacillus* spp. as well), which characteristic complicates formulation of those bacteria for commercial use (Weller 2007).

Pseudomonas strains produce one or more secondary metabolites involved in biocontrol. Some 5.7 % of the *Pseudomonas fluorescens* Pf-5 genome (~400 kb) has been estimated to be devoted to secondary metabolism (Paulsen et al. 2005). For example, this pseudomonad produces 2,4-diacetylphloroglucinol (2,4-DAPG), pyrrolnitrin (PRN), pyoluteorin (PLT), hydrogen cyanide (HCN), the siderophores pyochelin and pyoverdine, the cyclic lipopeptide orfamide, and rhizoxin derivatives (Paulsen et al. 2005; Brendel et al. 2007; Gross et al. 2007; Loper et al. 2008).

The polyketide antibiotic 2,4-DAPG is the main metabolite of *P. fluorescens* that is involved in the control of plant diseases caused by soilborne pathogens. Eight ORFs have been identified in the DAPG locus (*phlH, phlG, phlF, phlA, phlC, phlB, phlD,* and *phlE*). The ORFs *phlG, phlA, phlC, phlB, phlD,* and *phlE* have a common transcriptional orientation, while the loci *phlH* and *phlF* are oppositely oriented (Bangera and Thomashow 1999; Delany et al. 2000; Schnider-Keel et al. 2000; Abbas et al. 2002, 2004; Haas and Keel 2003; Bottiglieri and Keel 2006; Yang and Cao 2012). The key biosynthetic gene is *phlD*, which catalyzes the synthesis of phloroglucinol from three molecules of malonyl-CoA (Zha et al. 2006). On the basis of the genetic sequence of the conserved region of the *phlD* from different strains, several specific primers have been designed, evaluated, and shown to be useful for the isolation and identification of 2,4-DAPG-producing pseudomonads within the rhizosphere. In addition, these primers could be used to characterize the abundance and diversity of $phlD^+$ bacteria in the field (Raaijmakers et al. 1997; McSpadden Gardener et al. 2001).

Other pseudomonad antibiotics are the phenazines—pigmented, heterocyclic compounds. Those metabolic products have a broad antagonistic effect as a result of their redox properties (Mavrodi et al. 2006) and are produced from chorismate via a common intermediate, phenazine-1-carboxylate. Although the phenazine biosynthetic locus is highly conserved among *Pseudomonas* spp., individual strains differ in the range of phenazine compounds that they produce. The genes for the production of phenazine derivatives have been identified in different species of *Pseudomonas*—such as *P. fluorescens* 2-79, *P. aureofaciens* 30-84, and *P. chlororaphis* PCL1391. The phenazine biosynthetic cluster comprises the *phzABCDEFG* structural genes. Upstream from this operon are located the *phzI* and *phzR* genes, those encoding respectively the autoinducer synthase and the transcriptional activator, with *phzR* being oriented in the opposite orientation to the *phz* operon (Pierson et al. 1995; Mavrodi et al. 1998; Chin-A-Woeng et al. 2001a, 2003; Haas and Keel 2003; Khan et al. 2005).

PLT is a phenolic polyketide antibiotic that suppresses the plant diseases caused by the oomycete *P. ultimum*. The PLT biosynthesis gene cluster contains the structural operon *pltLABCDEFG* and the *pltM* gene, the ATP-binding cassette (ABC), the transport operon *pltHIJKNO*, and two regulatory genes, *pltR* and *pltZ*—these latter two genes being divergently transcribed from those two operons in both *P. fluorescens* Pf-5 and *P. aeruginosa* M18. The former gene encodes the transcriptional activator LysR required for the transcription of genes for the biosynthesis of PLT, while the latter encodes a transcriptional repressor that negatively regulates PLT production (Nowak-Thompson et al. 1999; Huang et al. 2004, 2006; Brodhagen et al. 2005).

PRN is a chlorinated phenylpyrrole antibiotic that has activity against several bacteria and soilborne fungi. The mode of action of PRN involved an inhibition of the fungal respiratory chain (Tripathi and Gottlieb 1969). In *P. fluorescens*, four genes are essential for PRN biosynthesis—*prnA, prnB, prnC,* and prnD—from the precursor tryptophan (Hammer et al. 1997). De Souza and Raaijmakers (2003) developed primers from conserved sequences within *prnD* and *pltC* for the respective detection of PRN- and PLT-producing pseudomonads. The availability of these primers enables a quick screening of large collections of *Pseudomonas* spp. for those with good biocontrol potential (De Souza and Raaijmakers 2003). For example, the *prnD* gene was identified with these primers in two native strains of *Pseudomonas* isolated from the rhizosphere of maize crops in Argentina (strains DGR22 and MGR39). These strains were considered strong antagonists since they inhibited a broad number of fungi belonging to the *Fusarium* genus (Cordero et al. 2012).

Other antifungal metabolites produced by *Pseudomonas* spp. are CLPs. These compounds have surfactant properties and are thus able to insert into fungal and bacterial membranes and perturb their function (Haas and Défago 2005). CLPs from the *Pseudomonas* genus have been classified in four groups (viscosin, amphisin, tolaasin, and syringomycin; Raaijmakers et al. 2006). The synthesis of these biosurfactants is nonribosomal and as such is catalyzed by large peptide synthetase complexes (Marahiel et al. 1997). The role of CLPs in biocontrol activity has been demonstrated by De Souza et al. (2003), who isolated surfactant-producing *Pseudomonas* strains from the rhizosphere of wheat.

A surfactant-deficient mutant from strain *P. fluorescens* SS101, for example, was not able to control *Pythium* root rot in the hyacinth, thus strongly suggesting that such biosurfactants are involved in the biocontrol against *Pythium intermedium*. Moreover, in vitro assays indicated that the viscosinamide cyclic lipopeptide produced by *P. fluorescens* DR54 adversely affected mycelial density, oospore formation, and intracellular activity of *P. ultimum* (Thrane et al. 2000).

In addition to diffusible antibiotics, *Pseudomonas* spp. produces volatile antifungal compounds: one of the best known is HCN. This metabolite is produced from glycine under essentially microaerophilic conditions. In *P. fluorescens*, the *hcnABC* genes encode for the HCN synthetase critical for HCN production (Haas and Défago 2005). A PCR-based assay targeting the *hcnAB* genes allowed a sensitive detection of HCN$^+$ pseudomonads (Svercel et al. 2007). In our laboratory, this protocol has been used within a collection of native isolates of *Pseudomonas* to detect HCN-producing strains that inhibited a wide range of phytopathogenic fungi. The majority of those strains were able to produce HCN, in addition to other secondary metabolites such as proteases, celluloses, and siderophores (Cordero et al. 2012). In recent years, VOCs have begun to receive special consideration within the biocontrol scenario. Different species of the *Pseudomonas* genus isolated from canola root and stubble and from soybean roots produced volatiles that were inhibitory for survival (sclerotia), infection (mycelia), and reproductive structures production (ascospores) of *S. sclerotiorum* (Fernando et al. 2005). From six VOCs, each with high fungicidal activity, two were aldehydes (nonanal and *n*-decanal), two were alcohols (cyclohexanol and 2-ethyl, 1-hexanol), and two were sulfur-based compounds (benzothiazole and dimethyl trisulfide; Fernando et al. 2005). In addition, *P. fluorescens* L13-6-12 and *P. trivialis* 3Re2-7 emit complex blends of volatiles that drastically inhibit the growth of *R. solani*. An analysis of the VOC spectra of these strains revealed that certain compounds are isolate specific, while others are emitted by the two antagonists. Mass spectroscopic analysis of these compounds and comparison with the 147,000 compounds comprising in the library of the National Institute of Standards and Technology allowed the identification of only a few compounds: undecene, undecadiene, and benzyloxybenzonitrile (Kai et al. 2007). Furthermore, *P. chlororaphis* O6, like the *Bacillus* isolates, has been shown to produce the volatile 2R, 3R-butanediol, and systemic resistance induced by *P. chlororaphis* O6 against the soft-rot pathogen of tobacco was correlated with the production of this same volatile (Han et al. 2006).

Finally, nonpathogenic rhizobacteria *Pseudomonas* spp. can reduce disease in plant tissues through ISR. For example, the ability of *P. fluorescens* CHA0, *P. fluorescens* WCS417, *P. putida* WCS358, *P. fluorescens* Q2-87, or *P. aeruginosa* 7NSK2 to induce or prime the defense responses and trigger the systemic resistance of grapevine against *B. cinerea* has been demonstrated. Depending on the bacterial strain, different compounds (such as SA, pyochelin, or 2,4-DAPG, among others) have likewise been found capable of inducing disease resistance (Verhagen et al. 2010).

The biosynthesis of antifungal metabolites is controlled by both regulatory genes and environmental conditions. The siderophores and antibiotics positively autoregulate their own biosynthesis at the transcriptional level (Haas and Défago 2005). A control is also provided by the relative concentrations of the sigma factors RpoD, RpoN, and RpoS in the bacterial cells. That a balance among these three sigma factors is essential in determining antibiotics levels has been suggested. For example, RpoN positively controls the expression of *plt* structural genes and PLT production in *P. fluorescens* CHA0 but exerts a strong negative effect on the expression of the *phl* structural genes and DAPG production. RpoS might interfere with the RpoN-mediated control of antibiotic production, since RpoS exerts opposite effect to that of RpoN on the PLT expression. The cellular levels of RpoN and RpoS could therefore be influencing the relative amounts of this antibiotic produced by *P. fluorescens* CHA0 (Haas and Keel 2003; Péchy-Tarr et al. 2005). The regulation by RpoS also varies depending on the microorganism and the antibiotic in question. An *rpoS* mutant of *P. fluorescens* Pf-5 exhibited a decreased PRN production along with increased levels of DAPG and PLT (Sarniguet et al. 1995), whereas in *P. chlororaphis* PA23 the expression of PRN, lipase, and protease was repressed, while PHZ was positively regulated by RpoS (Manuel et al. 2012). In addition, RpoS also modulates the production of other compounds: for example, in an *rpoS* mutant of strain Pf-5, the HCN production was found to be increased (Whistler et al. 1998).

The production of certain antifungal metabolites is also regulated by QS, e.g., the biosynthesis of phenazine. In this instance, the *phzI* gene encodes an N-AHL-synthase that produces AHL. After a threshold level of AHL accumulation, the molecules bind to and activate PhzR, whose activated conformation then induces the transcription of the phenazine biosynthetic operon (Pierson et al. 1998; Chin-A-Woeng et al. 2001b).

Of interest too is the possibility that PHZ may function as a signaling molecule and in that capacity affect *P. chlororaphis* biofilm development (Selin et al. 2010): Maddula et al. (2008) have demonstrated that a phenazine structural mutant, *phzB*, derived from *P. chlororaphis* 30-84 was defective in phenazine biosynthesis and impaired in biofilm formation as well.

Moreover, the production of PRN is also under the control by QS in the pseudomonads (Selin et al. 2012); but, unlike PHZ, PRN does not affect biofilm development, thus suggesting that this trait is not related to antibiotic production in general; rather the characteristic may be unique to PHZ (Selin et al. 2010).

Environmental conditions are also consequential in the regulation of biocontrol genes. For example, Lim et al. (2012) recently investigated the transcriptomic and proteomic impacts of iron limitation on *P. fluorescens* Pf-5. At low iron levels, the genes involved in the biosynthesis of the secondary metabolites 2,4-DAPG, orfamide A, PRN, and chitinase were overexpressed, but HCN-gene transcription was downregulated in an iron-depleted culture medium. Other compounds widely known to be induced under conditions of iron limitation are the siderophores. Minerals and carbon sources commonly found in root exudates have a differential influence on the production of antibiotics and siderophores. For example, 2,4-DAPG is stimulated by Zn^{2+}, ammonium molybdate, and glucose in *P. fluorescens* CHA0; but the production of PLT is increased by Zn^{2+}, Co^{2+}, and glycerol though at the same time repressed by glucose (Duffy and Défago 1999). Furthermore, low oxygen concentrations are necessary for the biosynthesis of HCN (Haas and Keel 2003).

At the posttranscriptional level, the GacS/GacA two-component system is involved in the regulation of extracellular enzymes and secondary metabolites. A myriad of poorly unknown signals is believed to stimulate the autophosphorylation of the GacS sensor kinase. Thereafter, a phosphate group is transferred to the GacA response regulator and this phosphorylation, in turn, activates GacA in response to those signals or stimuli. The phosphorylated GacA induces the transcription of small regulatory RNAs—such as RsmZ and Rsm—that, for their part, sequesters the small RNA-binding proteins RsmA and RsmE, thus relieving translational repression through those regulatory proteins. In *P. fluorescens*, the RsmA and RsmE function as posttranscriptional repressors of the structural genes for DAPG, PRN, HCN, or PLT biosynthesis (Heeb and Haas 2001; Valverde et al. 2003; Haas and Défago 2005).

9.5 The Commercialization of Biocontrol Agents

Biocontrol agents (BCAs) have several advantages compared to chemical pesticides: BCAs are safer, have a reduced environmental impact, and carry a lower risk to human and animal health. BCAs multiply by themselves but their proliferation is controlled by the plant as well as by the indigenous microbial community. Furthermore, BCAs decompose more rapidly than conventional chemical pesticides, while resistance development is reduced through several mechanisms (Berg 2009). Despite all the advantages of BCAs over chemical pesticides, the marketing and implementation of BCAs has been delayed.

Many biocontrol agents work well in the laboratory and under greenhouse conditions, but then fail in the field (Heydari and Pessarakli 2010). One of the reasons for such biocontrol failure is the lack of appropriate screening procedures to select those microorganisms that are most suitable for disease control in diverse soil environments (Pliego et al. 2011). Because that screening is a critical step in the development of field-effective biocontrol agents, the success of all subsequent stages depends on the ability to identify the appropriate candidate beforehand (McSpadden Gardener and Fravel 2002). In addition, a better understanding of the biotic and abiotic factors present in the rhizosphere is necessary since those conditions could affect the biocontrol microorganism itself and/or its production of the requisite metabolites for the desired form of control (Heydari and Pessarakli 2010).

In conclusion, the development of a biopesticide requires several steps. The first is the establishment of collections of bacterial isolates from selected sites. The rhizospheres of soils that are naturally suppressive to diseases can be good sources of BCAs because the plant roots are associated with microbial communities that have an overall beneficial effect on plant health (Mendes et al. 2011), although ordinary (conductive) soils also contain BCAs (Haas and Défago 2005). Once the bacterial collection is established, the choice of the screening procedure will dictate the selection of the different BCA candidates. This decision must be based on the type of pathogen to be controlled, the environmental parameters, the desired future biocontrol strategy (preventive or curative), the difficulties for future formulation, and other considerations (Pliego et al. 2011).

The BCAs must compete effectively with the native microflora, survive, become established in the plant root, and then function under the particular conditions of the ecosystem at hand. Therefore, the next step in the screening is a test under greenhouse conditions. A good correlation has been found between the efficacy of biocontrol microorganisms against soilborne pathogens and their ability to colonize the plant root, especially when the mode of action used by BCAs is antibiosis or competition for niches and nutrients (Lugtenberg and Kamilova 2009; Pliego et al. 2011). Ideally, greenhouse assays should be followed by field trials under different climatic conditions and diverse soil qualities. A determination of to what extent temperature, moisture, soil type, host cultivar, and other factors affect biocontrol performance is also essential. Notwithstanding, screening in the field may be difficult to perform because of inconsistent abiotic and biotic parameters in addition to the time and cost required for such trials (Fravel 2005; Pliego et al. 2011).

Once a good candidate is obtained, further studies such as a pathogenicity test must necessarily be performed and the biosafety of the biopesticide confirmed in order to avoid undesirable side effects on plants and/or animals as well as on the environment (Montesinos 2003).

For the commercial development of a biopesticide, the microorganism should be produced on an industrial scale (i.e., fermentation), preserved, stored, and formulated. One condition limiting commercial interest in biocontrol is the high cost of production for most biocontrol agents. This costliness may be attributed to the high cost of a required substrate or to a low biomass productivity. In general, bacteria are produced by liquid fermentation in continuously stirred bioreactor tanks, but apart from the method used for fermentation, the aim is always to achieve the highest possible yield at the lowest cost of the culture medium: that end is attained through the use of molasses, peptones, or industrial-grade protein hydrolysates (Montesinos 2003; Fravel 2005).

The need for storage and preservation of the biopesticide for commercialization requires a sufficiently long shelf life, which feature is one of the main limiting conditions.

Particular methods can affect certain aspects of biocontrol performance, shelf life, and other variables. For example, many products are obtained by spray-drying, but this method can produce a high loss of viability in certain microorganisms (e.g., Gram-negative bacteria) because of the high temperatures involved. Some biocontrol agents such as *Bacillus* spp. are capable of producing spores that are relatively simple to formulate (Montesinos 2003; Fravel 2005). The final product should be formulated before use by means of biocompatible additives that increase the microorganism's survival, improve its application, and stabilize it (Boyetchko et al. 1998).

The formulated product developed commercially as a biopesticide can be financially protected by means of a patent. Biopesticide patents are considered biotechnologic inventions because they include microbial products and processes. A patent, however, is not an authorization for commercial use. A given commercial product of this type must be registered as a biopesticide, where the appropriate registration depends on the specific laws and regulations within each country. Obstacles in that procedure are the cost and investment of time involved. In spite of the relatively high number of patents for biopesticides, only a few are registered for agricultural use. In the United States, several products have been registered as biopesticides. The majority of the commercial BCAs have a broad host range and can be applied to numerous crops (Berg 2009). Table 9.1 shows examples of biopesticides based on *Bacillus* spp. and *Pseudomonas* spp. that have been registered and are available on the market in some countries. Although, a few reports have been published on the microbial biocontrol of plant diseases in Argentina (Cavaglieri et al. 2005; Correa et al. 2009; Alvarez et al. 2012), commercial products have not yet been registered in that country.

Our current work is focussed on the application of the native strains of *B. amyloliquefaciens* to control phytopathogens. To that end, field trials, carried out during 2009–2010, have shown promising results. In cooperation with an Argentine company, a formulation based on *B. amyloliquefaciens* was developed and further tested under field conditions. The viability of the *Bacillus* strains as well as the activity of the metabolites present in the inoculant continued for 1 year. The formulation was used for foliar application on R5.3-stage soybean plants showing

9 Fighting Plant Diseases Through the Application of *Bacillus* and...

Table 9.1 Commercial biofungicides based on *Bacillus* spp. and *Pseudomonas* spp.

Product	Biocontrol agent	Target pathogen or disease	Crop	Manufacturer/distributor
Ballad plus	*B. pumilus* strain QST2808	Rust, powdery mildew, *Cercospora*	Soybean	AgraQuest, USA
HiStick N/T	*B. subtilis* MBI600 (rhizobia also in formulation)	*Fusarium* spp., *Rhizoctonia* spp., *Aspergillus*	Soybean, alfalfa, dry/snap beans, peanuts	Becker Underwood, USA
Companion	*B. subtilis* GB03	*Rhizoctonia, Pythium, Fusarium, Phytophthora, Sclerotinia*	Greenhouse and ornamental crops	Growth Products, Ltd., USA
Kodiak	*B. subtilis* GB03	*Rhizoctonia, Fusarium, Aspergillus*	Cotton, legumes, soybean	Bayer CropScience, USA
Serenade/Rhapsody/Serenade Garden	*B. subtilis* QST713	Powdery mildew, downy mildew, *Cercospora* leaf spot, early blight, late blight, brown rot, fire blight, and others	Cucurbits, grapes, hops, vegetables, peanuts, pome fruits, stone fruits, and others	AgraQuest, Inc., USA
Taegro	*B. subtilis* var. *amyloliquefaciens* FZB24	*Rhizoctonia, Fusarium*	Tree seedlings, ornamentals and shrubs	Novozymes, Denmark
Cedomon	*P. chlororaphis*	*Fusarium, Pyrenophora graminea, Helminthosporium teres* (*Pyrenophora teres*)	Barley net blotch, barley leaf stripe (*Drechslera graminea*) and *Fusarium* oat leaf spot	Bioagri, Sweden
Biosave	*P. syringae* strains ESC-10 and ESC-11	*Botrytis cinerea, Penicillium* spp., *Mucor pyroformis, Geotrichum candidum*	Citrus, stone and pome fruits, cherries, potatoes	JET Harvest Solutions, USA
BlightBan A506	*P. fluorescens* strain A506	Frost damage, *Erwinia amylovora*, and russet-inducing bacteria	Almond, apple, apricot, blueberry, cherry, peach, pear, potato, strawberry, tomato	Nufarm, Inc., USA
Spotless biofungicide	*P. aureofaciens* strain Tx-1	Dollar spot, anthracnose, *Pythium*, pink snow mold	Turfgrass	Turf Science Laboratories, Inc., USA

(continued)

Table 9.1 (continued)

Product	Biocontrol agent	Target pathogen or disease	Crop	Manufacturer/distributor
Cerall	P. chlororaphis strain MA 342	Seed-borne diseases common wheat bunt (*Tilletia caries*), wheat leaf spot (*Septoria nodorum*), and Fusarium (*Fusarium* spp.) in wheat, rye, and triticale	Cereal	Bioagri, Sweden
Proradix	Pseudomonas sp. strain DSMZ 13134	*Erwinia* spp., *Rhizoctonia solani*	Potatoes, tomatoes, pepper	Sourcon-Padena GmbH & Co. KG, Tübingen, Germany

typical symptoms of Septoria brown spots (at a 40 % of incidence) and traces of Cercospora leaf spot. A reduction in the severity of the symptoms and a delayed defoliation were observed after 20 days postapplication compared to nontreated plants. The average soybean yield following the *Bacillus* treatments increased between 101 and 180 kg/ha with respect to nontreated plants. Moreover, the foliar application of *Bacillus* strains resulted in an increased soybean yield compared to that obtained after treatment with the chemical fungicide Amistar Xtra®. Our data thus suggest that the foliar application of the lipopeptide-producing *B. amyloliquefaciens* strains could be a promising strategy for the management of soybean diseases.

9.6 Final Considerations

Because pathogenic fungi cause significant losses in the yield and quality of crops, the use of chemical pesticides to fight pests that adversely affect crops has increased in recent years. Unfortunately, agrochemicals also have a negative impact on human health and on the environment. Since an alternative, and more environmentally friendly, strategy is the use of microorganisms capable of improving the health of plants, many bacterial strains have been evaluated as biocontrol agents. Few of these attempts, however, were successful after testing in the field. One of the reasons for this failure is the method of screening used to select the candidate most capable of functioning under diverse environmental conditions. The development of a biopesticide, accordingly, requires several steps: namely, the isolation of bacterial candidates, screening in vitro, experiments under first greenhouse and then field conditions, and finally development of the methods of application. In addition, extensive research is also needed to elucidate the mode of action used by the biocontrol agents and to improve the scale-up and bioprocess development for microbial inoculants.

Many scientific challenges for research lie ahead in the field of biocontrol. Of particular relevance will be the exploitation of molecular techniques to study in situ the expression of bacterial factors necessary for efficient biocontrol activity. A special area of interest will be the molecular analysis of the mode of interaction between the bacteria and their host plant and between them and the pathogen, including the mechanism whereby they affect its regulation. For example, the so-called omics technologies—viz., proteomics and metabolomics—could prove useful in identifying the relevant genes that are expressed in the rhizosphere (Barea et al. 2005; Berg 2009).

Acknowledgment The authors wish to thank Dr. Donald F. Haggerty, a retired career investigator and native English speaker, for editing the final version of the manuscript.

References

Abbas A, Morrissey JP, Marquez PC, Sheehan MM, Delany IR, O'Gara F (2002) Characterization of interactions between the transcriptional repressor PhlF and its binding site at the *phlA* promoter in *Pseudomonas fluorescens* F113. J Bacteriol 184(11):3008–3016

Abbas A, McGuire JE, Crowley D, Baysse C, Dow M, O'Gara F (2004) The putative permease PhlE of *Pseudomonas fluorescens* F113 has a role in 2,4-diacetylphloroglucinol resistance and in general stress tolerance. Microbiology 150(7):2443–2450

Ahimou F, Jacques P, Deleu M (2000) Surfactin and iturin A effects on *Bacillus subtilis* surface hydrophobicity. Enzyme Microb Technol 27:749–754

Alvarez F, Castro M, Príncipe A, Borioli G, Fischer S, Mori G, Jofre E (2012) The plant-associated *Bacillus amyloliquefaciens* strains $MEP_{2}18$ and $ARP_{2}3$ capable of producing the cyclic lipopeptides iturin or surfactin and fengycin are effective in biocontrol of sclerotinia stem rot disease. J Appl Microbiol 112:159–174

Arrebola E, Jacobs R, Korsten L (2010) Iturin A is the principal inhibitor in the biocontrol activity of *Bacillus amyloliquefaciens* PPCB004 against postharvest fungal pathogens. J Appl Microbiol 108:386–395

Babalola OO (2010) Beneficial bacteria of agricultural importance. Biotechnol Lett 32:1559–1570

Bangera MG, Thomashow LS (1999) Identification and characterization of a gene cluster for synthesis of the polyketide antibiotic 2,4-diacetylphloroglucinol from *Pseudomonas fluorescens* Q2-87. J Bacteriol 181:3155–3163

Bardin SD, Huang HC (2001) Research on biology and control of *Sclerotinia* diseases in Canada. Can J Plant Pathol 23:88–98

Barea JM, Pozo MJ, Azcón R, Azcon-Aguilar C (2005) Microbial co-operation in the rhizosphere. J Exp Bot 56(417):1761–1778

Bari R, Jones JDG (2009) Role of plant hormones in plant defense responses. Plant Mol Biol 69:473–488

Bauer WD, Mathesius U (2004) Plant responses to bacterial quorum sensing signals. Curr Opin Plant Biol 7(4):429–433

Berg G (2009) Plant–microbe interactions promoting plant growth and health: perspectives for controlled use of microorganisms in agriculture. Appl Microbiol Biotechnol 84:11–18

Bottiglieri M, Keel C (2006) Characterization of PhlG, a hydrolase that specifically degrades the antifungal compound 2,4-diacetylphloroglucinol in the biocontrol agent *Pseudomonas fluorescens* CHA0. Appl Environ Microbiol 72:418–427

Boyetchko S, Pedersen E, Punja Z, Reddy M (1998) Formulations of biopesticides. In: Hall FR, Menn JJ (eds) Biopesticides: use and delivery. Humana Press, Totowa, NJ, pp 487–508

Brendel N, Partida-Martinez LP, Scherlach K, Hertweck C (2007) A cryptic PKS-NRPS gene locus in the plant commensal *Pseudomonas fluorescens* Pf-5 codes for the biosynthesis of an antimitotic rhizoxin complex. Org Biomol Chem 5:2211–2213

Brodhagen M, Paulsen I, Loper JE (2005) Reciprocal regulation of pyoluteorin production with membrane transporter gene expression in *Pseudomonas fluorescens* Pf-5. Appl Environ Microbiol 71:6900–6909

Bulgari D, Casati P, Brusetti L, Quaglino F, Brasca M, Daffonchio D, Bianco PA (2009) Endophytic bacterial diversity in grapevine (*Vitis vinifera* L.) leaves described by 16S rRNA gene sequence analysis and length heterogeneity-PCR. J Microbiol 47:393–401

Carrillo C, Teruel JA, Aranda FJ, Ortiz A (2003) Molecular mechanism of membrane permeabilization by the peptide antibiotic surfactin. Biochim Biophys Acta 1611:91–97

Cavaglieri LR, Andrés L, Ibáñez M, Etcheverry MG (2005) Rhizobacteria and their potential to control *Fusarium verticillioides*: effect of maize bacterisation and inoculum density. Antonie Van Leeuwenhoek 87:179–187

Cazorla FM, Romero D, Pérez-García A, Lugtenberg BJJ, de Vicente A, Bloemberg G (2007) Isolation and characterization of antagonistic *Bacillus subtilis* strains from the avocado rhizoplane displaying biocontrol activity. J Appl Microbiol 103:1950–1959

Chen C, Belanger RR, Benhamou N, Paulitz TC (2000) Defense enzymes induced in cucumber roots by treatment with plant growth promoting rhizobacteria (PGPR) and *Pythium aphanidermatum*. Physiol Mol Plant Pathol 56:13–23

Chen XH, Koumoutsi A, Scholz R, Schneider K, Vater J, Sussmuth R, Piel J, Borriss R (2009) Genome analysis of *Bacillus amyloliquefaciens* FZB42 reveals its potential for biocontrol of plant pathogens. J Biotechnol 140:27–37

Chin-A-Woeng TFC, Thomas-Oates JE, Lugtenberg BJJ, Bloemberg GV (2001a) Introduction of the *phzH* gene of *Pseudomonas chlororaphis* PCL1391 extends the range of biocontrol ability of phenazine-1-carboxylic acid-producing *Pseudomonas* spp. strains. Mol Plant Microbe Interact 14:1006–1015

Chin-A-Woeng TFC, van den Broek D, de Voer G, van der Drift K, Tuinman S, Thomas-Oates JE, Lugtenberg BJJ, Bloemberg GV (2001b) Phenazine-1-carboxamide production in the biocontrol strain *Pseudomonas chlororaphis* PCL1391 is regulated by multiple factors secreted into the growth medium. Mol Plant Microbe Interact 14:969–979

Chin-A-Woeng TFC, Bloemberg GB, Lugtenberg BJJ (2003) Phenazines and their role in biocontrol By *Pseudomonas* bacteria. New Phytol 157:503–523

Cho SJ, Lim WJ, Hong SY, Park SR, Yun HD (2003) Endophytic colonization of balloon flower by antifungal strain *Bacillus* sp. CY22. Biosci Biotechnol Biochem 67(10):2132–2138

Cho SM, Kang BR, Han SH, Anderson AJ, Park JY, Lee YH, Cho BH, Yang KY, Ryu CM, Kim YC (2008) 2R,3R-butanediol, a bacterial volatile produced by *Pseudomonas chlororaphis* O6, is involved in induction of systemic tolerance to drought in *Arabidopsis thaliana*. Mol Plant Microbe Interact 21:1067–1075

Choudary DK, Johri BN (2009) Interactions of *Bacillus* spp. and plants-with special reference to induced systemic resistance (ISR). Microbiol Res 164:493–513

Chulze SN, Ramírez ML, Farnochi MC, Pascale M, Visconti A, March G (1996) *Fusarium* and fumonisins occurrence in Argentinean corn at different ear maturity stages. J Agric Food Chem 44:2797–2801

Compant S, Duffy B, Nowak J, Clément C, Barka EA (2005) Use of plant growth-promoting bacteria for biocontrol of plant diseases: principles, mechanisms of action, and future prospects. Appl Environ Microbiol 71:4951–4959

Conrath U, Pieterse CMJ, Mauch-Mani B (2002) Priming in plant-pathogen interactions. Trends Plant Sci 7:210–216

Conrath U, Beckers GJ, Flors V, Garcia-Agustin P, Jakab G, Mauch F, Newman MA, Pieterse CM, Poinssot B, Pozo MJ, Pugin A, Schaffrath U, Ton J, Wendehenne D, Zimmerli L, Mauch-Mani B (2006) Priming: getting ready for battle. Mol Plant Microbe Interact 19:1062–1071

Cordero P, Cavigliasso A, Príncipe A, Godino A, Jofré E, Mori G, Fischer S (2012) Genetic diversity and antifungal activity of native *Pseudomonas* isolated from maize plants grown in a central region of Argentina. Syst Appl Microbiol 35:342–351

Correa OS, Montecchia MS, Berti MF, Fernández Ferrari MC, Pucheu NL, Kerber NL, García AF (2009) *Bacillus amyloliquefaciens* BNM122, a potential microbial biocontrol agent applied on soybean seeds, causes a minor impact on rhizosphere and soil microbial communities. Appl Soil Ecol 41:185–194

De Souza JT, Raaijmakers JM (2003) Polymorphisms within the *prnD* and *pltC* genes from pyrrolnitrin and pyoluteorin-producing *Pseudomonas* and *Burkholderia* spp. FEMS Microbiol Ecol 43:21–34

De Souza JT, De Boer M, De Waard P, Van Beek TA, Raaijmakers JM (2003) Biochemical, genetic, and zoosporicidal properties of cyclic lipopeptide surfactants produced by *Pseudomonas fluorescens*. Appl Environ Microbiol 69:7161–7172

Dean R, Van Kan JAL, Pretorius ZA, Hammond-Kosack KE, Di Pietro A, Spanu PD, Rudd JJ, Dickman M, Kahmann R, Ellis J, Foster GD (2012) The top 10 fungal pathogens in molecular plant pathology. Mol Plant Pathol 13(4):414–430

Delany I, Sheehan MM, Fenton A, Bardin S, Aarons S, O'Gara F (2000) Regulation of production of the antifungal metabolite 2,4-diacetylphloroglucinol in *Pseudomonas fluorescens* F113: genetic analysis of *phlF* as a transcriptional repressor. Microbiology 146:537–546

Dong YH, Wang LH, Xu JL, Zhang HB, Zhang XF, Zhang LH (2001) Quenching quorum sensing dependent bacterial infection by an N-acyl homoserine lactonase. Nature 411:813–817

Drogue B, Doré H, Borland S, Wisniewski-Dyé F, Prigent-Combaret C (2012) Which specificity in cooperation between phytostimulating rhizobacteria and plants? Res Microbiol 163:500–510

Duffy BK, Défago G (1999) Environmental factors modulating antibiotic and siderophore biosynthesis by *Pseudomonas fluorescens* biocontrol strains. Appl Environ Microbiol 65:2429–2438

Duijff BJ, Gianinazzi-Pearson V, Lemanceau P (1997) Involvement of the outer membrane lipopolysaccharides in the endophytic colonization of tomato roots by biocontrol *Pseudomonas fluorescens* strain WCS417r. New Phytol 135:325–334

Edwards SG (2004) Influence of agricultural practices on *Fusarium* infection of cereals and subsequent contamination of grain by trichothecene mycotoxins. Toxicol Lett 153(1):29–35

Fernando WGD, Ramarathnam R, Krishnamoorthy AS, Savchuk SC (2005) Identification and use of potential bacterial organic antifungal volatiles in biocontrol. Soil Biol Biochem 37:955–964

Francis I, Holsters M, Vereecke D (2010) The Gram-positive side of plant–microbe interactions. Environ Microbiol 12(1):1–12

Fravel DR (2005) Commercialization and implementation of biocontrol. Annu Rev Phytopathol 43:337–359

Fuqua WC, Winans SC, Greenberg EP (1994) Quorum sensing in bacteria: the LuxR-LuxI family of cell density-responsive transcriptional regulators. J Bacteriol 176:269–275

Fuqua WC, Parsek MR, Greenberg EP (2001) Regulation of gene expression by cell-to-cell communication: acyl-homoserine lactone quorum sensing. Annu Rev Genet 35:439–468

Galloway WRJD, Hodgkinson JT, Bowden S, Welch M, Spring DR (2012) Applications of small molecule activators and inhibitors of quorum sensing in Gram-negative bacteria. Trends Microbiol 20(9):449–458

Gross H, Stockwell VO, Henkels MD, Nowak-Thompson B, Loper JE, Gerwick WH (2007) The genomisotopic approach: a systematic method to isolate products of orphan biosynthetic gene clusters. Chem Biol 14:53–63

Haas D, Défago G (2005) Biological control of soil-borne pathogens by fluorescent pseudomonads. Nat Rev Microbiol 3:307–319

Haas D, Keel C (2003) Regulation of antibiotic production in root-colonizing *Pseudomonas* spp. and relevance for biological control of plant diseases. Annu Rev Phytopathol 41:117–153

Hammer PE, Hill DS, Lam ST, van Pée KH, Ligon JM (1997) Four genes from *Pseudomonas fluorescens* that encode the biosynthesis of pyrrolnitrin. Appl Environ Microbiol 63:2147–2154

Hamoen LW, Venema G, Kuipers OP (2003) Controlling competence in *Bacillus subtilis*: shared use of regulators. Microbiology 149:9–17

Han SH, Lee SJ, Moon JH, Park KH, Yang KY, Cho BH, Kim KY, Kim YW, Lee MC, Anderson AJ, Kim YC (2006) GacS-dependent production of 2R, 3R-butanediol by *Pseudomonas chlororaphis* O6 is a major determinant for eliciting systemic resistance against *Erwinia carotovora* but not against *Pseudomonas syringae* pv. *tabaci* in tobacco. Mol Plant Microbe Interact 19:924–930

Heeb S, Haas D (2001) Regulatory roles of the GacS/GacA two-component system in plant-associated and other Gram-negative bacteria. Mol Plant Microbe Interact 14:1351–1363

Heil M, Bostock RM (2002) Induced systemic resistance (ISR) against pathogens in the context of induced plant defenses. Ann Bot 89:503–512

Heydari A, Pessarakli M (2010) A review on biological control of fungal plant pathogens using microbial antagonist. J Biol Sci 10(4):273–290

Horst L, Locke JC, Krause CR, McMahon RW, Madden LV, Hoitink HA (2005) Suppression of *Botrytis* blight of begonia by *Trichoderma hamatum* 382 in peat and compost-amended potting mixes. Plant Dis 89:1195–1200

Hovmøller MS, Sørensen CK, Walter S, JusteseN AF (2011) Diversity of *Puccinia striiformis* on cereals and grasses. Annu Rev Phytopathol 49:197–217

Huang X, Zhu D, Ge Y, Hu H, Zhang X, Xu Y (2004) Identification and characterization of *pltZ*, a gene involved in the repression of pyoluteorin biosynthesis in *Pseudomonas* sp. M18. FEMS Microbiol Lett 232:197–202

Huang X, Yan A, Zhang X, Xu Y (2006) Identification and characterization of a putative ABC transporter PltHIJKN required for pyoluteorin production in *Pseudomonas* sp. M18. Gene 376:68–78

Jetiyanon K, Kloepper JW (2002) Mixtures of plant growth promoting rhizobacteria for induction of systemic resistance against multiple plant diseases. Biol Control 24:285–291

Kai M, Effmert U, Berg G, Piechulla B (2007) Volatiles of bacterial antagonists inhibit mycelial growth of the plant pathogen *Rhizoctonia solani*. Arch Microbiol 187:351–360

Kai M, Haustein M, Molina F, Petri A, Scholz B, Piechulla B (2009) Bacterial volatiles and their action potential. Appl Microbiol Biotechnol 81:1001–1012

Khan SR, Mavrodi DV, Jog GJ, Suga H, Thomashow LS, Farrand SK (2005) Activation of the *phz* operon of *Pseudomonas fluorescens* 2–79 requires the LuxR homolog PhzR, N-(3-OHhexanoyl)-L-homoserine lactone produced by the LuxI homolog PhzI, and a cis-acting phz box. J Bacteriol 187:6517–6527

Kloepper JW, Ryu CM, Zhang SA (2004) Induced systemic resistance and promotion of plant growth by *Bacillus* spp. Phytopathology 94:1259–1266

Krause MS, DeCeuster TJJ, Tiquia SM, Michel FC Jr, Madden LV, Hoitink HAJ (2003) Isolation and characterization of rhizobacteria from composts that suppress the severity of bacterial leaf spot of radish. Phytopathology 93:1130–1292

Lazazzera BA, Solomon JM, Grossman AD (1997) An exported peptide functions intracellularly to contribute to cell density signaling in *B. subtilis*. Cell 89(6):917–925

Lim CK, Hassan KA, Tetu SG, Loper JE, Paulsen IT (2012) The effect of iron limitation on the transcriptome and proteome of *Pseudomonas fluorescens* Pf-5. PLoS One 7(6):e39139

Liu B, Qiao H, Huang L, Buchenauer H, Han Q, Kang Z, Gong Y (2009) Biological control of take-all in wheat by endophytic *Bacillus subtilis* E1R-j and potential mode of action. Biol Control 49:277–285

Long HH, Schmidt DD, Baldwin IT (2008) Native bacterial endophytes promote host growth in a species-specific manner; phytohormone manipulations do not result in common growth responses. PLoS One 3:e2702

Loper JE, Henkels MD, Shaffer BT, Valeriote FA, Gross H (2008) Isolation and identification of rhizoxin analogs from *Pseudomonas fluorescens* Pf-5 using a genomic mining strategy. Appl Environ Microbiol 74(10):3085–3093

Lugtenberg B, Kamilova F (2009) Plant-growth-promoting rhizobacteria. Annu Rev Microbiol 63:541–556

Lugtenberg BJ, Dekkers L, Bloemberg GV (2001) Molecular determinants of rhizosphere colonization by *Pseudomonas*. Annu Rev Phytopathol 39:461–490

Maddula VSRK, Pierson EA, Pierson LS (2008) Altering the ratio of phenazines in *Pseudomonas chlororaphis* (aureofaciens) strain 30-84: effects on biofilm formation and pathogen inhibition. J Bacteriol 190:2759–2766

Magan N, Aldred D, Mylona K, Lambert RJW (2010) Limiting mycotoxins in stored wheat. Food Addit Contam 27:644–650

Magnin-Robert M, Trotel-Aziz P, Quantinet D, Biagianti S, Aziz A (2007) Biological control of *Botrytis cinerea* by selected grapevine-associated bacteria and stimulation of chitinase and β-1,3 glucanase activities under field conditions. Eur J Plant Pathol 118:43–57

Manuel J, Selin C, Fernando WGD, de Kievit TR (2012) Stringent response mutants of *Pseudomonas chlororaphis* PA23 exhibit enhanced antifungal activity against *Sclerotinia sclerotiorum* in vitro. Microbiology 158:207–216

Marahiel MA, Stacelhaus T, Mootz HD (1997) Modular peptide synthetases involved in nonribosomal peptide synthesis. Chem Rev 97:2651–2673

Mauch F, Mauch-Mani B, Boller T (1988) Antifungal hydrolases in pea tissue. II. Inhibition of fungal growth by combinations of chitinase and β-1,3-glucanase. Plant Physiol 88:936–942

Mavrodi DV, Ksenzenko VN, Bonsall RF, Cook RJ, Boronin AM, Thomashow LS (1998) A seven-gene locus for synthesis of phenazine-1-carboxylic acid by *Pseudomonas fluorescens* 2–79. J Bacteriol 180:2541–2548

Mavrodi DV, Blankenfeldt W, Thomashow LS (2006) Phenazine compounds in fluorescent *Pseudomonas* spp. biosynthesis and regulation. Annu Rev Phytopathol 44:417–445

Mavrodi DV, Mavrodi OV, Parejko JA, Bonsall RF, Kwak YS, Paulitz TC, Thomashow LS, Weller DM (2012) Accumulation of the antibiotic phenazine-1-carboxylic acid in the rhizosphere of dryland cereals. Appl Environ Microbiol 78(3):804–812

McSpadden Gardener BB, Fravel DR (2002) Biological control of plant pathogens: research, commercialization, and application in the USA. Plant Health Prog. doi:10.1094/PHP-2002-0510-01-RV

McSpadden Gardener BB, Mavrodi DV, Thomashow LS, Weller DM (2001) A rapid polymerase chain reaction-based assay characterizing rhizosphere populations of 2,4-diacetylphloroglucinol-producing bacteria. Phytopathology 91(1):44–54

Melnick RL, Zidack NK, Bailey BA, Maximova SN, Guiltinan M, Backman PA (2008) Bacterial endophytes: *Bacillus* spp. from annual crops as potential biological control agents of black pod rot of cacao. Biol Control 46:46–56

Mendes R, Kruijt M, Bruijn I, Dekkers E, van der Voort M, Schneider JHM, Piceno YM, DeSantis TZ, Andersen GL, Bakker PAHM, Raaijmakers JM (2011) Deciphering the rhizosphere microbiome for disease-suppressive bacteria. Science 332:1097–1100

Montesinos E (2003) Development, registration and commercialization of microbial pesticides for plant. Int Microbiol 6:245–252

Moyne AL, Cleveland TE, Tuzun S (2004) Molecular characterization and analysis of the operon encoding the antifungal lipopeptide bacillomycin D. FEMS Microbiol Lett 234:43–49

Niu DD, Liu HX, Jiang CH, Wang YP, Wang XY, Jin HL, Guo JH (2011) The plant growth promoting rhizobacterium *Bacillus cereus* AR156 induces systemic resistance in *Arabidopsis thaliana* by simultaneously activating salicylate- and jasmonate/ethylene dependent signaling pathway. Mol Plant Microbe Interact 24(5):533–542

Nowak-Thompson B, Chaney N, Gould SJ, Loper JE (1999) Characterization of the pyoluteorin biosynthetic gene cluster of *Pseudomonas fluorescens* Pf-5. J Bacteriol 181:2166–2274

Ongena M, Jacques P (2008) *Bacillus* lipopeptides: versatile weapons for plant disease biocontrol. Trends Microbiol 16:115–125

Ongena M, Jourdan E, Schafer M, Kech C, Budzikiewicz H, Luxen A, Thonart P (2005) Isolation of an N-alkylated benzylamine derivative from *Pseudomonas putida* BTP1 as elicitor of induced systemic resistance in bean. Mol Plant Microbe Interact 18:562–569

Ongena M, Jourdan E, Adam A, Paquot M, Brans A, Joris B, Arpigny JL, Thonart P (2007) Surfactin and fengycin lipopeptides of *Bacillus subtilis* as elicitors of induced systemic resistance in plants. Environ Microbiol 9:1084–1090

Paulsen IT, Press CM, Ravel J, Kobayashi DY, Myers GSA, Mavrodi DV, DeBoy RT, Seshadri R, Ren Q, Madupu R, Dodson RJ, Durkin AS, Brinkac LM, Daugherty SC, Sullivan SA, Rosovitz MJ, Gwinn ML, Zhou L, Schneider DJ, Cartinhour SW, Nelson WC, Weidman J, Watkins K, Tran K, Khouri H, Pierson EA, Pierson LS III, Thomashow LS, Loper JE (2005) Complete genome sequence of the plant commensal *Pseudomonas fluorescens* Pf-5. Nat Biotechnol 23:873–878

Péchy-Tarr M, Bottiglieri M, Mathys S, Bang Lejbølle K, Schnider-Keel U, Maurhofer M, Keel C (2005) RpoN (σ54) controls production of antifungal compounds and biocontrol activity in *Pseudomonas fluorescens* CHA0. Mol Plant Microbe Interact 18:260–272

Pérez-García A, Romero D, de Vicente A (2011) Plant protection and growth stimulation by microorganisms: biotechnological applications of Bacilli in agriculture. Curr Opin Biotechnol 22:187–193

Pierson LS, Gaffney T, Lam S, Gong FC (1995) Molecular analysis of genes encoding phenazine biosynthesis in the biological control bacterium *Pseudomonas aureofaciens* 30-84. FEMS Microbiol Lett 134:299–307

Pierson LS, Wood DW, Pierson EA (1998) Homoserine lactone-mediated gene regulation in plant-associated bacteria. Annu Rev Phytopathol 36:207–225

Pieterse CMJ, Van Wees SCM, van Pelt JA, Knoester M, Laan R, Gerrits H, Weisbeek PJ, Van Loon LC (1998) A novel signaling pathway controlling induced systemic resistance in *Arabidopsis*. Plant Cell 10:1571–1580

Pieterse CMJ, van der Ent S, van Pelt JA, van Loon LC (2007) The role of ethylene in rhizobacteria-induced systemic resistance (ISR). In: Ramina A, Chang C, Giovannoni J, Klee H, Perata P, Wolterings (eds) Advances in plant ethylene research. Proceeding of the 7th international symposium on plant hormone ethylene. Springer, Dordrecht, pp 325–331

Pliego C, Ramos C, de Vicente A, Cazorla F (2011) Screening for candidate bacterial biocontrol agents against soilborne fungal plant pathogens. Plant Soil 340:505–520

Prieto P, Mercado-Blanco J (2008) Endophytic colonization of olive roots by the biocontrol strain *Pseudomonas fluorescens* PICF7. FEMS Microbiol Ecol 64(2):297–306

Prieto P, Schiliró E, Maldonado-González MM, Valderrama R, Barroso-Albarracín JB, Mercado-Blanco J (2011) Root hairs play a key role in the endophytic colonization of olive roots by *Pseudomonas* spp. with biocontrol activity. Microb Ecol 62(2):435–445

Príncipe A, Alvarez F, Castro M, Zacchi L, Fischer S, Mori G, Jofré E (2007) Biocontrol and PGPR features in native strains isolated from saline soils of Argentina. Curr Microbiol 55:314–322

Purdy LH (1979) *Sclerotinia sclerotiorum*: history, diseases and symptomatology, host range, geographic distribution, and impact. Phytopathology 69:875–880

Raaijmakers J, Weller DM, Thomashow LS (1997) Frequency of antibiotic-producing *Pseudomonas* spp. in natural environments. Appl Environ Microbiol 63:881–887

Raaijmakers JM, de Bruijn I, de Kock MJD (2006) Cyclic lipopeptide production by plant-associated *Pseudomonas* spp.: diversity, activity, biosynthesis, and regulation. Mol Plant Microbe Interact 19:699–710

Raaijmakers JM, Paulitz TC, Steinberg C, Alabouvette C, Moënne-Loccoz Y (2009) The rhizosphere: a playground and battlefield for soilborne pathogens and beneficial microorganisms. Plant Soil 321:341–361

Raudales RE, Stone E, McSpadden Gardener BB (2009) Seed treatment with 2,4-diacetylphloroglucinol-producing pseudomonads improves crop health in low pH soils by altering patterns of nutrient uptake. Phytopathology 99:506–511

Reading NC, Sperandio V (2006) Quorum sensing: the many languages of bacteria. FEMS Microbiol Lett 254(1):1–11

Romero D, de Vicente A, Rakotoaly RH, Dufour SE, Veening J-W, Arrebola E, Cazorla FM, Kuipers OP, Paquot M, Pérez-García A (2007) The iturin and fengycin families of lipopeptides are key factors in antagonism of *Bacillus subtilis* toward *Podosphaera fusca*. Mol Plant Microbe Interact 20:430–440

Rudrappa T, Biedrzycki ML, Kunjeti SG, Donofrio NF, Czymmek KJ, Paré PW, Bais HP (2010) The rhizobacterial elicitor acetoin induces systemic resistance in *Arabidopsis thaliana*. Commun Integr Biol 3:130–138

Ryan RP, Germaine K, Franks A, Ryan DJ, Dowling DN (2008) Bacterial endophytes: recent developments and applications. FEMS Microbiol Lett 278:1–9

Ryu CM, Farag MA, Hu CH, Reddy MS, Wei HX, Paré PW, Kloepper JW (2003) Bacterial volatiles promote growth in *Arabidopsis*. Proc Natl Acad Sci USA 100:4927–4932

Ryu CM, Murphy JF, Mysore KS, Kloepper JW (2004) Plant growth-promoting rhizobacteria systemically protect *Arabidopsis thaliana* against cucumber mosaic virus by a salicylic acid and NPR1-independent and jasmonic acid-dependent signaling pathway. Plant J 39:381–392

Saravanakumar D, Samiyappan R (2007) ACC deaminase from *Pseudomonas fluorescens* mediated saline resistance in groundnut (*Arachis hypogea*) plants. J Appl Microbiol 102 (5):1283–1292

Sarniguet A, Kraus J, Henkels MD, Muehlchen AM, Loper JE (1995) The sigma factor σ^s affects antibiotic production and biological control activity of *Pseudomonas fluorescens* Pf-5. Proc Natl Acad Sci USA 92:12255–12259

Schallmey M, Singh A, Ward OP (2004) Developments in the use of *Bacillus* species for industrial production. Can J Microbiol 50:1–17

Schnider-Keel U, Seematter A, Maurhofer M, Blumer C, Duffy B, Gigot-Bonnefoy C, Reimmann C, Notz R, Défago G, Haas D, Keel C (2000) Autoinduction of 2,4-diacetylphloroglucinol biosynthesis in the biocontrol agent *Pseudomonas fluorescens* CHA0 and repression by the bacterial metabolites salicylate and pyoluteorin. J Bacteriol 182:1215–1225

Segura A, Ramos JL (2013) Plant-bacteria interactions in the removal of pollutants. Curr Opin Biotechnol 24:467–473, http://www.dx.doi.org/10.1016/j.copbio.2012.09.011

Selin C, Habibian R, Poritsanos N, Athukorala SNP, Fernando D, de Kievit TR (2010) Phenazines are not essential for *Pseudomonas chlororaphis* PA23 biocontrol of *Sclerotinia sclerotiorum*, but do play a role in biofilm formation. FEMS Microbiol Ecol 71:73–83

Selin C, Fernando WGD, de Kievit T (2012) The PhzI/PhzR quorum-sensing system is required for pyrrolnitrin and phenazine production, and exhibits cross-regulation with RpoS in *Pseudomonas chlororaphis* PA23. Microbiology 158:896–907

Shahraki M, Heydari A, Hassanzadeh N (2009) Investigation of antibiotic, siderophore and volatile metabolites production by *Bacillus* and *Pseudomonas* bacteria. Iran J Biol 22:71–85

Shoda M (2000) Bacterial control of plant diseases. J Biosci Bioeng 89:515–521

Steller S, Vollenbroich D, Leenders F, Stein T, Conrad B, Hofmeister J, Jacques P, Thonart P (1999) Structural and functional organization of the fengycin synthetases multienzyme system from *Bacillus subtilis* b213 and A1/3. Chem Biol 6:31–41

Svercel M, Duffy B, Défago G (2007) PCR amplification of hydrogen cyanide biosynthetic locus *hcnAB in Pseudomonas* spp. J Microbiol Methods 70:209–213

Tendulkar SR, Saikumari YK, Patel V, Raghotama S, Munshi TK, Balaram P, Chatoo BB (2007) Isolation, purification and characterization of an antifungal molecule produced by *Bacillus licheniformis* BC98, and its effect on phytopathogen *Magnaporthe grisea*. J Appl Microbiol 103:2331–2339

Thrane C, Harder Nielsen T, Neiendam Nielsen M, Sorensen J, Olsson S (2000) Viscosinamide-producing *Pseudomonas fluorescens* DR54 exerts a biocontrol effect on *Pythium ultimum* in sugar beet rhizosphere. FEMS Microbiol Ecol 33:139–146

Tripathi RK, Gottlieb D (1969) Mechanism of action of the antifungal antibiotic pyrrolnitrin. J Bacteriol 100:310–318

Tsuge K, Akiyama T, Shoda M (2001) Cloning, sequencing and characterization of the iturin A operon. J Bacteriol 183:6265–6273

Valverde C, Heeb S, Keel C, Haas D (2003) RsmY, a small regulatory RNA, is required in concert with RsmZ for GacA-dependent expression of biocontrol traits in *Pseudomonas fluorescens* CHA0. Mol Microbiol 50:1361–1379

van Loon LC, Bakker PAHM (2006) Root-associated bacteria inducing systemic resistance. In: Gnanamanickam SS (ed) Plant-associated bacteria. Springer, Dordrecht, pp 269–316

van Loon LC, Van Strien EA (1999) The families of pathogenesis related proteins, their activities, and comparative analysis of PR-1 type proteins. Physiol Mol Plant Pathol 55:85–97

Van Oosten VR, Bodenhausen N, Reymond P, Van Pelt JA, Van Loon LC, Dicke M, Pieterse CM (2008) Differential effectiveness of microbially induced resistance against herbivorous insects in *Arabidopsis*. Mol Plant Microbe Interact 21:919–930

Van Peer R, Niemann GJ, Schippers B (1991) Induced resistance and phytoalexin accumulation in biological control of *Fusarium* wilt of carnation by *Pseudomonas* sp. WCS417r. Phytopathology 81:728–734

Venturi V (2006) Regulation of quorum sensing in *Pseudomonas*. FEMS Microbiol Rev 30 (2):274–291

Verhagen BWM, Glazebrook J, Zhu T, Chang HS, van Loon LC, Pieterse CMJ (2004) The transcriptome of rhizobacteria-induced systemic resistance in *Arabidopsis*. Mol Plant Microbe Interact 17(8):895–908

Verhagen BWM, Trotel-Aziz P, Couderchet M, Höfte M, Aziz A (2010) *Pseudomonas* spp.-induced systemic resistance to *Botrytis cinerea* is associated with induction and priming of defense responses in grapevine. J Exp Bot 61:249–260

Vermeiren H, Willems A, Schoofs G, de Mot R, Keijers V, Hai W, Vanderleyden J (1999) The rice inoculant strain *Alcaligenes faecalis* A15 is a nitrogen-fixing *Pseudomonas stutzeri*. Syst Appl Microbiol 22:215–224

Ward ER, Uknes SJ, Williams SC, Dincher SS, Wiederhold DL, Alexander DC, Ahl-Goy P, Métraux JP, Ryals JA (1991) Coordinate gene activity in response to agents that induce systemic acquired resistance. Plant Cell 3:1085–1094

Weller DM (2007) *Pseudomonas* biocontrol agents of soilborne pathogens: looking back over 30 years. Phytopathology 97:250–256

Weyens N, Truyens S, Dupae J, Newman L, Taghavi S, van der Lelie D, Carleer R, Vangronsveld J (2010) Potential of the TCE degrading endophyte *Pseudomonas putida* W619-TCE to improve plant growth and reduce TCE phytotoxicity and evapotranspiration in poplar cuttings. Environ Pollut 158:2915–2919

Wheatley RE (2002) The consequences of volatile organic compound mediated bacterial and fungal interactions. Antonie Van Leeuwenhoek 81:357–364

Whistler CA, Corbell NA, Sarniguet A, Ream W, Loper JE (1998) The two component regulators GacS and GacA influence accumulation of the stationary phase sigma factor σ^S and the stress response in *Pseudomonas fluorescens* Pf-5. J Bacteriol 180:6635–6641

Williamson B, Tudzynski B, Tudzynski P, van Kan JAL (2007) *Botrytis cinerea*: the cause of grey mould disease. Mol Plant Pathol 8:561–580

Wrather JA, Anderson TR, Arsyad DM, Tan Y, Ploper LD, Porta-Puglia A, Ram HH, Yorinori JT (2001) Soybean disease loss estimates for the top ten soybean-producing countries in 1998. Can J Plant Pathol 23:115–121

Yang F, Cao Y (2012) Biosynthesis of phloroglucinol compounds in microorganisms – review. Appl Microbiol Biotechnol 93:487–495

Zha WJ, Rubin-Pitel SB, Zhao HM (2006) Characterization of the substrate specificity of PhlD, a type III polyketide synthase from *Pseudomonas fluorescens*. J Biol Chem 281 (42):32036–32047

Zhang H, Kim MS, Sun Y, Dowd SE, Shi H, Paré PW (2008) Soil bacteria confer plant salt tolerance by tissue-specific regulation of the sodium transporter HKT1. Mol Plant Microbe Interact 21:737–744

Chapter 10
Functional Diversity of Endophytic Bacteria

Lucía Ferrando and Ana Fernández-Scavino

10.1 Introduction

Endophytic bacteria have deserved the interest of researchers based on their putative properties for the protection against biotic or abiotic stress as well as for the growth promotion of the plant (Sturz and Nowak 2000). These properties have been tested by inoculation of the bacteria, isolated from soil or plants, in seeds or new plants that, as a result, increase the yields or the tolerance toward stress conditions. Many of these properties are associated to secondary metabolism of the microorganisms rather than to their central metabolic pathways, though these metabolic pathways may explain the interactions and their contributions to the flow nutrients inside the plant.

The strong interest in PGP (plant growth promoting) endophytic bacteria, a fraction of the whole community, limited the information available to evaluate the diversity of endophytes. With the exception of diazotrophs, to our knowledge there are no studies focusing on the description of specific functional groups of endophytes such as methanogens, methanotrophs, or ammonia oxidizers, which are able to metabolize substrates that may be abundant in the endosphere.

Recent reports expanded the information available and revealed the identity and phylogenetic affiliation of endophytes that until now have an unknown role.

In this chapter, the diversity of bacterial endophytes is presented with special emphasis in two functional groups, methanotrophs and diazotrophs.

L. Ferrando
Cátedra de Microbiología, Departamento de Biociencias, Facultad de Química, Universidad de la República, Montevideo, Uruguay

A. Fernández-Scavino (✉)
Cátedra de Microbiología, Departamento de Biociencias, Facultad de Química, Universidad de la República, Gral. Flores 2124, Casilla de Correo, 1157, Montevideo, Uruguay
e-mail: afernand@fq.edu.uy

10.2 General Properties of Endophytes

Microbial endophytes have been defined as the microorganisms that have the ability to colonize and survive in the internal plant tissues without causing damage on the host (Hallmann et al. 1997). Thus, they have been obtained after surface sterilization of vegetal tissues of healthy plants.

The endophytes originate from the seed and from the surrounding environment, with the contribution of each source to the whole community in mature plants depending on environmental conditions. Moreover, the endophytic community is an assemblage rather than a fortuitous combination of partial bacterial communities from soil and seed. Gottel et al. (2011) observed that the endophytic bacterial richness was tenfold lower than rhizosphere samples associated to the roots of poplar trees and that rhizospheric community was dominated by *Acidobacteria* (31 %) and *Alphaproteobacteria* (30 %), whereas most endophytes were associated to *Gammaproteobacteria* (54 %) and *Alphaproteobacteria* (23 %). Ringelberg et al. (2012) proposed that mature plants of wheatgrass were inhabited by *Actinobacteria*, *Firmicutes*, and *Gammaproteobacteria* translocated from seeds and by exogenous organisms able to colonize the plant tissues. Furthermore, the authors suggested that the plant-growing media influenced the endophytic community of mature plants, which were richer in *Firmicutes* and *Cyanobacteria* after propagation in sand, whereas *Betaproteobacteria* increased in plants propagated in a peat-based growing mix.

The phenotype of endophytic is unpredictable since different strains of one bacterial species do not have the same ability to colonize the internal tissues of the plant. This has been well illustrated with *Azospirillum brasilense*, a widely studied microorganism due to its PGP properties, for which rhizospheric and endophytic strains have been described (Baldani et al. 1997a). In contrast with mycorrhiza, the endophytic bacteria can live as free organisms in soil or other environments. Moreover, the high degree of transcriptional regulators detected by metagenomic analysis of the endophytic community of rice indicates that the endophytic lifestyle may require active mechanisms of adaptation (Sessitsch et al. 2012).

Other interesting feature is that the endophytic community is sensitive to the environmental conditions where the plant is growing, although is not clear if this is a direct effect over the endophytic community or reflects a change in the rhizospheric community that is in close interaction with the endophytes. Long-term application of herbicides or different fertilizers (Seghers et al. 2004), as well as differences in the chemical composition of environmental contaminants from the petroleum industry (Phillips et al. 2010), influences the endophytic community.

The association among the endophytes and plants seems to be ancestral and quite conserved. Johnston-Monje and Raizada (2011) investigated the microbial endophytic community of seeds from the wild ancestor to domesticated maize harvested in different regions. There was no significant difference in the endophytic diversity in wild versus domesticated *Zea* species, but bacterial communities varied with *Zea* phylogeny. A core of endophytes composed by three culturable genera, *Enterobacter*,

Pantoea, and *Pseudomonas*, and two genera detected by molecular methods, *Clostridium* and *Paenibacillus*, was found to significantly contribute to the most conserved endophytic traits across *Zea*.

10.3 Diversity of Endophytic Populations

Most of the diversity studies of the total endophytic bacterial community do not show a dominance pattern of few species indicating that there is not a unique microorganism interacting with the plant. Furthermore, after inoculation with endophytic PGPB, the microorganism is recovered from the plant, but often its dominance is not confirmed. The allocation in patches of different microorganisms with spatial and niche separation could explain such species distribution and diversity. The high diversity of genes involved in universal quorum-sensing signals detected in metagenomic analysis of the root rice (Sessitsch et al. 2012) could reinforce the importance of the concerted activities between the bacterial community to survive or remain active inside the plant.

Endophytic bacteria have been detected in all tissues of very diverse host plants. A few recent reports and results of endophytic bacteria are summarized here, and some examples are shown in Table 10.1.

10.3.1 Cultivated Endophytic Bacteria

Endophytic bacteria isolated from ginseng plants were grouped in four clusters: *Firmicutes*, *Actinobacteria*, *Alphaproteobacteria*, and *Gammaproteobacteria*, with *Bacillus* and *Staphylococcus* being the predominant genera in 1-year-old and 4-year-old plants, respectively (Vendan et al. 2010). The isolates of young radish plants belonged to four major phylogenetic groups: high G + C Gram-positive bacteria, low G + C Gram-positive bacteria, *Bacteroidetes*, and *Proteobacteria*, with this phylum being predominant in leaves and roots (Seo et al. 2010).

Besides, endophytic bacteria are not rare in symbiotic legumes nodules colonized by rhizobia. The nonsymbiotic endophytic bacterial strains isolated from soybean root nodules were classified into six genera: *Pantoea*, *Serratia*, *Acinetobacter*, *Bacillus*, *Agrobacterium*, and *Burkholderia* with strong dominance of *Pantoea agglomerans* (Li et al. 2008). Although most of these endophytes had PGP properties, their inoculation had no significant effects on the growth and nodulation of soybean. However, the horizontal transfer of *nifH* genes between *Bradyrhizobium japonicum* and the endophytic *Bacillus* strains was proposed by these authors. Diverse bacterial groups of endophytes with multiple PGP properties were isolated from nodules of the herbaceous legume *Lespedeza* sp. colonized by *Rhizobium* and *Bradyrhizobium* bacteria. Interestingly, the main genera recovered, *Arthrobacter*, *Bacillus*, *Burkholderia*, *Dyella*, *Methylobacterium*, *Microbacterium*, and *Staphylococcus*, also correspond to the same phylogenetic groups present in nonlegumes plants (Palaniappan et al. 2010).

Table 10.1 Non-diazotrophic endophytic bacteria found in several plant tissues

Genera or phyla	Plant and tissue	Relative abundance[a]	Detection method[b]	Reference
Bacillus, Staphylococcus	Ginseng roots	D	C	Vendan et al. (2010)
Pantoea	Soybean root nodules	D	C	Li et al. (2008)
Pantoea	Rice seeds, roots, leaves	D	C	Loaces et al. (2011)
Pseudomonas	Rice leaves	D	C	Ferrando et al. (2012)
Brevundimonas	Rice leaves	D	M (cl)	Ferrando et al. (2012)
Stenotrophomonas	Rice roots	D	M (cl)	Sun et al. (2008)
Methanospirillum, Methanoregula	Rice roots	R	M (cl)	Sun et al. (2008)
Methylosinus, Methylocystis	Rice leaves	D	M (cl *pmoA*)	This study
Enterobacter	Rice roots	D	M (m)	Sessitsch et al. (2012)
Planctomycetes, Verrucomicrobia	Rice roots	R	M (m)	Sessitsch et al. (2012)
Methylocella, Methylocapsa	*Sphagnum* mosses stems	D	M (F)	Raghoebarsing et al. (2005)
Methylocystis	*Sphagnum* mosses stems	D	M (ma *pmoA*)	Kip et al. (2012)
Curtobacterium, Pseudomonas	Poplar trees aerial parts	D	C	Ulrich et al. (2008)
Sphingomonas	Poplar trees aerial parts	D	M (cl)	Ulrich et al. (2008)
Pseudomonas	Poplar trees roots	D	M (p)	Gottel et al. (2011)
Beijerinckia, Staphylococcus,	Micropropagated *Atriplex*, callus	D	M (p)	Lucero et al. (2011)
Geobacillus, Caulobacter	Micropropagated *Atriplex*, callus	R	M (p)	Lucero et al. (2011)

[a] *D* dominant, *R* rare
[b] *C* cultivation method, *M* molecular methods based on the 16S rRNA or *pmoA* (when indicated) gene analysis, (*cl*) cloning, (*m*) metagenomic, (*p*) pyrosequencing, (*F*) fluorescent in situ hybridization, (*ma*) microarray

The endophytic bacterial community associated with copper-tolerant plant species seems to be composed by the same main groups found in legumes and nonlegumes plants. The Cu-resistant isolates belonged to three major phyla: *Firmicutes*, *Actinobacteria*, and *Proteobacteria* (Sun et al. 2010).

Irrigated rice is an interesting system to study the influence of the oxic–anoxic transition zone on the plant–endophyte interactions. Loaces et al. (2011) showed that the ability to produce siderophores, though common among soil bacteria, becomes relevant to persist associated to roots and leaves as the plant grows, particularly after flooding. The predominant genus isolated from roots and leaves during the crop cycle was *Pantoea*, which coexisted with bacteria of the genera *Sphingomonas*, *Pseudomonas*, *Burkholderia*, and *Enterobacter* that fluctuated during the plant growth. Interestingly, the species *Pantoea ananatis*, permanently associated to any plant tissue, comprised many different strains that were randomly distributed along time and tissue, suggesting that a common trait of the species *P. ananatis* determined the interaction with the rice plant.

10.3.2 Molecular-Detected Endophytic Bacteria

Usually, the molecular approach, even with the most recently improvements, confirms that the main phylogenetic groups found by cultivation are present in the internal tissues of the plants although the group distribution assessed by each method may be different.

A multiphasic approach was used to study the endophytes present in leaves of three rice varieties along two consecutive crop seasons (Ferrando et al. 2012). *P. ananatis* and *Pseudomonas syringae* constituted 51 % of the total isolates and were always present regardless of the variety or the season. The bacterial communities established in the three varieties were at least 74 % similar according to the 16S rRNA gene fingerprinting clustering analysis. *Brevundimonas*, the dominant genus in the clone library, was not recovered by cultivation. Conversely, bacteria from the genus *Pseudomonas* were not detected in the clone library. These results indicate that communities established in leaves of physiologically different rice varieties were highly similar and composed by a reduced group of strongly associated and persistent bacteria that were partially recovered by cultivation. Also, bacterial endophytes of the aerial parts of poplar trees displayed 53 different genera comprised into four major phyla: *Proteobacteria*, *Actinobacteria*, *Firmicutes*, and *Bacteroidetes*. Members of the family *Microbacteriaceae* made up the largest fraction of the isolates, whereas *Alpha-* and *Betaproteobacteria* dominated the 16S rRNA clone library. The most abundant genera among the isolates were *Pseudomonas* and *Curtobacterium*, while *Sphingomonas* prevailed among the clones (Ulrich et al. 2008).

Frequently, molecular methods revealed the presence of endophytes that cannot be recovered by traditional culturing methods. Endophytes of rice roots analyzed by cloning of the 16S rRNA genes revealed the presence of 52 different phylotypes distributed in six main phylogenetic groups: *Proteobacteria*, *Cytophaga/Flexibacter/Bacteroides* (CFB), low G + C Gram-positive bacteria, *Deinococcus-Thermus*, *Acidobacteria*, and *Archaea*. The clone library was dominated by the genus *Stenotrophomonas* (*Betaproteobacteria*), but a considerable percentage of the clones showed high similarity to uncultured bacteria. Also, sequences of *Archaea* related to two genera of the hydrogenotrophic methanogens, *Methanospirillum* and *Methanoregula*, were detected (Sun et al. 2008). However, their presence is not surprising in the anoxic roots given by flooding conditions, and it remains to be established if methane is produced in the inner tissues of the rice plant.

Modified methods were developed to overcome the limitations imposed by the simultaneous amplification of the 16S rRNA gene of plant organelles. A method designed to enrich in bacterial cells recalcitrant to cultivation from aerial parts of soybean showed that the main groups recovered by 16S rRNA gene cloning were affiliated to the *Proteobacteria*, *Actinobacteria*, *Firmicutes*, and *Bacteroidetes* (Ikeda et al. 2009). Also, detached bacterial cells from rice roots analyzed by a metagenomic approach (not biased by the PCR reaction) revealed that the main members of the bacterial endophytic community belonged to the genus *Enterobacter*, followed by *Alphaproteobacteria* associated to rhizobia and by *Firmicutes*, with a few members of other lineages such as *Verrucomicrobia*, *Planctomycetes*, and *Fusobacterium* that were common in soil but not detected previously as endophytes (Sessitsch et al. 2012). Although it was not surprising to find denitrifying bacteria because many endophytes belong to the *Gammaproteobacteria* phylum that comprises most of denitrifying bacteria, it is noteworthy to detect the expression of denitrifying genes inside the roots of rice plants (Sessitsch et al. 2012). These authors also detected the expression of nitrification and diazotrophy genes. Despite *nifH* genes were abundant, only a small fraction of the root-associated bacteria, mainly related to *Bradyrhizobium*, was active.

High throughput methods explored deeply the structure of the community revealing the identity of the rare members. In spite of the restricted information given by the short sequences obtained, the analysis of higher number of sequences reveals the presence of genera that have not been detected by cloning before. The massive sequencing analysis of roots of 12 potato cultivars allowed the identification of a total of 238 known genera from 15 phyla. Although almost 90 % of the sequences belonged to the same five main groups observed for other plants before (*Bacteroidetes*, *Gammaproteobacteria*, *Alphaproteobacteria*, *Betaproteobacteria*, and *Actinobacteria*), five of the ten most common genera (*Rheinheimera*, *Dyadobacter*, *Devosia*, *Pedobacter*, and *Pseudoxanthomonas*) had not been previously reported as endophytes of potato (Manter et al. 2010). Microproppagated plants offer a scenario where bacteria are transferred with the callus without the contribution of surrounding soil to the endophytic community. In *Atriplex* species, a globally distributed halophyte, this methodology allowed the analysis of more than 9,000 sequences in two plant species. Although the endophytic community was composed by three common phyla, *Bacteroidetes*, *Firmicutes*, and *Proteobacteria*, sequences homologous to a few rare genera were detected. *Firmicutes* were represented, among others, by sequences homologous to *Geobacillus* (a genus associated with thermophiles), *Clostridium*, and *Sporobacter* (two genera forming spores and known for sulfur-reducing activity). Within the *Proteobacteria* the main genera included *Beijerinckia*, *Rhizobium*, *Sphingomonas*, and *Caulobacter* (Lucero et al. 2011).

10.4 Endophytic Methanotrophic Bacteria

Environmental conditions present in plant tissues, which inhabit flooded ecosystems, allow methanotrophic bacteria to grow. As wetland plants possess tissues that facilitate gas exchange (O_2, CO_2, CH_4, etc.) from soil to the atmosphere and vice versa, they have been intensely studied in order to elucidate the presence of methanotrophs as endophytes.

Aerobic methanotrophic bacteria are divided into two major groups according to different characteristics of their intracytoplasmic membrane structure, carbon assimilation pathway, and phospholipid fatty acids. Types I and II methanotrophs belong to *Gammaproteobacteria* and *Alphaproteobacteria* class, respectively (Bowman 2006). Verrucomicrobial methanotrophs were recently isolated although their presence seems to be limited to extreme environments (Dunfield et al. 2007).

Many methanotrophs are also nitrogen-fixing bacteria mostly, but not exclusively, type II methanotrophs. This capability is the main PGP property tested for this physiological group. Other properties that could contribute to the plant growth remain unknown since no methanotrophic bacteria have been isolated as endophyte so far and their existence is an issue scarcely addressed. Several studies reported the detection of methanotrophs associated to different plant roots (Calhoun and King 1998; Gilbert et al. 1998; Takeda et al. 2008; Lüke et al. 2011), but most of these studies did not differentiate between endophytic and epiphytic bacteria.

Few studies are focused on culturing methanotrophic bacteria as endophytes from different plants, and mainly because of the well-known difficulties of cultivation, most of these studies have assessed their diversity using molecular approaches.

Methanotrophs associated to *Sphagnum* mosses in peat bogs have been the most studied. Raghoebarsing et al. (2005) determined that the endophytic methanotrophic bacteria play an important role on diminishing methane emissions from these ecosystems. They showed that methanotrophic symbionts associated to submerged *Sphagnum* mosses provide carbon for photosynthesis in peat bogs through a highly effective in situ methane recycling. The presence of acidophilic type II methanotrophs related to *Methylocella palustris* and *Methylocapsa acidiphila* (alphaproteobacteria previously isolated from *Sphagnum* bogs) was revealed by fluorescence in situ hybridization (FISH) in the hyaline cells of plants and on stem leaves. Parmentier et al. (2011) also confirmed that endophytic methanotrophs mitigated the methane emissions in *Sphagnum* mosses. Additionally, alpha- and gammaproteobacterial methanotrophs were present in all *Sphagnum* mosses from South American peat bogs (Kip et al. 2012). The *pmoA* microarray used revealed that the most abundant species in the most active mosses belonged to *Methylocystis*, an alphaproteobacteria.

On the other hand, the presence of methanotrophs as endophytes of maize (*Zea mays* L.) was revealed by Seghers et al. (2004). Type II methanotrophs were not detected by PCR amplification, while the DGGE analysis for type I methanotrophs allowed to successfully differentiate the mineral-fertilized maize plants from those cultivated using organic fertilizers.

Paddy fields are one of the most important anthropogenic sources of CH_4, contributing greatly to the greenhouse gas effect (Le Mer and Roger 2001). Aerobic methanotrophic bacteria have been extensively studied in order to understand their behavior and ecology and to accomplish suitable strategies for mitigation of methane emissions in this ecosystem. The aerobic zones present in irrigated rice fields, rhizosphere, and soil–water interface layer, where O_2 and CH_4 are available for

Table 10.2 Methanotrophic bacteria detected in different rice plant tissues

Rice cultivation conditions			Methanotrophs detection			
			Plant tissue[a]		Methanotrophic enrichment[b]	
Rice variety	Plant tissue	Nitrogen fertilization	Methane oxidation activity	$pmoA$ gene detection	Methane oxidation activity	$pmoA$ gene detection
INIA El Paso 144	Leaf	Fertilized	nd	nd	+	+
INIA El Paso 144	Root	Fertilized	nd	nd	+	+
INIA Olimar	Leaf	Fertilized	−	−	+	+
INIA Olimar	Leaf	Not fertilized	−	−	+	+
INIA Olimar	Root	Fertilized	−	+	+	+
INIA Olimar	Root	Not fertilized	−	+	+	+

[a]nd not determined
[b]Different nitrogen sources were used for methanotrophic enrichments in nitrate mineral salt medium (NMS): NH_4Cl (INIA Olimar), no nitrogen source or NH_4NO_3 (INIA El Paso 144). Positive results were obtained for the three media

methanotrophy, sustain high abundance and activity of this physiological group (Eller and Frenzel 2001; Ferrando and Tarlera 2009). But little is known about the presence and activity of methanotrophs inside the rice plant. Recently, a metagenomic approach conducted by Sessitsch et al. (2012) revealed the presence of various genes required for methane oxidation (*pmo*, *mmo*, and *mxa* genes) in rice roots.

The results from our work on elucidating the presence of methanotrophs as rice endophytes are presented in the next paragraphs of this chapter. The searching for endophytic methanotrophs was performed on different surface disinfected rice tissues (seeds, roots, and leaves) corresponding to two rice varieties (INIA Olimar and INIA El Paso 144) and in different nitrogen fertilization conditions (with or without usual fertilizer application). Methanotrophic enrichments (NMS medium, Whittenbury et al. 1970, with different N sources) and the molecular detection of the methanotrophy gene marker *pmoA* were performed from the different rice tissues and the enrichments as previously described (Ferrando and Tarlera 2009).

Methanotrophs were only detected when soil was used for rice cultivation. Seeds and rice tissues from a hydroponic experiment did not show the presence of methanotrophs. Neither *pmoA* gene direct detection nor the enrichments gave positive results. These findings suggest that methanotrophs could colonize rice roots from the surrounding soil in early stages of rice culture and that they do not seem to have a seed borne origin. Nevertheless, it is important to take into account that some methanotrophic bacteria could be undetectable using this approach by the limitation of cultivation methods and by the absence of *pmoA* gene in some strains [*M. palustris* and *Methyloferula stellata*; Dedysh et al. (2000) and Vorobev et al. (2010), respectively].

The results from the enrichments performed from tissues of rice plants grown on soil are summarized in Table 10.2. In spite of the undetectable methanotrophic activity in plants, the presence of *pmoA* genes was confirmed. Thus, the fact that genes responsible for methane oxidation are present despite the undetectable activity could be due to the method sensitivity or measurement conditions distant from the optimal ones.

On the other hand, enrichments showed both methane oxidation activity and *pmoA* gene amplification. Type II methanotrophs were recovered from the two rice varieties at any condition assayed.

The methanotrophs were detected by sequencing *pmoA* in the enrichments from leaves of fertilized rice, variety INIA El Paso 144, even after five successive subcultures. Two methanotrophic sequences were retrieved from duplicated enrichments without nitrogen source. This fact suggests that these bacteria could act as growth promoters for the rice plants. Both sequences were associated to type II methanotrophs (accession numbers JX481157 to JX481159). One of them was very similar to Methanotroph C9 (EU359001), which was previously isolated from rice rhizosphere of Uruguayan paddy fields by Ferrando and Tarlera (2009) (99 % identity retrieved through the BLAST tool), and the other was similar to *Methylocystis hirsuta* strain CSC1 (DQ364434, 99 % identity). Despite *M. hirsuta* has been isolated by ordinary strategies (Lindner et al. 2007), our attempts to obtain isolates by dilution to extinction were unsuccessful which could indicate that these particular strains could have special requirements for growth.

On the other hand, methanotrophs were also detected in enrichments from roots and leaves from Olimar rice variety. Methane oxidation and *pmoA* gene amplification were verified for both enrichments from fertilized and not fertilized rice plants.

Coincidentally, *pmoA* clone libraries from leaf enrichments allowed the detection of sequences (accession numbers JX481160 to JX481165) also related to *Methylosinus trichosporium* (AJ459037 and AJ459021, >95 % identity) and Methanotroph C9 (99 % identity). The *pmoA* sequences from Methanotroph C9 and *M. trichosporium* strains are closely related (>96 % identity) which would indicate that the diversity of methanotrophic endophytes inhabiting rice leaves is low and that nitrogen fertilization would not affect the composition of this particular community.

These results show that closely related methanotrophs were detected as endophytes of leaves from two different rice varieties (INIA Olimar, INIA El Paso 144) using different soils and fertilization conditions. This finding suggests that *M. trichosporium* present in soil can colonize rice plants translocating to their leaves during the plant growth. Further studies must be performed to understand the adaptations that the soil bacterial strains undergo to become endophytes and make them more difficult to cultivate. Thus, the ecological role of methanotrophic endophytes should be further explored. The presence of oxygen and methane could not be the major factor favoring the methanotrophy in the inner tissues but also the use of other substrates from the plant, such as methanol, or eventually by the presence of yet-unknown facultative methanotrophs (Dedysh et al. 2005).

10.5 Diazotrophic Endophytic Bacteria

Several bacteria have been isolated and characterized as nitrogen-fixing endophytes from nonleguminous plants in the last decades. Rice and sugarcane have been the most widely explored grasses in order to find out endophytic bacteria able to supply part of the nitrogen required for these two important crops.

A variety of diazotrophic bacteria, including *Pseudomonas*, *Stenotrophomonas*, *Xanthomonas*, *Acinetobacter*, *Rhanella*, *Enterobacter*, *Pantoea*, *Shinella*, *Agrobacterium*, and *Achromobacter* (Taulé et al. 2012), have been isolated from sugarcane plants, but the most prominent is *Gluconacetobacter diazotrophicus*, described for the first time by Cavalcante and Dobereiner (1988). The ability of *G. diazotrophicus* to grow and fix nitrogen in sugarcane inner tissues has been demonstrated (Sevilla et al. 2001), and its great contribution to the plant nitrogen content, particularly in poor soils, has been evaluated (Boddey et al. 1995). As few PGP traits were identified in *G. diazotrophicus*, it was proposed that during the endophytic colonization, this bacterium could influence the sugarcane gene expression. It has been speculated that the strong effect of the bacterium on the plant physiology depends on the bacterial density and, then, bacterial signal transduction may be relevant for the PGP properties of *G. diazotrophicus*. More recently, the occurrence of *Gluconacetobacter*-like diazotrophs has been reported in diverse plants like sweet potato, coffee, tea, banana, carrot, radish, and rice, with a wide geographical distribution (Saravanan et al. 2008). Moreover, an increasing number of species of the genera *Swaminathania* and *Acetobacter* have been reported as diazotrophs within the family *Acetobacteraceae*.

Rice harbors a wide range of endophytic diazotrophs (Barraquio et al. 1997), and it seems that the diversity of this group of bacteria is still underexplored. The complexity of diazotrophic endophytic community of rice has been reported by Prakamhang et al. (2010). The genera *Herbaspirillum* (Baldani et al. 1986) and *Burkholderia* (Baldani et al. 1997b) were early described as the main diazotrophic endophytes isolated from rice. *Azoarcus* also earned much attention because, though was originally isolated from Kallar grass (Reinhold-Hurek et al. 1993), it can colonize rice in laboratory experiments (Hurek et al. 1994). But this model deeply studied to understand the interaction between the rice plant and diazotrophic bacteria is quite restricted to the strain *Azoarcus* sp. BH72. The ability of certain strains of *Azospirillum* to colonize the intercellular spaces between the cortex and the endodermis cells as well as vascular tissue of maize (Patriquin and Dobereiner 1978) and rice (James and Olivares 1998) roots was also early recognized.

Other bacteria, mostly *Enterobacteriaceae*, have been isolated and characterized as nitrogen-fixing endophytes in nonleguminous plants including *Serratia* spp. (Gyaneshwar et al. 2001), *Klebsiella oxytoca* (Palus et al. 1996), and *P. agglomerans* YS19 (Feng et al. 2006). The endophytic diazotrophic bacteria found in plants from relatively extreme environments, like dune grasses, also belong to the *Proteobacteria*, being *Stenotrophomonas*, *Pseudomonas*, and *Burkholderia* the main genera detected (Dalton et al. 2004).

Many diazotrophic endophytes have a broad host range. This is particularly common in Gramineae plants but not exclusive. *Herbaspirillum seropedicae* has been found in a variety of crops, including maize, sorghum, sugarcane, and other Gramineae plants (Baldani et al. 1986; Olivares et al. 1996). Examples of cross-colonization have also been illustrated. A *Herbaspirillum* strain isolated from rice could colonize sugarcane tissues (Njoloma et al. 2006), and *Burkholderia* sp. isolated from onion was able to grow in grapes (Compant et al. 2005). Also, both *Pantoea*, isolated from sweet potato, and *Enterobacter*, isolated from sugarcane, could colonize wild and cultivated rice (Zakria et al. 2008).

This lack of host specificity and the relatively restricted incubation conditions may have conditioned the diversity of diazotrophic bacteria recovered from the inner tissues of different plants around the world. The traditional method of isolation of diazotrophic bacteria in media without nitrogen incubated in microaerophilic conditions may hide other nitrogen-fixing bacteria that could be important inhabitants in plant tissues.

There are few isolated diazotrophic endophytes that do not belong to the *Proteobacteria* already described. It has been shown that the strain of *Microbacterium* sp. SH16 colonized intercellular spaces of root cortices in sugarcane. The inoculation with this diazotrophic actinobacteria increased significantly the biomass and nitrogen content of micropropagated sugarcane (Lin et al. 2012). In addition, anaerobic nitrogen-fixing consortia consisting of diazotrophic clostridia and diverse non-diazotrophic oxygen-consuming bacteria (*Burkholderia*, *Enterobacter*, *Bacillus*, and *Microbacterium*) have been reported by Minamisawa et al. (2004). These consortia are widespread in the aerial part of wild rice and pioneer plants, which are able to grow in unfavorable locations. The authors suggest that the efficient nitrogen fixation by clostridia depends on the association with non-diazotrophic endophytes that guarantee the required anoxic conditions.

The *nifH* gene has been employed as functional marker to explore the diversity and activity of diazotrophic bacteria. Diazotrophs are a diverse physiological group comprising members of the domains *Bacteria* and *Archaea* (Martínez Romero 2006). As a consequence there are a variety of designed primers that only targeted a fraction of the total diazotrophs (Demba Diallo et al. 2008) and conditioned the results obtained by different studies.

Several studies assessed the molecular diversity of *nifH* genes and their expression in sugarcane. The cloning of *nifH* genes from stems of sugarcane allowed Ando et al. (2005) to detect sequences related to *Bradyrhizobium*, *Serratia*, and *Klebsiella*. Noticeably, although cultivation of diazotrophic endophytes usually recovers *G. diazotrophicus*, this species was not detected by molecular methods based on the *nifH* gene amplification or expression. *Bradyrhizobium* spp. or their closest relatives appear as dominant members in most of these studies. In addition, *Rhizobium rosettiformans* was found as the dominant active DGGE phylotype in field-grown sugarcane plants from Africa and America (Burbano et al. 2011).

Bradyrhizobium was also detected as one of the most active diazotrophs in other crops. Terakado-Tonooka et al. (2008) found that rhizobial species were the most abundant in stems and roots of field-grown sweet potatoes. Other alphaproteobacteria, including the methanotroph *M. trichosporium* and *Beijerinckia derxii*, were present in lower proportion as well as the betaproteobacteria *H. seropedicae*, *Burkholderia* spp., *Pelomonas saccharophila* and *Azohydromonas australica*, and cyanobacteria. However, the expression analyses revealed that the active diazotrophic bacteria on sweet potatoes were mostly affiliated to *Bradyrhizobium*, *Pelomonas*, and *Bacillus*.

Other proteobacteria were dominant diazotrophic endophytes in many plants. Flores-Mireles et al. (2007) reported that in mangrove roots most of *nifH* sequences retrieved by cloning belonged to *Gamma-* (*Pseudomonas*-related sequences) and *Deltaproteobacteria* (*Aeromonas*, *Oceanimonas*, *Vibrio*). Rare sequences were affiliated to *Alphaproteobacteria* (*Paracoccus* sp.), *Actinobacteria* (*Corynebacterium* sp., which was previously isolated from mangrove roots), and sulfate-reducing bacteria related to *Desulfonemalimicola*. Coincidentally, *Gammaproteobacteria* (*Pseudomonas pseudoalcaligenes*) dominated the diazotrophic community inhabiting roots of the drought-tolerant perennial grass *Lasiurus sindicus* (Chowdhury et al. 2009). Conversely, the dominant *nifH* sequences in maize stems belonged to the *Alpha-* and *Betaproteobacteria* (Roesch et al. 2008). The dominant genera found in roots and stems of maize were *Herbaspirillum*, *Ideonella*, and *Klebsiella*, while the presence of certain genera was restricted to particular tissues (*Dechloromonas* to roots and *Methylosinus*, *Raoultella*, and *Rhizobium* to stems).

Endophytic nitrogen-fixing bacteria have been extensively studied in rice (*Oryza sativa*) roots, and the dominant genera belonged mostly to the *Proteobacteria* class. Dominance of betaproteobacterial *nifH* sequences (50 % of clone library) from roots of modern rice cultivars was reported by Wu et al. (2009) with *Azoarcus* spp. as their closest relatives. But other interesting community members were found as methylotrophs and methane oxidizers that may also play an important role in fixing nitrogen in rice roots. Ueda et al. (1995) reported that mainly the sulfate-reducing deltaproteobacteria *Desulfovibrio gigas* was the dominant diazotroph associated to rice roots. Also, sequences of *nifH* belonging to *Gammaproteobacteria* (*Bradyrhizobium* sp. and *Methylocystis parvus*) were abundant. A *nifH* microarray analysis revealed that *Betaproteobacteria* and Gram-positive bacteria were present in wild rice roots (Zhang et al. 2007). However, the authors observed that only a small subset of the whole endophytic community of diazotrophs was actually active, being the Gram-positive bacteria the dominant *nifH* genes expressed. Conversely, the active diazotrophic bacteria in flooded rice were dominated by *Deltaproteobacteria* with around 70 % of the *nifH* transcripts phylogenetically distant from any sequence of databases although they gathered with *Geobacter sulfurreducens* (Elbeltagy and Ando 2008).

10.6 Conclusions

Endophytic bacteria are much more diverse than the extensively studied PGP bacteria. Even in the restrictive environment imposed by the inner plant tissues and despite the limitations of the cultivation and cultivation-independent methods, the universe of the endophytic bacteria is permanently increasing.

Although the main endophytes detected in different plants as trees, pioneer plants, metal-tolerant plants, nodules of legumes, or flooded plants belong to the soil predominant lineages (*Proteobacteria*, *Firmicutes*, and *Actinobacteria*), endophytic bacteria are a selected group of the soil bacteria. Moreover, the endophytes are closely and ancestrally related to the plant and constituted a community of adapted bacteria with concerted activities.

The cultivable endophytic bacterial community partially overlaps with the bacteria detected by molecular methods, but the predominant genera depend on the method used to study the community. Also, the molecular methods allow the detection of rare community members such as methanogens and methanotrophs, which have a restricted range of substrates and growing conditions.

Though with different intensities, diazotrophs and methanotrophs are two of the functional groups of endophytes more studied due to their biotechnological and ecological implications. Albeit no methanotrophs have been isolated as endophytes, their presence was evidenced in several plants by enrichments and molecular detection, and its diversity seems to be restricted to a few genera in rice plants.

Endophytic diazotrophs have deserved much attention as an alternative to the nitrogen fertilization in crops. Most of the isolated diazotrophs with PGP properties are affiliated to the *Proteobacteria*. However, bacteria from other lineages or consortia of several bacteria may be relevant for the plant. The molecular methods evidence not just a wider diversity of *nifH* sequences compared with the cultivable diazotrophs but a remarkable shift in the composition of the active members of this group.

In summary, the improvement of the cultivation conditions and the development of better molecular tools will contribute to understand how the endophytic bacterial community is composed and will be useful to reveal new endophytes with potential PGP properties. Furthermore, the interactions among endophytes must be taken into account to elucidate the influence of the whole community over the PGP bacteria eventually inoculated in plants.

Acknowledgments We are very grateful to ANII (Agencia Nacional de Investigación e Innovación), CSIC (Comisión Sectorial de Investigación Científica), and PEDECIBA (Programa de Desarrollo de las Ciencias Básicas) for their finantial support.

References

Ando S, Goto M, Meunchang S, Thongra-ar P, Fujiwara T, Hayashi H, Yoneyama T (2005) Detection of *nifH* sequences in sugarcane (*Saccharum officinarum* L.) and pineapple (*Ananas comosus* [L.] Merr.). Soil Sci Plant Nutr 51:303–308

Baldani JI, Baldani VLD, Seldin L, Dobereiner J (1986) Characterization of *Herbaspirillum seropedicae* gen. nov., sp. nov., a root-associated nitrogen-fixing bacterium. Int J Syst Bacteriol 36:86–93

Baldani J, Caruso L, Baldani VLD, Goi SR, Dobereiner J (1997a) Recent advances in BNF with non-legume plants. Soil Biol Biochem 29:911–922

Baldani VLD, Oliveira E, Balota E, Baldani JI, Kirchhof G, Dobereiner J (1997b) *Burkholderia brasilensis* sp. nov., uma nova especie de bacteria diazotrofica endofitica. An Acad Bras Cienc 69:116

Barraquio WL, Revilla L, Ladha JK (1997) Isolation of endophytic diazotrophic bacteria from wetland rice. Plant Soil 194:15–24

Boddey RM, de Oliveira OC, Urquiaga S, Reis VM, Olivares FL, Baldani VLD, Dobereiner J (1995) Biological nitrogen fixation associated with sugar cane and rice: contributions and prospects for improvement. Plant Soil 174:195–209

Bowman J (2006) The methanotrophs—the families *Methylococcaceae* and *Methylocystaceae*. In: Dworkin M, Falkow S, Rosenberg E, Schleifer KH, Strackebrandt E (eds) The prokaryotes. Springer, New York, NY, pp 266–289

Burbano CS, Liu Y, Rösner KL, Massena Reis B, Caballero-Mellado J, Reinhold-Hurek B, Hurek T (2011) Predominant *nifH* transcript phylotypes related to *Rhizobium rosettiformans* in field-grown sugarcane plants and in Norway spruce. Environ Microbiol Rep 3:383–389

Calhoun A, King GM (1998) Characterization of root-associated methanotrophs from three freshwater macrophytes: *Pontederia cordata*, *Sparganium eurycarpum*, and *Sagittaria latifolia*. Appl Environ Microbiol 64:1099

Cavalcante V, Dobereiner J (1988) A new acid-tolerant nitrogen-fixing bacterium associated with the sugarcane. Plant Soil 108:23–31

Chowdhury SP, Schmid M, Hartmann A, Tripathi AK (2009) Diversity of 16S-rRNA and *nifH* genes derived from rhizosphere soil and roots of an endemic drought tolerant grass, *Lasiurus sindicus*. Eur J Soil Biol 45:114–122

Compant S, Reiter B, Sessitsch A, Nowak J, Clement C, Barka EA (2005) Endophytic colonization of *Vitis vinifera* L. by plant growth-promoting bacterium *Burkholderia* sp. strain PsJN. Appl Environ Microbiol 71:1685–1693

Dalton DA, Kramer S, Azios N, Fusaro S, Cahill E, Kennedy C (2004) Endophytic nitrogen fixation in dune grasses (*Ammophila arenaria* and *Elymus mollis*) from Oregon. FEMS Microbiol Ecol 49:469–479

Dedysh SN, Liesack W, Khmelenina VN, Suzina NE, Trotsenko YA, Semrau JD, Bares AM, Panikov NS, Tiedje JM (2000) *Methylocella palustris* gen. nov., sp nov., a new methane-oxidizing acidophilic bacterium from peat bogs, representing a novel subtype of serine-pathway methanotrophs. Int J Syst Evol Microbiol 50:955–969

Dedysh SN, Knief C, Dunfield PF (2005) *Methylocella* species are facultatively methanotrophic. J Bacteriol 187:4665–4670

Demba Diallo M, Reinhold-Hurek B, Hurek T (2008) Evaluation of PCR primers for universal *nifH* gene targeting and for assessment of transcribed nifH pools in roots of *Oryza longistaminata* with and without low nitrogen input. FEMS Microbiol Ecol 65:220–228

Dunfield PF, Yuryev A, Senin P, Smirnova AV, Stott MB, Hou SB et al (2007) Methane oxidation by an extremely acidophilic bacterium of the phylum *Verrucomicrobia*. Nature 450:879–882

Elbeltagy A, Ando Y (2008) Expression of nitrogenase gene (*nif*H) in roots and stems of rice, *Oryza sativa*, by endophytic nitrogen-fixing communities. Afr J Biotechnol 7:1950–1957

Eller G, Frenzel P (2001) Changes in activity and community structure of methane oxidizing bacteria over the growth period of rice. Appl Environ Microbiol 67:2395–2403

Feng Y, Shen D, Song W (2006) Rice endophyte *Pantoea agglomerans* YS19 promotes host plant growth and affects allocations of host photosynthates. J Appl Microbiol 100:938–945

Ferrando L, Tarlera S (2009) Activity and diversity of methanotrophs in the soil water interface and rhizospheric soil from a flooded temperate rice field. J Appl Microbiol 106:306–316

Ferrando L, Fernandez Manay J, Fernandez Scavino A (2012) Molecular and culture-dependent analyses revealed similarities in the endophytic bacterial community composition of leaves from three rice (*Oryza sativa*) varieties. FEMS Microbiol Ecol 80:696–708

Flores-Mireles AL, Winans SC, Holguin G (2007) Molecular characterization of diazotrophic and denitrifying bacteria associated to mangrove roots. Appl Environ Microbiol 73:7308–7321

Gilbert B, Abmus B, Hartman A, Frenzel P (1998) In situ localization of two methanotrophic strains in the rhizosphere of rice plants. FEMS Microbiol Ecol 25:117–128

Gottel NR, Castro HF, Kerley M, Yang Z et al (2011) Distinct microbial communities within the endosphere and rhizosphere of *Populus deltoides* roots across contrasting soil types. Appl Environ Microbiol 77:5934–5944

Gyaneshwar P, James EK, Mathan N, Reddy PM, Reinhold-Hurek B, Ladha JK (2001) Endophytic colonization of rice by a diazotrophic strain of *Serratia marcescens*. J Biotechnol 183:2634–2645

Hallmann JA, Mahaffee WF, Kloepper JW (1997) Bacterial endophytes in agricultural crops. Can J Microbiol 43:895–914

Hurek T, Reinhold-Hurek B, Van Montagu M, Kellenberger E (1994) Root colonization and systemic spreading of *Azoarcus* sp. strain BH72 in grasses. J Bacteriol 176:1913–1923

Ikeda S, Kaneko T, Okubo T, Rallos LEE, Eda S, Mitsui H, Sato S, Nakamura Y, Tabata S, Minamisawa K (2009) Development of a bacterial cell enrichment method and its application to the community analysis in soybean stems. Microbial Ecol 58:703–714

James EK, Olivares FL (1998) Infection and colonization of sugarcane and other Graminaceous plants by endophytic diazotrophs. Crit Rev Plant Sci 17:77–119

Johnston-Monje D, Raizada MN (2011) Conservation and diversity of seed associated endophytes in *Zea* across boundaries of evolution, ethnography and ecology. PLoS One 6:e20396

Kip N, Fritz C, Langelaan ES, Pan Y, Bodrossy L, Pancotto V, Jetten MSM, Smolders AJP, Op den Camp HJM (2012) Methanotrophic activity and diversity in different *Sphagnum magellanicum* dominated habitats in the southernmost peat bogs of Patagonia. Biogeosciences 9:47–55

Le Mer J, Roger P (2001) Production, oxidation, emission and consumption of methane by soils: a review. Eur J Soil Biol 37:25–50

Li JH, Wang ET, Chen WF, Chen WX (2008) Genetic diversity and potential for promotion of plant growth detected in nodule endophytic bacteria of soybean grown in Heilongjiang province of China. Soil Biol Biochem 40:238–246

Lin L, Guo W, Xing Y, Zhang X, Li Z, Hu C, Li S, Li Y, An Q (2012) The actinobacterium *Microbacterium* sp. 16SH accepts pBBR1-based pPROBE vectors, forms biofilms, invades roots, and fixes N_2 associated with micropropagated sugarcane plants. Appl Microbiol Biotechnol 93:1185–1195

Lindner AS, Pacheco A, Aldrich HC, Costello A, Uz I, Hodson DJ (2007) *Methylocystis hirsuta* sp. nov., a novel methanotroph isolated from a groundwater aquifer. Int J Syst Evol Microbiol 57:1891–1900

Loaces I, Ferrando L, Fernández Scavino A (2011) Dynamics, diversity and function of endophytic siderophore-producing bacteria in rice. Microb Ecol 61:606–618

Lucero ME, Unc A, Cooke P, Dowd S, Sun S (2011) Endophyte microbiome diversity in micropropagated *Atriplex canescens* and *Atriplex torreyi* var *griffithsii*. PLoS One 6:e17693

Lüke C, Bodrossy L, Lupotto E, Frenzel P (2011) Methanotrophic bacteria associated to rice roots: the cultivar effect assessed by T-RFLP and microarray analysis. Environ Microbiol Rep 3:518–525

Manter DK, Delgado JÁ, Holm DG, Stong RA (2010) Pyrosequencing reveals a highly diverse and cultivar-specific bacterial endophyte community in potato roots. Microb Ecol 60:157–166

Martínez Romero E (2006) Dinitrogen-fixing prokaryotes. In: Dworkin M, Falkow S, Rosenberg E, Schleifer KH, Strackebrandt E (eds) The prokaryotes. Springer, New York, NY, pp 793–817

Minamisawa K, Nishioka K, Miyaki T, Ye B, Miyamoto T, You M, Saito A, Saito M, Barraquio WL, Teaumroong N, Sein T, Sato T (2004) Anaerobic nitrogen-fixing consortia consisting of clostridia isolated from *Gramineous* plants. Appl Environ Microbiol 70:3096–3102

Njoloma J, Tanaka K, Shimizu T, Nishiguchi T, Zakria M, Akashi R, Oota M, Akao S (2006) Infection and colonization of aseptically micropropagated sugarcane seedlings by nitrogen-fixing endophytic bacterium, *Herbaspirillum* sp. B501gfp1. Biol Fertil Soils 43:137–143

Olivares FL, Baldani VLD, Reis VM, Baldani JI, Dobereiner J (1996) Occurrence of the endophytic diazotrophs *Herbaspirillum* spp. in root, stems, and leaves, predominantly of *Gramineae*. Biol Fertil Soils 21:197–200

Palaniappan P, Chauhan PS, Saravanan VS, Anandham R, Sa T (2010) Isolation and characterization of plant growth promoting endophytic bacterial isolates from root nodule of *Lespedeza* sp. Biol Fertil Soils 46:807–816

Palus JA, Borneman J, Ludden PW, Triplett EW (1996) A diazotrophic bacterial endophyte isolated from stems of *Zea mays* L. and *Zea luxurians* Iltis and Doebley. Plant Soil 186: 135–142

Parmentier FJW, van Huissteden J, Kip N, Op den Camp HJM, Jetten MSM, Maximov TC, Dolman AJ (2011) The role of endophytic methane-oxidizing bacteria in submerged *Sphagnum* in determining methane emissions of Northeastern Siberian tundra. Biogeosciences 8: 1267–1278

Patriquin DG, Dobereiner J (1978) Light microscopy observations of tetrazolium-reducing bacteria in the endorhizosphere of maize and other grasses in Brazil. Can J Microbiol 24:734–742

Phillips LA, Armstrong SA, Headley JV, Greer CW, Germida JJ (2010) Shifts in root-associated microbial communities of *Typha latifolia* growing in naphthenic acids and relationship to plant health. Int J Phytoremediation 12:745–760

Prakamhang J, Boonkerd N, Teaumroong N (2010) Rice endophytic diazotrophic bacteria. In: Maheshwari DK (ed) Plant growth and health promoting bacteria. Springer, Berlin, pp 317–322

Raghoebarsing AA, Smolders AJP, Schmid MC, Rijpstra IC, Wolters-Arts M, Derksen J, Jetten MSM, Schouten S et al (2005) Methanotrophic symbionts provide carbon for photosynthesis in peat bogs. Nature 436:25

Reinhold-Hurek B, Hurek T, Gillis M, Hoste B, Vancanneyt M, Kersters K, De-Ley J (1993) *Azoarcus* gen. nov., nitrogen-fixing proteobacteria associated with roots of Kallar grass (*Leptochloa fusca* (L.) Kunth), and description of two species, *Azoarcus indigenes* sp. nov. and *Azoarcus communis* sp. nov. Int J Syst Bacteriol 43:574–584

Ringelberg D, Foley K, Reynolds CM (2012) Bacterial endophyte communities of two wheatgrass varieties following propagation in different growing media. Can J Microbiol 58:67–80

Roesch LFW, Camargo FAO, Bento FM, Triplett EW (2008) Biodiversity of diazotrophic bacteria within the soil, root and stem of field-grown maize. Plant Soil 302:91–104

Saravanan VS, Madhaiyan M, Osborne J, Thangaraju M, Sa TM (2008) Ecological occurrence of *Gluconacetobacter diazotrophicus* and nitrogen-fixing *Acetobacteraceae* members: their possible role in plant growth promotion. Microb Ecol 55:130–140

Seghers D, Wittebolle L, Top EM, Verstraete W, Siciliano SD (2004) Impact of agricultural practices on the *Zea mays* L. endophytic community. Appl Environ Microbiol 70:1475–1482

Seo WT, Lim WJ, Kim EJ, Yun HD, Lee YH, Cho KM (2010) Endophytic bacterial diversity in the young radish and their antimicrobial activity against pathogens. J Appl Biol Chem 53: 493–503

Sessitsch A, Hardoim P, Döring J, Weilharter A, Krause A, Woyke T, Mitter B et al (2012) Functional characteristics of an endophyte community colonizing rice roots as revealed by metagenomic analysis. Mol Plant Microbe Interact 25:28–36

Sevilla M, Burris RH, Gunapala N, Kennedy C (2001) Comparison of benefit to sugarcane plant growth and $^{15}N_2$ incorporation following inoculation of sterile plants with *Acetobacter diazotrophicus* wild-type and nif⁻ mutant strains. Mol Plant Microbe Interact 14:358–366

Sturz AV, Nowak J (2000) Endophytic communities of rhizobacteria and the strategies required to create yield enhancing associations with crops. Appl Soil Ecol 15:183–190

Sun L, Qiu F, Zhang X, Dai X, Dong X, Song W (2008) Endophytic bacterial diversity in rice (*Oryza sativa* L.) roots estimated by 16S rDNA sequence analysis. Microb Ecol 55:415–424

Sun LN, Zhang YF, He LY, Chen ZJ, Wang QY, Qian M, Sheng XF (2010) Genetic diversity and characterization of heavy metal-resistant-endophytic bacteria from two copper-tolerant plant species on copper mine wasteland. Bioresour Technol 101:501–509

Takeda K, Tonouchi A, Takada M, Suko T, Suzuki S, Kimura Y, Matsuyama N, Fujita T (2008) Characterization of cultivable methanotrophs from paddy soils and rice roots. Soil Sci Plant Nutr 54:876–885

Taulé C, Mareque C, Barlocco C, Hackembruch F, Reis VM, Sicardi M, Battistoni F (2012) The contribution of nitrogen fixation to sugarcane (*Saccharum officinarum* L.), and the identification and characterization of part of the associated diazotrophic bacterial community. Plant Soil 356:35–49

Terakado-Tonooka J, Ohwaki Y, Yamakawa H, Tanaka F, Yoneyama T, Fujihara S (2008) Expressed *nifH* genes of endophytic bacteria detected in field-grown sweet potatoes (*Ipomoea batatas* L.). Microbes Environ 23:89–93

Ueda T, Suga Y, Yahiro N, Matsuguchi T (1995) Remarkable N_2-fixing bacterial diversity detected in rice roots by molecular evolutionary analysis of *nifH* gene sequences. J Bacteriol 177:1414–1417

Ulrich K, Ulrich A, Ewald D (2008) Diversity of endophytic bacterial communities in poplar grown under field conditions. FEMS Microbiol Ecol 63:169–180

Vendan RT, Yu YJ, Lee SH, Rhee YH (2010) Diversity of endophytic bacteria in ginseng and their potential for plant growth promotion. J Microbiol 48:559–565

Vorobev AV, Baani M, Doronina NV, Brady AL, Liesack W, Dunfield PF, Dedysh SN (2010) *Methyloferula stellata* gen. nov., sp. nov., an acidophilic, obligately methanotrophic bacterium possessing only a soluble methane monooxygenase. Int J Syst Evol Microbiol 61:2456–2463. doi:10.1099/ijs.0.028118-0

Whittenbury R, Phillips KC, Wilkinson JF (1970) Enrichment, isolation and some properties of methane-utilizing bacteria. J Gen Microbiol 61:205–218

Wu L, Ma K, Lu Y (2009) Prevalence of betaproteobacterial sequences in *nifH* gene pools associated with roots of modern rice cultivars. Microb Ecol 57:58–68

Zakria M, Udonishi K, Ogawa T, Yamamoto A, Saeki Y, Akao S (2008) Influence of inoculation technique on the endophytic colonization of rice by *Pantoea* sp. isolated from sweet potato and by *Enterobacter* sp. isolated from sugarcane. Soil Sci Plant Nutr 54:224–236

Zhang L, Hurek T, Reinhold-Hurek B (2007) A *nifH*-based oligonucleotide microarray for functional diagnostics of nitrogen-fixing microorganisms. Microb Ecol 53:456–470

Part IV
Arbuscular Mycorrhizal Symbiosis

Chapter 11
Chemical Signalling in the Arbuscular Mycorrhizal Symbiosis: Biotechnological Applications

Juan A. López-Ráez and María J. Pozo

11.1 Introduction

Plants are living organisms that continuously communicate with other organisms, including microorganisms, present in their environment. Unlike animals, plants are sessile organisms that largely rely on chemicals as signalling molecules to perceive and respond to environmental changes. For instance, plants use molecules to recognize potential pathogens and defend themselves as well as to establish mutualistic beneficial associations with certain microorganisms in the rhizosphere. These beneficial associations belowground can affect plant growth and development, change nutrient dynamics, susceptibility to disease, tolerance to heavy metals and can help plants in the degradation of xenobiotics (Morgan et al. 2005). As a result, these plant–microorganism interactions have considerable potential for biotechnological exploitation.

The rhizosphere is the narrow soil zone surrounding plant roots and constitutes a very dynamic environment. It harbours many different microorganisms and is highly influenced by the root and the root exudates (Bais et al. 2006; Badri et al. 2009). Plants produce and exude through the roots a large variety of chemicals including sugars, amino acids, fatty acids, enzymes, plant growth regulators and secondary metabolites into the rhizosphere, some of which are used to communicate with their environment (Siegler 1998; Bertin et al. 2003; Bais et al. 2006). Moreover, the release of root exudates together with decaying plant material provides carbon sources for the heterotrophic soil biota. On the other hand, microbial activity in the rhizosphere affects rooting patterns and the supply of available nutrients to plants, thereby modifying the quantity and quality of root exudates (Barea et al. 2005). Of special interest for rhizosphere communication are the so-called secondary metabolites from root exudates, which received this name

J.A. López-Ráez (✉) • M.J. Pozo
Department of Soil Microbiology and Symbiotic Systems, Estación Experimental del Zaidín (CSIC), Profesor Albareda 1, 18008 Granada, Spain
e-mail: juan.lopezraez@eez.csic.es; mariajose.pozo@eez.csic.es

because of their presumed secondary importance in plant growth and survival (Siegler 1998). These metabolites include compounds from different biosynthetic origins that are of ecological significance because they act as signals in multiple pathogenic and mutualistic plant–microorganism interactions, including the arbuscular mycorrhizal (AM) symbiosis (Bertin et al. 2003; Bais et al. 2006).

11.2 The Arbuscular Mycorrhizal Symbiosis

The AM symbiosis is one of the best known beneficial plant–microorganism associations that take place in the rhizosphere (Barea et al. 2005; Smith and Read 2008; Bonfante and Genre 2010). This interaction is the most widely distributed symbiosis in the world (Parniske 2008; Smith and Read 2008). The AM symbiosis is based on the mutualistic association between certain soil fungi of the monophyletic phylum Glomeromycota and the vast majority of land plants, including most agricultural and horticultural crop species. This association dates back to more than 400 million years and it has been postulated to be a key step in the evolution of terrestrial plants (Smith and Read 2008). AM fungi are obligate biotrophs and depend entirely on the plant to obtain carbon and complete their life cycle. They colonize the root cortex of the host plant and form specialized tree-shaped subcellular structures called arbuscules, which are involved in nutrient exchange between the two symbiotic partners. The AM symbiosis gives rise to the formation of extensive hyphal networks in the soil that facilitate the acquisition of nutrients beyond the area of nutrient depletion, thereby assisting the plant in the acquisition of mineral nutrients (mainly phosphorous) and water (Smith and Read 2008). Therefore, the AM symbiosis positively affects plant growth and enhances the recycling of nutrients (Morgan et al. 2005; Parniske 2008; Bonfante and Genre 2010). Moreover, the AM symbiosis has non-nutritional effects such as the improvement of soil quality by stabilizing soil aggregates and preventing erosion (Gianinazzi et al. 2010). It also provides tolerance against different types of abiotic stresses such as drought, salinity or heavy metals (Smith and Read 2008; Ruiz-Lozano et al. 2012) and enhances the ability of the host plant to cope with biotic stresses (Pozo and Azcón-Aguilar 2007; Parniske 2008; Jung et al. 2012). Altogether, these benefits highlight the "true helper" character of this symbiosis and envisage the potential use of the AM symbiosis as biofertilizers and bioprotection agents for the sustainable management of agricultural ecosystems (Gianinazzi et al. 2010).

11.3 Signalling During the AM Symbiosis

AM symbiosis is a complex and very dynamic interaction which requires a high degree of coordination between the two partners. Symbiosis establishment implies a signal exchange between both partners that leads to mutual recognition and development of symbiotic structures (Siegler 1998; Bonfante and Genre 2010). Noteworthy, this molecular dialogue must be very precise in order to avoid opportunities for malevolent organisms (Bouwmeester et al. 2007). The plant–AM fungus communication occurs beyond the early stages of the interaction as it is necessary for the proper maintenance and functioning of the symbiosis.

11.3.1 The Pre-symbiotic Stage

The chemical signalling between the two symbiotic partners takes place even in the absence of direct physical contact and involves molecules that are produced by either the host plant or the AM fungus. Among them, strigolactones and Myc factors play an essential role.

11.3.1.1 Plant-Derived Signals: Strigolactones

Plant–AM fungus communication starts in the rhizosphere with the production and exudation of signalling molecules by the host plant that are recognized by AM fungi. These cues stimulate hyphal growth and favour the establishment of the first physical contact between the partners (Fig. 11.1). Spores of AM fungi can germinate spontaneously and undergo an initial asymbiotic stage of hyphal germ tube growth, which is limited by the amount of carbon storage in the spore. However, if a partner is in the vicinity, the hyphal germ tube grows and ramifies intensively through the soil towards the host root (López-Ráez et al. 2012). Although other compounds such as flavonoids and hydroxy fatty acids have been reported as hyphal growth stimulators (Scervino et al. 2005; Nagahashi and Douds 2011), strigolactones have been shown to be crucial for a successful root colonization by the AM fungi (Akiyama et al. 2005; Bouwmeester et al. 2007; López-Ráez et al. 2012). Experimental evidence of strigolactone relevance in the AM symbiosis establishment has been provided by the reduction in mycorrhizal colonization of mutant plants affected in strigolactone biosynthesis (Gómez-Roldán et al. 2008; Vogel et al. 2010; Kohlen et al. 2012). In agreement with their role as signalling molecules, strigolactones are short-lived in the rhizosphere (Akiyama et al. 2010). When perceived by the AM fungus, the hyphal germ tube grows and branches, increasing the possibility of contact with the host root (Bouwmeester et al. 2007; López-Ráez et al. 2012). Although they were initially identified as molecular cues in the rhizosphere, strigolactones are multifunctional molecules that have been

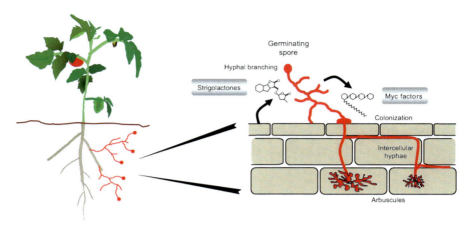

Fig. 11.1 Arbuscular mycorrhizal (AM) fungi–host plant symbiosis model. The process of AM symbiosis establishment begins with the production and exudation of strigolactones into the rhizosphere by the plant roots under deficient nutrient conditions (mainly phosphate shortage). Strigolactones induce hyphal branching of AM fungi-germinating spores, thus facilitating the contact with the host root. Parallel, AM fungus produces and releases the so-called Myc factors, which act on the plant roots inducing molecular responses required for a successful colonization. After contact, fungal hyphae grow intracellularly to the inner cortex, where the characteristic arbuscule is formed for nutrient exchange

classified as a new class of plant hormones regulating above- and belowground plant architecture (Bouwmeester et al. 2007; Gómez-Roldán et al. 2008; Umehara et al. 2008; Kapulnik et al. 2011a; Ruyter-Spira et al. 2011; López-Ráez et al. 2012). Strigolactones are present in the root exudates of a wide range of plants, and it has been shown that each plant produces a blend of different strigolactones, which suggest their broad spectrum action and importance in nature (Xie et al. 2010). They are derived from the carotenoids through sequential oxidative cleavage by *c*arotenoid *c*leavage *d*ioxygenases (CCD7 and CCD8) (Matusova et al. 2005; López-Ráez et al. 2008), thus belonging to the apocarotenoid class as the phytohormone abscisic acid (ABA) (Ohmiya 2009). All so far isolated and characterized strigolactones show a similar chemical structure (Fig. 11.2a), with a structural core consisting of a tricyclic lactone (the ABC-rings) connected via an enol ether bridge to a butyrolactone group (the D-ring) (Yoneyama et al. 2009; Zwanenburg et al. 2009). However, they can present different substituents on the AB-rings that make them different. It has been suggested that the biological activity of strigolactones resides in the enol ether bridge, which can be rapidly cleaved in aqueous and/or alkaline environments, indicating their short-lived and signalling character (Yoneyama et al. 2009; Zwanenburg et al. 2009; Akiyama et al. 2010). It was shown that phosphate and nitrogen deficiencies have a significant stimulatory effect on the production and exudation of strigolactones by plants (Yoneyama et al. 2007; López-Ráez et al. 2008), which will recruit AM fungi. Similarly, it has been recently shown that under salt stress conditions, plants also increase

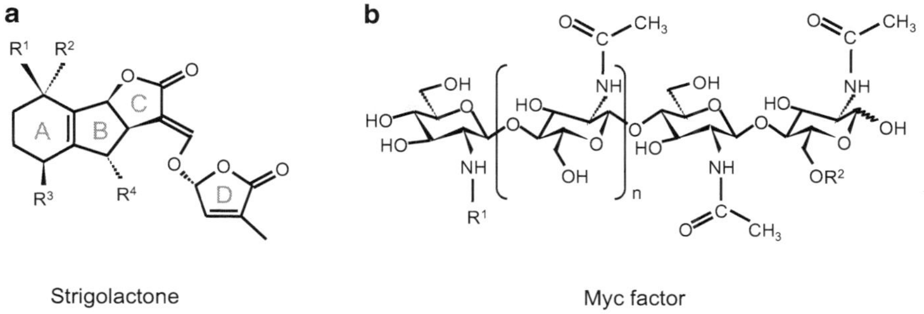

Fig. 11.2 Chemical structure of signalling molecules involved in AM symbiosis establishment. (**a**) General structure of host-derived strigolactones. (**b**) General Myc factor (lipochitooligosaccharide) structure

strigolactone production to promote AM symbiosis establishment (Aroca et al. 2013). According to the role of AM fungi, this increase in strigolactones is considered a cry for help under nutrient limiting and other stress conditions. Novel biological functions for strigolactones are being discovered continuously, indicating the relevance of these compounds in plant physiology and communication with other organisms from the rhizosphere (Kohlen et al. 2012).

Strigolactones present in the rhizosphere are perceived by the fungus through a so far uncharacterized receptor, thus inducing the so-called pre-symbiotic stage (Fig. 11.1). This stage is characterized by a profuse hyphal branching of the germinating spores, increasing the probability of contact with the root and that of establishing symbiosis (Akiyama et al. 2005; Besserer et al. 2006). Besides their role as hyphal branching factors, strigolactones stimulate spore germination in certain AM fungi (Besserer et al. 2006). Even, it has been suggested that they might act as the chemoattractant that directs the growth of the AM hyphae to the roots (Sbrana and Giovannetti 2005). Interestingly, strigolactones trigger a response only in AM fungi, but not in other beneficial fungal species such as *Trichoderma* and *Piriformospora indica* or in soilborne pathogens (Steinkellner et al. 2007), further supporting that they are rather specific signalling molecules.

11.3.1.2 AM Fungi-Derived Signals: Myc Factors

As in the nodulation process, another ecologically important mutualistic association in the rhizosphere established between legumes and rhizobia (Sprent 2009), a fungal factor analogous to the rhizobial Nod factor, has been recently characterized (Maillet et al. 2011) (Fig. 11.1). This Myc factor is produced by the metabolically active AM fungus, although it is not clear yet whether its production is induced by strigolactones. This fungal signal was previously described as a diffusible molecule, which is perceived by the plant inducing molecular responses in the host root required for a successful colonization (Parniske 2008). The chemical nature of the elusive Myc factor remained unknown for a long time. Maillet and co-workers recently showed that AM fungi produce and secrete a mixture of sulphated and non-sulphated simple lipochitooligosaccharides (LCOs) (Fig. 11.2b) that have structural similarities with rhizobial Nod factors (Maillet et al. 2011). Both consist of an N-acetylglucosamine backbone with various substitutions, which are the basis for their diversity. Since nodulation is more recent than AM symbiosis (it appeared about 60 million years ago), the authors suggest that the Nod factor-derived signalling pathway likely evolved from the AM association. Interestingly, Maillet and co-workers showed that Myc factors are not only symbiotic signals that stimulate the AM establishment, but they also act as plant growth regulators affecting root development (Maillet et al. 2011). Thus, AM fungi would trigger alterations in root architecture to increase the number of colonization sites. The discovery and characterization of these pre-symbiotic signals pave the way for the development of new environmentally friendly agricultural strategies, as we will discuss later in Sect. 11.5.

11.3.2 Communication During Symbiosis Establishment and Maintenance

After perception of the Myc factor by the host plant through a so far uncharacterized receptor, the plant cell actively prepares the intracellular environment for colonization by the AM fungus (Fig. 11.1). The plant responses comprise an extensive and specific reprogramming of root tissues for the establishment of the AM symbiosis under the control of the so-called common symbiosis (SYM) pathway, since it is common for a successful mycorrhization and nodulation in legumes (Parniske 2008; Bonfante and Genre 2010). The common SYM pathway seems to be primarily involved in controlling early events in AM establishment. As a consequence of sequential chemical and mechanical stimulation and reorganization, plant cells produce an AM-specific structure that is crucial for a successful fungal penetration known as the "prepenetration apparatus" (Genre et al. 2005). This transcellular path is structurally related to the pre-infection thread formed in legumes during nodulation (Parniske 2008). By this subcellular structure, the plant guides the fungal growth through root cells towards the cortex. Once in the inner root cortex, a perifungal membrane is assembled and it fuses with the plant plasma membrane to produce an invagination. Then, the intracellular hyphae branch repeatedly to form the characteristic arbuscules (Fig. 11.3) (Bonfante and Genre 2010).

Conversely to the pre-symbiotic stage, the information regarding the nature of the chemical cues involved in the molecular dialogue during AM establishment and maintenance is scarce. The central factor of this signalling pathway is calcium spiking. Thus, Myc factor perception triggers a rapid transient elevation of cytosolic calcium (Bonfante and Genre 2010). A receptor-like kinase has the potential to directly or indirectly perceive the fungal signals and transduce the signal to the

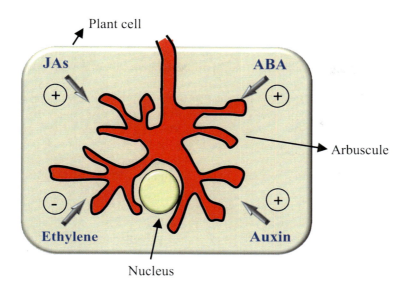

Fig. 11.3 Simplified model of the involvement of phytohormones in AM symbiosis. The picture gives a model of a root cortical cell containing an arbuscule, with the fungus in *red*. *Encircled plus* and *Encircled minus* indicates a positive and negative effect, respectively, of plant hormones on the establishment of AM symbiosis. *JAs* jasmonates, *ABA* abscisic acid

cytoplasm by phosphorylating a so far unknown substrate (Bonfante and Genre 2010). Since all the downstream elements of the SYM pathway are located in the nucleus, it has been suggested that the signal should be rapidly transduced to the nucleus. Additional chemical cues are probably needed to induce the establishment of the symbiosis and to provide certain specificity (Parniske 2008; Bonfante and Genre 2010). Unfortunately, the mechanisms controlling colonization of the host root tissue and arbuscule development are also largely unknown, although a role for phosphorous and carbohydrates has been proposed (Javot et al. 2007; Balzergue et al. 2011; Helber et al. 2011). Despite the importance of strigolactones in the initiation of AM symbiosis, it is unclear whether they also play a role in subsequent steps of the symbiosis. In this sense, it has been recently shown that strigolactones are also regulators of root nodulation in legumes (Soto et al. 2010; Foo and Davies 2011). Unlike AM symbiosis, strigolactones here would not act as host detection cues, but they would be required for optimal nodule number. This finding suggests that strigolactones might be also involved in further stages of AM symbiosis establishment and/or maintenance. Indeed, it has been proposed that they would promote sustained intercellular root colonization (Kretzschmar et al. 2012). However, further research is needed to clarify this aspect.

11.3.2.1 Involvement of Phytohormones

Mycorrhizal colonization is a very dynamic process in which the plant must control fungal growth and development. In this process, phytohormones orchestrate the modifications that occur in the host plant (Hause et al. 2007; López-Ráez et al. 2011b). Among them, jasmonic acid and its derivatives, known as jasmonates (JAs), are believed to play a major role in the AM symbiosis (Isayenkov et al. 2005; Hause et al. 2007; Herrera-Medina et al. 2008; López-Ráez et al. 2010b). JAs have been proposed to be involved in the reorganization of cytoskeleton and in the alteration of sink status of mycorrhizal roots (Hause et al. 2007). In this sense, they could stimulate carbohydrate biosynthesis in the shoots and transport into the roots for nutrient exchange in the arbuscules. JAs might also increase fitness in mycorrhizal plants by interacting with other phytohormones such as cytokinins (Hause et al. 2007). Abscisic acid (ABA) was also recently shown to be a key component in the establishment of the symbiosis in tomato plants. This hormone is necessary to complete the arbuscule formation process, enhance its functionality and promote sustained colonization of the root (Herrera-Medina et al. 2007; Martín-Rodríguez et al. 2010). In addition, it has been recently suggested that ABA plays a role in the regulation of strigolactone biosynthesis (López-Ráez et al. 2010a). Thus, ABA would be involved both in the early stages of the plant–fungus interaction and in the maintenance of the symbiosis once established. A role for auxin signalling within the host root during the early stages of AM formation has also been described (Hanlon and Coenen 2011). Since auxins are involved in regulating plant architecture, they could be responsible for lateral root formation on the plant, favouring host plant–AM fungus contact and therefore symbiosis establishment. Ethylene has also arisen as an important regulator of the symbiosis. Several findings suggest that ethylene would act as a negative regulator of mycorrhizal intensity (Zsogon et al. 2008; Martín-Rodríguez et al. 2011; Mukherjee and Ane 2011). Interestingly, a cross talk between ethylene and ABA was also shown (Herrera-Medina et al. 2007). The authors showed that ABA deficiency induced ethylene production and proposed that one of the mechanisms by which ABA regulates fungal colonization might be through the negative modulation of ethylene biosynthesis. In summary, mycorrhizal symbiosis establishment and maintenance implies a fine regulation of phytohormone levels in the host plant, whose understanding is still in its infancy.

11.3.2.2 Autoregulation of AM Symbiosis

A plant regulatory mechanism to control excessive root colonization has been proposed, a phenomenon known as autoregulation (García-Garrido et al. 2009; Staehelin et al. 2011). This mechanism shares some similarities with control of nodulation in legumes and allows the host plant to suppress further colonization once a certain level of mycorrhization has been reached, presumably to prevent costs of the symbiosis exceeding the benefits. It is known that mycorrhizal plants with a well-established colonization produce fewer strigolactones than non-mycorrhizal ones (López-Ráez et al. 2011a). Therefore, it has been suggested that the reduction in strigolactone biosynthesis might be involved in the autoregulation process (García-Garrido et al. 2009; López-Ráez et al. 2011b). However, the mechanism by which AM symbiosis reduces strigolactone production is unknown. As aforementioned, strigolactone biosynthesis is negatively affected by phosphate levels (Yoneyama et al. 2007; López-Ráez et al. 2008). Moreover, phosphate acts systemically to repress AM colonization by affecting essential symbiotic genes, in particular genes coding enzymes of carotenoid and strigolactone biosynthesis, and symbiosis-associated phosphate transporters (Breuillin et al. 2010). Therefore, it is plausible that changes in phosphate levels within the plant are responsible of autoregulation in mycorrhizal plants (Breuillin et al. 2010).

More recently, a different mechanism has been proposed for autoregulation of mycorrhization. Staehelin and co-workers (2011) suggested that long-distance transport of chemical signals would be related to autoregulation. The authors proposed that fungal Myc factors induce expression or post-translation processing of root-derived short peptides belonging to the CLE peptide family, which likely function as ascending long-distance signals to the shoot. Extracellular CLE peptides show hormone-like activities and are involved in the regulation of several processes in plant development. In legumes, some of these short peptides have been shown to be involved in autoregulation of rhizobial colonization (Mortier et al. 2010). Somehow, these CLE peptides would induce the production of a descending signal (shoot-derived inhibitor) that inhibits mycorrhizal root colonization (Staehelin et al. 2011). In this scenario, it would be logical to think that this unknown shoot-derived signal inhibits strigolactone production in the roots. A third mechanism proposed to mediate autoregulation has been suggested. Microarray analysis revealed that mycorrhization in tomato induced the expression of key genes in the 9-LOX branch of the oxylipin pathway (López-Ráez et al. 2010b; León-Morcillo et al. 2012). Oxylipins are a diverse class of lipid metabolites that include biologically active molecules as the phytohormone JAs (Mosblech et al. 2009). The 9-LOX pathway gives rise to the biosynthesis of, among others, colnelenic and colneleic acids for which a role in defence against plant pathogens has been proposed (Mosblech et al. 2009). Therefore, these compounds might act as inhibitors to avoid an excessive colonization (López-Ráez et al. 2010b). In any case, none of the above-described hypothesis are exclusive, and most likely they act in concert. Further research is needed to elucidate how the phenomenon of autoregulation during AM symbiosis is regulated and what are the chemical molecules involved.

11.4 Strigolactones: Ecological Significance in the Rhizosphere

In addition to their role as signalling molecules in AM symbiosis, strigolactones have been recognized as a new class of plant hormones that inhibits shoot branching and hence controls above-ground architecture (Gómez-Roldán et al. 2008; Umehara et al. 2008). More recently, it was shown that they also affect root growth, root hair elongation and adventitious rooting (Kapulnik et al. 2011a; Ruyter-Spira et al. 2011; Kohlen et al. 2012; Rasmussen et al. 2012), which shows they are even more important components in the regulation of plant architecture than already postulated. Root system architecture (RSA) is of great importance for plants. In addition to a site of interaction with symbiotic microorganisms, roots are essential for different plant functions such as uptake of nutrients and water and anchorage in the substrate (Den Herder et al. 2010). RSA is altered under nutrient-deficient conditions such as phosphorous starvation, resulting in a shorter primary root, more and longer lateral roots (which are the preferred colonization sites for AM fungi) and a greater density of root hairs, thus expanding the exploratory capacity of the root system (Sánchez-Calderón et al. 2005; Rouached et al. 2010). These responses are regulated through a complex network of interconnected signalling pathways in which plant hormones, including strigolactones, play a key role (Kapulnik et al. 2011b; Ruyter-Spira et al. 2011). According to this positive effect on root development, strigolactone biosynthesis is promoted under limited Pi conditions (Yoneyama et al. 2007; López-Ráez et al. 2008). Consequently, it has been proposed that they play a pivotal role in plants as modulators of the coordinated development of roots and shoots and acting as a "cry for help" signal in the rhizosphere in response to Pi starvation (López-Ráez et al. 2011b; Ruyter-Spira et al. 2011).

11.4.1 Strigolactones Are Germination Stimulants of Root Parasitic Plants

Long before the discovery of their function as plant hormones and signalling molecules for mycorrhization, strigolactones were described as germination stimulants for the seeds of root parasitic plants of the family Orobanchaceae, including *Striga*, *Orobanche* and *Phelipanche* genera (Cook et al. 1972; Bouwmeester et al. 2003; López-Ráez et al. 2012). Strigolactones are produced and exuded into the rhizosphere by plants in very low amounts, being able to stimulate the germination of these parasitic plants in nano- and pico-molar concentrations. These parasitic weeds are some of the most damaging agricultural pests, causing large crop losses worldwide in cultivars such as rice, maize, tomato and legumes (Joel et al. 2007; Parker 2009). They attach to the roots and acquire nutrients and water from their host through a specialized organ called haustorium

(Bouwmeester et al. 2003). The life cycle of these parasitic weeds involves germination in response to strigolactones, radicle growth towards the host root, attachment and penetration. After emergence from the soil, the parasitic plants flower and produce enormous amount of seeds that are scattered in the soil, increasing the seed bank (Bouwmeester et al. 2003; López-Ráez et al. 2009). Parasitic weeds are difficult to control because most of their life cycle occurs underground, making the diagnosis of infection difficult and possible only when irreversible damage has already been caused to the crop. Therefore new control strategies focused on the initial steps in the host–parasite interaction, and especially those targeting seed germination induced by strigolactones are required (López-Ráez et al. 2009).

11.5 Biotechnological Applications of Signalling Communication in the AM Symbiosis: The Green Technology

Biotechnological tools are defined as those processes of biological interest that use chemistry of living organisms to develop new and more effective ways of producing traditional products while maintaining the natural environment. Even though the mechanistic bases of chemical signalling and regulation of plant responses during the AM symbiosis are not fully elucidated, it is worthy to eavesdrop this signalling process and use it for crop protection and yield increase. Thus, the AM symbiosis can greatly contribute to crop productivity and ecosystem sustainability (Gianinazzi et al. 2010). They not only improve plant growth through increased uptake of water and nutrients but also have "non-nutritional" effects such as stabilizing soil aggregates, preventing erosion as well as alleviating abiotic and biotic stress (Parniske 2008; Smith and Read 2008). Since the identity and chemical structure of the key molecules involved in AM symbiosis establishment belowground is now being elucidated, manipulating their production to promote AM symbiosis, and its benefits, seems of great potential for a "green agriculture."

11.5.1 AM Symbiosis as Bioprotective Agent Against Root Parasitic Plants

The fact that strigolactones play a dual role in the rhizosphere as signalling molecules for AM fungi and root parasitic plants makes them a suitable candidate to develop environmentally friendly control methods against these parasitic weeds. Remarkably, it was shown that AM fungal inoculation of maize and sorghum led to a reduction in *Striga hermonthica* infection in the field (Lendzemo et al. 2005). It was proposed that this reduced infection was caused, at least partially, by a reduction in the production of strigolactones in mycorrhizal plants (Lendzemo et al. 2007). Similarly, AM colonization in pea and tomato induced less germination of seeds of *Orobanche* and *Phelipanche* species compared to non-colonized plants (Fernández-Aparicio et al. 2010; López-Ráez et al. 2011a). Moreover, in the case of tomato, it was demonstrated that such reduction was caused by a decrease in the production of strigolactones (López-Ráez et al. 2011a). This decrease was associated to a fully established symbiosis, likely related to the phenomenon mycorrhizal autoregulation.

These observations in different plant systems suggest that the reduction in strigolactone production induced by AM symbiosis is conserved across the plant kingdom. Therefore, since AM fungi colonize roots of most agricultural and horticultural species and are widely distributed around the globe, AM symbiosis could be used as a biocontrol strategy for economically important crops that suffer from these root parasitic weeds. In addition, these crops would also take advantage of all the other well-known benefits of the symbiosis, such as the positive effect on plant fitness and higher tolerance/resistance against biotic and abiotic stresses. Addressing the possible specificity of different strigolactones in plant–parasitic plant and plant–AM fungus interactions is a major research challenge to further develop biotechnological strategies that favour one against another. In this line, by generating tomato transgenic plants blocked at the CCD8 enzyme (biosynthesis of strigolactones), we have recently shown that a mild reduction in strigolactone exudation is sufficient to reduce parasitic infection about 90 % without severely compromising AM symbiosis (Kohlen et al. 2012).

11.5.2 AM Symbiosis as Biofertilizer

AM fungi are already being used as biofertilizers for enhancing plant growth and biomass production in agriculture, although their use is still rather low compared with conventional methods (Barea et al. 2005; Gianinazzi et al. 2010). Remarkably, the benefits of the association for the host plant depend on the plant and AM fungus combination. Therefore, selecting appropriate AM fungal isolates adapted to local conditions and functionally optimal for the plant of interest is essential. Understanding the mechanisms regulating colonization, symbiosis efficiency and competence is crucial to optimize mycorrhizal benefits. Elucidation of the molecules involved in the interaction of host plant–AM fungus will definitively contribute to a better implementation of the "mycorrhizal technology" agrosystems.

11.5.2.1 Use of Strigolactones to Promote AM Symbiosis

Strigolactones have been shown to be essential for AM symbiosis establishment through the induction of hyphal branching, thus increasing the possibility of contact with the host plant (Akiyama et al. 2005; Bouwmeester et al. 2007). Therefore, breeding for high strigolactone production cultivars appears as an attractive strategy to improve mycorrhization in agronomical conditions. On the other hand, this improved mycorrhizal colonization could be achieved by the exogenous application of strigolactones or strigolactone analogues. However, due to their stimulatory role of germination of root parasitic plants in the rhizosphere, a higher plant infection by these parasitic weeds would also take place. Moreover, strigolactones as phytohormones are involved in multiple physiological functions, which could give rise to undesired side effects. Therefore, a better understanding in the biology of strigolactones is essential prior to its application to agriculture. As mentioned above, one single plant species produces a mixture of different strigolactones and in different quantities (Xie et al. 2010). In addition, structural differences between strigolactones on the AM fungal branching response and for parasitic plant seed germination have been reported (Yoneyama et al. 2009; Akiyama et al. 2010). Similarly, specificity in strigolactone transport *in* and *ex planta* has been recently reported (Kohlen et al. 2012). Kohlen and co-workers showed that certain strigolactones are mainly exuded into the rhizosphere, while others are preferentially loaded into the xylem and transported to the shoot. Thus, new strategies for breeding or strigolactone application should be focused on strigolactone specificity in order to favour mycorrhization without affecting other functions in the plant.

11.5.2.2 Application of Myc Factors

With regard to the Myc factors, they are the fungal signals that stimulate AM symbiosis establishment in leguminous and other mycotrophic plant species. Besides stimulating AM symbiosis, they act as plant growth regulators by stimulating root branching and development (Maillet et al. 2011). Elucidation of the chemical nature of these Myc factors (Maillet et al. 2011) makes Myc factors as the ideal candidates for their use as biofertilizers. Indeed, this "green technology" has currently been applied for the treatment of leguminous seeds with Nod factors and yielded increase in crops such as soybean, alfalfa and pea. The fact that more than 80 % of land plants are mycotrophic suggests that Myc factors might have a broader spectrum of activity than Nod factors. Besides their beneficial effect on plant growth through an improved mycorrhization, they would facilitate the development of the root system, with the consequent improvement in nutrient uptake and resistance to abiotic stresses. In the search of methods for efficient synthesis of Myc factors, its production by bacteria has been already developed (Maillet et al. 2011), allowing their production on a large scale. Still further research is needed on the biological activity and specificity of Myc factors and the optimal conditions for application before its implementation in agriculture.

11.6 Conclusions

Chemical fertilizers and pesticides are being used to promote plant growth and to prevent plant diseases. However, the environmental pollution caused by excessive use and misuse of agrochemicals has led to public concerns about the use of these chemicals in agriculture. Therefore, there is a need to find more environmentally friendly alternatives for fertilization and disease control. The key to achieve successful biological control is the knowledge on plant interactions in an ecological context. We emphasize here the importance of the chemical communication that occurs in the rhizosphere between plants and beneficial microorganisms, especially AM fungi, and its potential use as biofertilizers and biocontrol agents. Further research on the mechanisms regulating the production and release of these cues and about their specificities requirements will expand our knowledge on biological processes occurring underground. The information generated may result in the development of new green technologies for sustainable agriculture in the near future.

Acknowledgments Our research is supported by grant AGL2009-07691 from the National R&D Plan of the MINCIN.

References

Akiyama K, Matsuzaki K, Hayashi H (2005) Plant sesquiterpenes induce hyphal branching in arbuscular mycorrhizal fungi. Nature 435:824–827

Akiyama K, Ogasawara S, Ito S, Hayashi H (2010) Structural requirements of strigolactones for hyphal branching in AM fungi. Plant Cell Physiol 51:1104–1117

Aroca R, Ruiz-Lozano JM, Zamarreño AM, Paz JA, García-Mina JM, Pozo MJ, López-Ráez JA (2013) Arbuscular mycorrhizal symbiosis influences strigolactone production under salinity and alleviates salt stress in lettuce plants. J Plant Physiol 170(1):47–55. doi:10.1016/j.jplph.2012.08.020

Badri DV, Weir TL, van der Lelie D, Vivanco JM (2009) Rhizosphere chemical dialogues: plant-microbe interactions. Curr Opin Biotechnol 20:642–650

Bais HP, Weir TL, Perry LG, Gilroy S, Vivanco JM (2006) The role of root exudates in rhizosphere interactions with plants and other organisms. Annu Rev Plant Biol 57:233–266

Balzergue C, Puech-Pages V, Becard G, Rochange SF (2011) The regulation of arbuscular mycorrhizal symbiosis by phosphate in pea involves early and systemic signalling events. J Exp Bot 62:1049–1060

Barea JM, Pozo MJ, Azcon R, Azcon-Aguilar C (2005) Microbial co-operation in the rhizosphere. J Exp Bot 56:1761–1778

Bertin C, Yang XH, Weston LA (2003) The role of root exudates and allelochemicals in the rhizosphere. Plant Soil 256:67–83

Besserer A, Puech-Pages V, Kiefer P, Gomez-Roldan V, Jauneau A, Roy S, Portais JC, Roux C, Becard G, Sejalon-Delmas N (2006) Strigolactones stimulate arbuscular mycorrhizal fungi by activating mitochondria. PLoS Biol 4:1239–1247

Bonfante P, Genre A (2010) Mechanisms underlying beneficial plant-fungus interactions in mycorrhizal symbiosis. Nat Commun 1:1–11

Bouwmeester HJ, Matusova R, Zhongkui S, Beale MH (2003) Secondary metabolite signalling in host-parasitic plant interactions. Curr Opin Plant Biol 6:358–364

Bouwmeester HJ, Roux C, López-Ráez JA, Bécard G (2007) Rhizosphere communication of plants, parasitic plants and AM fungi. Trends Plant Sci 12:224–230

Breuillin F, Schramm J, Hajirezaei M, Ahkami A, Favre P, Druege U, Hause B, Bucher M, Kretzschmar T, Bossolini E, Kuhlemeier C, Martinoia E, Franken P, Scholz U, Reinhardt D (2010) Phosphate systemically inhibits development of arbuscular mycorrhiza in *Petunia hybrida* and represses genes involved in mycorrhizal functioning. Plant J 64:1002–1017

Cook CE, Whichard LP, Wall ME, Egley GH, Coggon P, Luhan PA, McPhail AT (1972) Germination stimulants. 2. The structure of strigol-a potent seed germination stimulant for witchweed (*Striga lutea* Lour.). J Am Chem Soc 94:6198–6199

Den Herder G, Van Isterdael G, Beeckman T, De Smet I (2010) The roots of a new green revolution. Trends Plant Sci 15:600–607

Fernández-Aparicio M, García-Garrido JM, Ocampo JA, Rubiales D (2010) Colonisation of field pea roots by arbuscular mycorrhizal fungi reduces *Orobanche* and *Phelipanche* species seed germination. Weed Res 50:262–268

Foo E, Davies NW (2011) Strigolactones promote nodulation in pea. Planta 243:1073–1081

García-Garrido JM, Lendzemo V, Castellanos-Morales V, Steinkellner S, Vierheilig H (2009) Strigolactones, signals for parasitic plants and arbuscular mycorrhizal fungi. Mycorrhiza 19:449–459

Genre A, Chabaud M, Timmers T, Bonfante P, Barker DG (2005) Arbuscular mycorrhizal fungi elicit a novel intracellular apparatus in *Medicago truncatula* root epidermal cells before infection. Plant Cell 17:3489–3499

Gianinazzi S, Gollotte A, Binet MN, van Tuinen D, Redecker D, Wipf D (2010) Agroecology: the key role of arbuscular mycorrhizas in ecosystem services. Mycorrhiza 20:519–530

Gómez-Roldán V, Fermas S, Brewer PB, Puech-Pagés V, Dun EA, Pillot JP, Letisse F, Matusova R, Danoun S, Portais JC, Bouwmeester H, Bécard G, Beveridge CA, Rameau C, Rochange SF (2008) Strigolactone inhibition of shoot branching. Nature 455:189–194

Hanlon MT, Coenen C (2011) Genetic evidence for auxin involvement in arbuscular mycorrhiza initiation. New Phytol 189:701–709

Hause B, Mrosk C, Isayenkov S, Strack D (2007) Jasmonates in arbuscular mycorrhizal interactions. Phytochemistry 68:101–110

Helber N, Wippel K, Sauer N, Schaarschmidt S, Hause B, Requena N (2011) A versatile monosaccharide transporter that operates in the arbuscular mycorrhizal fungus *Glomus* sp. is crucial for the symbiotic relationship with plants. Plant Cell 23:3812–3823

Herrera-Medina MJ, Steinkellner S, Vierheilig H, Ocampo-Bote JA, García-Garrido JM (2007) Abscisic acid determines arbuscule development and functionality in the tomato arbuscular mycorrhiza. New Phytol 175:554–564

Herrera-Medina MJ, Tamayo MI, Vierheilig H, Ocampo JA, García-Garrido JM (2008) The jasmonic acid signalling pathway restricts the development of the arbuscular mycorrhizal association in tomato. J Plant Growth Regul 27:221–230

Isayenkov S, Mrosk C, Stenzel I, Strack D, Hause B (2005) Suppression of allene oxide cyclase in hairy roots of *Medicago truncatula* reduces jasmonate levels and the degree of mycorrhization with *Glomus intraradices*. Plant Physiol 139:1401–1410

Javot H, Pumplin N, Harrison MJ (2007) Phosphate in the arbuscular mycorrhizal symbiosis: transport properties and regulatory roles. Plant Cell Environ 30:310–322

Joel DM, Hershenhom Y, Eizenberg H, Aly R, Ejeta G, Rich JP, Ransom JK, Sauerborn J, Rubiales D (2007) Biology and management of weedy root parasites. Hort Rev 33:267–349

Jung SC, Martínez-Medina A, López-Ráez JA, Pozo MJ (2012) Mycorrhiza-induced resistance and priming of plant defenses. J Chem Ecol 38:651–664

Kapulnik Y, Delaux PM, Resnick N, Mayzlish-Gati E, Wininger S, Bhattacharya C, Sejalon-Delmas N, Combier JP, Bécard G, Belausov E, Beeckman T, Dor E, Hershenhorn J, Koltai H

(2011a) Strigolactones affect lateral root formation and root-hair elongation in Arabidopsis. Planta 233:209–216

Kapulnik Y, Resnick N, Mayzlish-Gati E, Kaplan Y, Wininger S, Hershenhorn J, Koltai H (2011b) Strigolactones interact with ethylene and auxin in regulating root-hair elongation in Arabidopsis. J Exp Bot 62:2915–2924

Kohlen W, Charnikhova T, Lammers M, Pollina T, Tóth P, Haider I, Pozo MJ, de Maagd RA, Ruyter-Spira C, Bouwmeester HJ, López-Ráez JA (2012) The tomato *CAROTENOID CLEAVAGE DIOXYGENASE8* (SlCCD8) regulates rhizosphere signaling, plant architecture and affects reproductive development through strigolactone biosynthesis. New Phytol 196:535–547

Kretzschmar T, Kohlen W, Sasse J, Borghi L, Schlegel M, Bachelier JB, Reinhardt D, Bours R, Bouwmeester HJ, Martinoia E (2012) A petunia ABC protein controls strigolactone-dependent symbiotic signalling and branching. Nature 483:341–344

Lendzemo VW, Kuyper TW, Kropff MJ, van Ast A (2005) Field inoculation with arbuscular mycorrhizal fungi reduces *Striga hermonthica* performance on cereal crops and has the potential to contribute to integrated *Striga* management. Field Crop Res 91:51–61

Lendzemo VW, Kuyper TW, Matusova R, Bouwmeester HJ, van Ast A (2007) Colonization by arbuscular mycorrhizal fungi of sorghum leads to reduced germination and subsequent attachment and emergence of *Striga hermonthica*. Plant Signal Behav 2:58–62

León-Morcillo RJ, Ángel J, Martín R, Vierheilig H, Ocampo JA, García-Garrido JM (2012) Late activation of the 9-oxylipin pathway during arbuscular mycorrhiza formation in tomato and its regulation by jasmonate signalling. J Exp Bot 63:3545–3558

López-Ráez JA, Charnikhova T, Gómez-Roldán V, Matusova R, Kohlen W, De Vos R, Verstappen F, Puech-Pages V, Bécard G, Mulder P, Bouwmeester H (2008) Tomato strigolactones are derived from carotenoids and their biosynthesis is promoted by phosphate starvation. New Phytol 178:863–874

López-Ráez JA, Matusova R, Cardoso C, Jamil M, Charnikhova T, Kohlen W, Ruyter-Spira C, Verstappen F, Bouwmeester H (2009) Strigolactones: ecological significance and use as a target for parasitic plant control. Pest Manag Sci 64:471–477

López-Ráez JA, Kohlen W, Charnikhova T, Mulder P, Undas AK, Sergeant MJ, Verstappen F, Bugg TDH, Thompson AJ, Ruyter-Spira C, Bouwmeester H (2010a) Does abscisic acid affect strigolactone biosynthesis? New Phytol 187:343–354

López-Ráez JA, Verhage A, Fernández I, García JM, Azcón-Aguilar C, Flors V, Pozo MJ (2010b) Hormonal and transcriptional profiles highlight common and differential host responses to arbuscular mycorrhizal fungi and the regulation of the oxylipin pathway. J Exp Bot 61:2589–2601

López-Ráez JA, Charnikhova T, Fernández I, Bouwmeester H, Pozo MJ (2011a) Arbuscular mycorrhizal symbiosis decreases strigolactone production in tomato. J Plant Physiol 168:294–297

López-Ráez JA, Pozo MJ, García-Garrido JM (2011b) Strigolactones: a cry for help in the rhizosphere. Botany 89:513–522

López-Ráez JA, Bouwmeester H, Pozo MJ (2012) Communication in the rhizosphere, a target for pest management. In: Lichtfouse E (ed) Agroecology and strategies for climate change. Springer, Dordrecht, pp 109–133

Maillet F, Poinsot V, Andre O, Puech-Pages V, Haouy A, Gueunier M, Cromer L, Giraudet D, Formey D, Niebel A, Martinez EA, Driguez H, Becard G, Denarie J (2011) Fungal lipochitooligosaccharide symbiotic signals in arbuscular mycorrhiza. Nature 469:58–63

Martín-Rodríguez JA, León-Morcillo R, Vierheilig H, Ocampo JA, Ludwig-Muller J, García-Garrido JM (2010) Mycorrhization of the *notabilis* and *sitiens* tomato mutants in relation to abscisic acid and ethylene contents. J Plant Physiol 167:606–613

Martín-Rodríguez JA, León-Morcillo R, Vierheilig H, Ocampo JA, Ludwig-Muller J, García-Garrido JM (2011) Ethylene-dependent/ethylene-independent ABA regulation of tomato plants colonized by arbuscular mycorrhiza fungi. New Phytol 190:193–205

Matusova R, Rani K, Verstappen FWA, Franssen MCR, Beale MH, Bouwmeester HJ (2005) The strigolactone germination stimulants of the plant-parasitic *Striga* and *Orobanche* spp. are derived from the carotenoid pathway. Plant Physiol 139:920–934

Morgan JAW, Bending GD, White PJ (2005) Biological costs and benefits to plant-microbe interactions in the rhizosphere. J Exp Bot 56:1729–1739

Mortier V, den Herder G, Whitford R, van de Velde W, Rombauts S, D'Haeseleer K, Holsters M, Goormachtig S (2010) CLE peptides control *Medicago truncatula* nodulation locally and systemically. Plant Physiol 153:222–237

Mosblech A, Feussner I, Heilmann I (2009) Oxylipins: structurally diverse metabolites from fatty acid oxidation. Plant Physiol Biochem 47:511–517

Mukherjee A, Ane JM (2011) Germinating spore exudates from arbuscular mycorrhizal fungi: molecular and developmental responses in plants and their regulation by ethylene. Mol Plant Microbe Interact 24:260–270

Nagahashi G, Douds DD (2011) The effects of hydroxy fatty acids on the hyphal branching of germinated spores of AM fungi. Fungal Biol 115:351–358

Ohmiya A (2009) Carotenoid cleavage dioxygenases and their apocarotenoid products in plants. Plant Biotechnol 26:351–358

Parker C (2009) Observations on the current status of *Orobanche* and *Striga* problems worldwide. Pest Manag Sci 65:453–459

Parniske M (2008) Arbuscular mycorrhiza: the mother of plant root endosymbioses. Nat Rev Microbiol 6:763–775

Pozo MJ, Azcón-Aguilar C (2007) Unravelling mycorrhiza-induced resistance. Curr Opin Plant Biol 10:393–398

Rasmussen A, Mason MG, De Cuyper C, Brewer PB, Herold S, Agusti J, Geelen D, Greb T, Goormachtig S, Beeckman T, Beveridge CA (2012) Strigolactones suppress adventitious rooting in Arabidopsis and pea. Plant Physiol 158:1976–1987

Rouached H, Arpat AB, Poirier Y (2010) Regulation of phosphate starvation responses in plants: signaling players and cross-talks. Mol Plant 3:288–299

Ruiz-Lozano JM, Porcel R, Azcón C, Aroca R (2012) Regulation by arbuscular mycorrhizae of the integrated physiological response to salinity in plants: new challenges in physiological and molecular studies. J Exp Bot 63:4033–4044

Ruyter-Spira C, Kohlen W, Charnikhova T, van Zeijl A, van Bezouwen L, de Ruijter N, Cardoso C, López-Ráez JA, Matusova R, Bours R, Verstappen F, Bouwmeester H (2011) Physiological effects of the synthetic strigolactone analog GR24 on root system architecture in Arabidopsis: another belowground role for strigolactones? Plant Physiol 155:721–734

Sánchez-Calderón L, López-Bucio J, Chacón-López A, Cruz-Ramirez A, Nieto-Jacobo F, Dubrovsky JG, Herrera-Estrella L (2005) Phosphate starvation induces a determinate developmental program in the roots of *Arabidopsis thaliana*. Plant Cell Physiol 46:174–184

Sbrana C, Giovannetti M (2005) Chemotropism in the arbuscular mycorrhizal fungus *Glomus mosseae*. Mycorrhiza 15:539–545

Scervino JM, Ponce MA, Erra-Bassells R, Vierheilig H, Ocampo JA, Godeas A (2005) Arbuscular mycorrhizal colonization of tomato by *Gigaspora* and *Glomus* species in the presence of root flavonoids. J Plant Physiol 162:625–633

Siegler DS (1998) Plant secondary metabolism. Kluwer Academic, Boston, MA

Smith SE, Read DJ (2008) Mycorrhizal symbiosis. Academic, London

Soto MJ, Fernández-Aparicio M, Castellanos-Morales V, García-Garrido JM, Ocampo JA, Delgado MJ, Vierheilig H (2010) First indications for the involvement of strigolactones on nodule formation in alfalfa (*Medicago sativa*). Soil Biol Biochem 42:383–385

Sprent JI (2009) Legume nodulation. A global perspective. Wiley-Blackwell, Chichester

Staehelin C, Xie ZP, Illana A, Vierheilig H (2011) Long-distance transport of signals during symbiosis: are nodule formation and mycorrhization autoregulated in a similar way? Plant Signal Behav 6:372–377

Steinkellner S, Lendzemo V, Langer I, Schweiger P, Khaosaad T, Toussaint JP, Vierheilig H (2007) Flavonoids and strigolactones in root exudates as signals in symbiotic and pathogenic plant-fungus interactions. Molecules 12:1290–1306

Umehara M, Hanada A, Yoshida S, Akiyama K, Arite T, Takeda-Kamiya N, Magome H, Kamiya Y, Shirasu K, Yoneyama K, Kyozuka J, Yamaguchi S (2008) Inhibition of shoot branching by new terpenoid plant hormones. Nature 455:195–200

Vogel JT, Walter MH, Giavalisco P, Lytovchenko A, Kohlen W, Charnikhova T, Simkin AJ, Goulet C, Strack D, Bouwmeester HJ, Fernie AR, Klee HJ (2010) *SlCCD7* controls strigolactone biosynthesis, shoot branching and mycorrhiza-induced apocarotenoid formation in tomato. Plant J 61:300–311

Xie XN, Yoneyama K, Yoneyama K (2010) The strigolactone story. Annu Rev Phytopathol 48:93–117

Yoneyama K, Xie X, Kusumoto D, Sekimoto H, Sugimoto Y, Takeuchi Y, Yoneyama K (2007) Nitrogen deficiency as well as phosphorous deficiency in sorghum promotes the production and exudation of 5-deoxystrigol, the host recognition signal for arbuscular mycorrhizal fungi and root parasites. Planta 227:125–132

Yoneyama K, Xie X, Yoneyama K, Takeuchi Y (2009) Strigolactones: structures and biological activities. Pest Manag Sci 65:467–470

Zsogon A, Lambais MR, Benedito VA, Figueira AVD, Peres LEP (2008) Reduced arbuscular mycorrhizal colonization in tomato ethylene mutants. Scientia Agricola 65:259–267

Zwanenburg B, Mwakaboko AS, Reizelman A, Anilkumar G, Sethumadhavan D (2009) Structure and function of natural and synthetic signalling molecules in parasitic weed germination. Pest Manag Sci 65:478–491

Chapter 12
Carbon Metabolism and Costs of Arbuscular Mycorrhizal Associations to Host Roots

Alex J. Valentine, Peter E. Mortimer, Aleysia Kleinert, Yun Kang, and Vagner A. Benedito

12.1 Introduction

The symbiotic relationship of plants with mycorrhizal fungi is cited as being a driving force behind the shift of plants to land, with the hyphal network enabling improved uptake by poorly developed roots in nutrient-poor conditions. This 470-million-year-old collaboration is a widespread phenomenon with 70–90 % of terrestrial plant species colonized by arbuscular mycorrhizal fungi. The evolutionary history of this relationship is supported by both fossil records and the fungal recognition mechanism that are conserved among plants. The premise of the symbiosis is the reliance of the fungi on photosynthetically derived carbohydrates supplied by the plant in exchange for improved nutrient (predominantly phosphate) and water uptake mediated by the extensive hyphal network. Recent studies have shown that the flow of carbon to the fungus can be downregulated under sufficient nutrient regimes and that the fungus is also able to control the transfer of nutrients to

---

A.J. Valentine (✉) • A. Kleinert
Botany and Zoology Department, Faculty of Science, University of Stellenbosch, Private Bag X1, Matieland 7602, South Africa
e-mail: alexvalentine@mac.com; kleinert@sun.ac.za

P.E. Mortimer
World Agroforestry Centre (ICRAF), China and East Asia Node, c/o: Kunming Institute of Botany, Kunming 650204, China
e-mail: peter@mail.kib.ac.cn

Y. Kang
Plant Biology Division, Samuel Roberts Noble Foundation, 2510 Sam Noble Parkway, Ardmore, OK 73401, USA
e-mail: ykang@noble.org

V.A. Benedito
Genetics and Developmental Biology Program, Division of Plant and Soil Sciences, West Virginia University, 2090 Agricultural Science Building, Morgantown, WV 26506, USA
e-mail: vagner.benedito@mail.wvu.edu

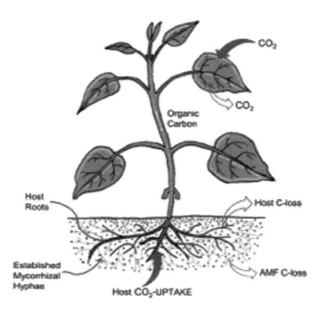

Fig. 12.1 The carbon costs of the arbuscular mycorrhizal root system, expressed as C inputs and losses above and below ground. Aboveground C inputs are via photosynthetic CO_2 assimilation and the losses are via respiratory CO_2 release. Belowground costs include CO_2 losses from host root and fungal symbiont respiration, due to tissue growth and nutrient acquisition. The refixation of CO_2 from the rhizosphere may replenish some of the respiratory CO_2 losses. Additionally, the increased translocation of organic C from the aboveground to belowground tissues may also incur a C cost (Figure provided by Dr. Yun Kang)

less than beneficial hosts (Selosse and Rousset 2011; Maillet et al. 2011; Kiers et al. 2011). This points to a biological "market" with reciprocal rewards to stabilize mycorrhizal symbiosis.

12.2 The C Cost of Arbuscular Mycorrhizae

AM fungi are dependent on the host plant as a C source and therefore act as a C sink. Once established, the fungal metabolic activities and growth can promote carbon sink stimulation of photosynthesis from the host plant (Fig. 12.1). It has been estimated that the fungus receives between 10 and 23 % of the plant's photosynthetically fixed carbon (Snellgrove et al. 1982; Kucey and Paul 1982; Koch and Johnson 1984; Harris et al. 1985; Jakobsen and Rosendahl 1990). Black et al. (2000) showed that mycorrhizal plants have a higher photosynthetic rate than non-mycorrhizal plants. This may be because of either an increased level of phosphate in the leaves due to the mycorrhizae (Azcon et al. 1992; Black et al. 2000) or because the AM fungus acts as a carbon sink (Snellgrove et al. 1982; Kucey and Paul 1982; Koch and Johnson 1984; Jakobsen and Rosendahl 1990). Both explanations have been found to be true but under different conditions and for different plants. Therefore it may result from a combination of both, depending on the growing conditions and the developmental stage of both the fungus and the plant.

In particular, the developmental stage of the symbiosis may warrant further exploration in order to gain a more accurate account of the C costs involved. Although the percentage AM colonization may be used as a means of estimating the potential mycorrhizal sink, there is conflicting evidence as to whether or not the percentage colonization of the root by AM fungi is related to the soluble carbohydrate content of the root. Pearson and Schweiger (1993) found that colonization was negatively correlated with the soluble carbohydrate content of the root, while Thompson et al. (1990) found a positive correlation. This may be because the conflicting experiments were carried out during different developmental stages of colonization. The three stages of colonization are the lag phase, the phase of rapid development, and the plateau phase (Smith and Read 1997). Pearson and Schweiger (1993) carried out their experimental work towards the end of the phase of rapid development, when the colonization period starts to decline and therefore is a subsequent decline in the demand for C by the fungus. Thompson et al. (1990) experimented during the end of the lag phase and the start of the phase of rapid development, when the demand for C is high. It appears that the process of colonization does depend on carbohydrates from the root during the initial phases and then reaches equilibrium as the process of colonization comes to an end and a stable symbiotic relationship develops. The carbon that is taken up by the fungus is incorporated into the growth and development of new fungal structures and spores. More than 90 % of the root can be colonized by an AM fungus (Motosugi et al. 2002) and can constitute up to 20 % of the root dry mass (Harris and Paul 1987).

Respiration of colonized roots was found to be between 6.6 and 16.5 % (depending on fungal species) higher than non-colonized roots in cucumber plants (Pearson and Jakobsen 1993). The increased respiration rate contributes to the sink effect of the fungus and indicates that colonized roots have a higher metabolic activity than non-colonized roots. This may in part be due to the losses incurred by the symbiont itself. There are three main ways that organic C is lost from the host via the fungus (Fig. 12.1): firstly via the loss of sloughed off fungal material, secondly through the release of fungal spores into the soil, or thirdly via the exudation of organic acids and phosphatase enzymes by the fungus. Fungal mycelia are constantly being replaced because of older material either breaking off as the root pushes through the soil or dying and being released into the soil. Bethlenfalvay et al. (1982a, b) found that as much as 88 % of the fungal biomass was external of the root for soybean; similarly Olsson and Johansen (2000) found that 70 % of the fungal biomass was external mycelium on cucumber roots. This will account for a large portion of C lost into the soil considering that at some stage the external hyphae will be released into the soil. The release of spores from the external mycelium accounts for a high percentage of lost organic C. In a study done by

Sieverding (1989), it was estimated that 919 kg ha^{-1} of plant C went into the production of spores, which are subsequently released into the soil by the fungus. Furlan and Fortin (1977) found spore production was influenced by the amount of C that is available to the fungus. The third means of organic C loss is through the exudation of organic acids and phosphatase enzymes by the fungal hyphae in order to aid in the uptake of nutrients such as phosphate. However, the main body of evidence supporting this has been found in ectomycorrhizae (Bolan et al. 1987). Although the release of organic acids is not thought to be the primary means of P uptake (Bolan 1991), it does constitute a loss of organic C derived from the host.

12.3 Arbuscular Mycorrhizal Efficiency and C-Flux Estimation

The transfer of fixed C from the host to the symbiont has a direct effect on the host plant and thus it is important to quantify this process. Koide and Elliott (1989) described this relationship mathematically using various models to describe both the gross benefit of the mycorrhizal colonization and the net benefit of colonization.

Gross benefit was defined as the difference between the quantity of gross C assimilation (mole C) in mycorrhizal and non-mycorrhizal plants over a given period of time (Koide and Elliott 1989):

$$\Delta A_m^g - \Delta A_{nm}^g$$

where ΔA_m^g and ΔA_{nm}^g are the gross C assimilation of the mycorrhizal and non-mycorrhizal plants during that time interval, respectively.

The net benefit of colonization for the same time period was described as the difference between mycorrhizal and non-mycorrhizal C accumulation (moles C) in the whole plant over the given time period (Koide and Elliott 1989):

$$\Delta C_m^w - \Delta C_{nm}^w$$

where ΔC_m^w and ΔC_{nm}^w represent the amount of C accumulated in the mycorrhizal and non-mycorrhizal plants over the given time period.

Koide and Elliott (1989) also described the efficiency of the relationship in terms of P acquisition, P utilization, and belowground C utilization. The efficiency of the P acquisition was defined as

$$\frac{\Delta P^w}{\Delta C^b}$$

where ΔP^w is the total P that has accumulated in the plant during the given time interval and ΔC^b is the total belowground C expenditure over the same time period

(Koide and Elliott 1989). This describes the efficiency of the relationship in terms of the amount of P taken up compared to the amount of C used for the uptake of P. C^b can be calculated as follows (Koide and Elliott 1989):

$$C^b = C^r + C^o + C^n$$

where C^r is the C that is allocated to the root tissue, C^o is the C lost via root belowground respiration, and C^n is the nonrespiratory, belowground C loss.

The efficiency of P utilization was defined by the following equation (Koide and Elliott 1989):

$$\frac{\Delta C^w}{\Delta P^w}$$

where ΔC^w is the total amount of C accumulated, in the whole plant, over the same period (Koide and Elliott 1989). This efficiency can be applied to any of the respective plant components. The final model proposed by Koide and Elliott (1989) was used to define the efficiency of belowground C utilization and was expressed as the ratio $\Delta C^w : \Delta C^b$. This ratio is the product of the previous two models:

$$\frac{\Delta C^w}{\Delta C^b} = \frac{\Delta C^w}{\Delta P^w} \times \frac{\Delta P^w}{\Delta C^b}$$

Koide and Elliott (1989) defined C^b (see above) as the total belowground C expenditure, which included all the C in the living tissue of the root system and the C lost from the root, via exudation, leaching, respiration, cell death, and direct transport to the fungus. Jones et al. (1991) went one step further and formulated two models that defined C^b in terms of the factors influencing the changes in C^b.

(a) The first model expressed C^b as a function of the C fixed via photosynthesis:

$$C_{b(Pn)} = Pn \frac{\%C_{BG}}{100 - \%C_{SR}} t$$

where $C_{b(Pn)}$ is the amount of photosynthetically fixed C that is allocated below ground in a given period of time. Pn is the net photosynthetic rate as mmol C s^{-1} for the whole shoot system; $\%C_{BG}$ is the percentage of the total fixed C which is allocated below ground, over a given period of time; $\%C_{SR}$ is the percentage of the fixed C which was released via respiration in the shoot; and t is the length of the daily light period, measured in seconds. The term $100 - \%C_{SR}$ represents the total amount of C left after respiration.

(b) Their second model expressed C^b as a function of the change in shoot mass, which will give an indication of C fluxes within the shoot:

$$\Delta C_{b(Wt)} = \Delta W_s \frac{\%C_{BG}}{\%C_{ST}}$$

where ΔW_s is the mean increase in shoot weight over a given time period and $\%C_{ST}$ and $\%C_{BG}$ are the mean percentages of the C fixed and allocated to the shoot tissue and to the belowground components, respectively.

The work of Koide and Elliott (1989) forms the backbone of mycorrhizal efficiency modeling, but they never tested their models experimentally. Therefore they have not defined the influencing factors that affected each of the parameters involved in the different models. The models proposed by Jones et al. (1991) elaborated on those of Koide and Elliott (1989) by defining C^b as a function of its influencing factors, not just its components.

However, the expression of C^b in terms of photosynthetically fixed C can be misleading. It assumes that photosynthetic C is the only source of C available to the plant. It does not include structural and nonstructural C that is already stored in the plant, which may be used and transported below ground or anywhere else in the plant for that matter. Similarly the expression of C^b in terms of the changes shoot mass assumes that the shoots are the only structures that will have an influence on belowground C, again ignoring other preexisting sources of C within the plant. This also neglects to take into account that AM and non-AM plants may allocate photosynthetic C in different proportions to different organs (Smith 1980; Koide 1985). Furthermore, these models do not include the stimulation of photosynthesis by mycorrhizal symbioses and, for this reason, may result in a misrepresented estimation of the carbon costs of mycorrhizal colonization.

12.4 Tissue Construction Cost and Belowground Respiration

Williams et al. (1987) proposed a model that can be used to determine the construction cost of various tissues within a plant. They defined construction cost as the amount of glucose required to provide C skeletons, reductant, and ATP for synthesizing the organic compounds in a tissue via standard biochemical pathways. They calculated tissue construction cost as

$$C_w = \left\{ (0.06968 \times \Delta H_c - 0.065)(1 - A) + \frac{kN}{14.0067} \times \frac{180.15}{24} \right\} \frac{1}{E_g}$$

where C_w is the construction cost of the tissue (g glucose g DW^{-1}) and ΔH_c is the ash-free heat of combustion of the sample (kJ g^{-1}). A is the ash content of the sample (g ash g DW^{-1}); k is the reduction state of the N substrate (NO_3 was used, therefore k is +5) and E_g is the deviation of growth efficiency from 100 %. E_g represents the fraction of the construction cost that provides reductant that is not incorporated into biomass. Williams et al. (1987) determined the value of E_g to be 0.89.

Peng et al. (1993) slightly modified this equation and converted the g glucose into mmol C:

$$C_w = \left\{(0.06968 \times \Delta H_c - 0.065)(1 - A) + \frac{kN}{14.0067} \times \frac{180.15}{24}\right\} \frac{1}{0.89} \times \frac{6,000}{180}$$

The units of construction cost are now mmol C g DW^{-1}. However, the tissue construction cost equation was further modified by Mortimer et al. (2005):

$$C_w = [C + kN/14 \times 180/24]\,(1/0.89)(6,000/180)$$

where C_w is the construction cost of the tissue (mmol C g DW^{-1}), C is the carbon concentration (mmol C g^{-1}), k is the reduction state of the N substrate, and N is the organic nitrogen content of the tissue (g/g DW) (Williams et al. 1987). The constant (1/0.89) represents the fraction of the construction cost which provides reductant that is not incorporated into biomass (Williams et al. 1987; Peng et al. 1993) and (6,000/180) converts units of g glucose/g DW to mmol C g DW^{-1}.

Peng et al. (1993) use the construction cost to determine the growth respiration, which was defined as the respired C associated with the biosynthesis of new tissue:

$$R_{G(t)} = C_t - \Delta W_c$$

where $R_{G(t)}$ is the growth respiration (îmol CO_2 d^{-1}); C_t (îmol CO_2 d^{-1}) is the C required for daily construction of new tissue. C_t was calculated by multiplying the root growth rate (ΔW_w, mg DW d^{-1}) by tissue construction cost (C_w). ΔW_c (îmol d^{-1}) is the change in root C content and was calculated by multiplying the root C content and the root growth rate (ΔW_w, mg DW d^{-1}).

12.5 Arbuscular Mycorrhizal Carbon Costs in Dual-Symbiotic Roots

Arbuscular mycorrhizal (AM) hosts plants are also able to form dual root symbioses with AM and nitrogen-fixing symbionts. In particular, the ability of legumes to form tripartite associations with AM fungi and rhizobia gives them access to sources of P and N that would normally not be available to the plant. Rhizobia are able to fix atmospheric N and convert it into an organic form which is subsequently made available to the host (Lodwig and Poole 2003). The AM are not only more efficient in the uptake of P from the soil but can also access pockets of soil P that would ordinarily not be available to the host. This allows for improved plant growth, especially in nutrient-poor soils. The ability to form these associations makes legumes valuable as a crop plant as it provides both the legumes and the subsequent crops with a renewable source of N and the ability to grow in low P soils (Frey and Schuepp 1992; Udvardi et al. 2005).

However, these nutrients provided to the host plant come at a cost; in exchange the host plant supplies the two symbionts with photosynthetically derived sugars (Vessey and Layzell 1987; Smith and Read 1997; Vance 2002). In spite of the C costs of maintaining a dual symbiosis, the cumulative benefits of dual inoculation are greater than those of singular inoculation to both the host plant and to the respective symbionts (Daft and El-Giahmi 1974; Cluett and Boucher 1983; Kawai and Yamamoto 1986; Pacovsky et al. 1986; Chaturvedi and Singh 1989). However the legume is able to balance these costs by increasing its photosynthetic rate, thus producing more sugars for the growth and maintenance of both itself and the two symbionts.

12.5.1 Symbiotic P and N Nutrition

One of the main functions of AM is the provision of soil P to the host plant; therefore, under low soil P conditions, the dependency of the host on the AM fungi increases (Smith and Read 1997). Fredeen and Terry (1988) found that AM-colonized legumes growing under low soil P had higher shoot P as well as greater shoot and nodule dry weights. This indicates the role that AM play in both the P nutrition and the growth of the host as well as the nodules. The enhancement of host growth is attributed to an increase in the production of photosynthate by the host, resulting from of the improved nutrition, obtained from the root symbioses (Fredeen and Terry 1988; Jia et al. 2004).

Rhizobia, found in the root nodules, play a crucial role in the N nutrition of their legume hosts. The bacteria are able to provide the host with reduced N, which is derived from atmospheric N_2. Atmospheric N is fixed into ammonia, with the aid of the enzyme nitrogenase, which is consequently exchanged with the host for C (Thorneley 1992; Lodwig and Poole 2003). Nodules are strong P sinks and the process of N fixation is energy intensive, resulting in the nodules requiring more energy and P than the host roots (Sa and Israel 1991; Al-Niemi et al. 1998; Almeida et al. 2000). Vadez et al. (1997) reported that nodules had a threefold greater concentration of P than other plant tissues, which gives an indication of the nodular sink strength for P. It has also been reported that a deficiency of P can lead to a reduction in both nodulation and symbiotic N fixation (Othman et al. 1991; Drevon and Hartwig 1997). Alternatively, P availability has been found to increase the ratio of nodule to total plant mass; nodule mass appears to be more influenced by the availability of P than nodule number (Othman et al. 1991; Almeida et al. 2000). This is confirmed by the work of Olivera et al. (2004), who reported that an increase in the P supplied to host plants led to a fourfold increase in nodule mass. This dependency on P by the nodule bacteria will also create a strengthened dependency on AM fungi by the host in order to supply the high amounts of P required by the rhizobia.

Although legumes rely on the N contribution of the rhizobia for growth and development, the plant can access other sources of N. Numerous studies have shown that AM play an important role, both directly and indirectly, in the uptake of N and the subsequent supply of N to the host plant (Marschner and Dell 1994; Constable et al. 2001; Toussaint et al. 2004; Govindarajulu et al. 2005; Mortimer et al. 2009). This indirect effect of AM on host N nutrition is apparent in legumes. Mycorrhizal legumes have been reported to have a greater number of nodules, increased nodular weight, and improved N fixation rates, thereby enhancing the N nutrition of the host (Carling et al. 1978; Kawai and Yamamoto 1986; Luis and Lim 1988, Vesjsadova et al. 1993; Goss and de Varennes 2002; Mortimer et al. 2008). Another indirect means of AM influencing the N nutrition of host legumes can be seen in the study carried out by Mortimer et al. (2012); this study found that AM can counter the asparagine-induced inhibition of biologic nitrogen fixation (BNF; under N nutrition), thus allowing for the continued function nodules when exposed to an external source of N.

12.5.2 C Costs of the Dual Symbiosis

Root symbionts require C from the host plant in exchange for the nutrients provided, thus acting as C sinks. Therefore, by default they both compete for the same source of photosynthate from the host plant (Harris et al. 1985, Mortimer et al. 2008). This double sink created by the root symbionts can impose a considerable drain on host C resources, altering rates of photosynthesis and host growth. Numerous studies have revealed that the AM fungus receives between 10 and 23 % of host photosynthate (Snellgrove et al. 1982; Koch and Johnson 1984; Kucey and Paul 1982; Jakobsen and Rosendahl 1990) and the nodules between 6 and 30 % (Kucey and Paul 1981, 1982; Harris et al. 1985; Provorov and Tikhonovich 2003). Nodule number as well as BNF is influenced by the availability of photosynthate; therefore, any factors which influence the rates of photosynthesis will in turn influence the BNF process (Bethlenfalvay and Phillips 1977; Murphy 1986; Atkins et al. 1989; Sussanna and Hartwig 1996; Schortemeyer et al. 1999). As has been shown in the studies of Fredeen and Terry (1988) and Peng et al. (1993), when soil P and N are not limiting to plant development, a growth depression in the host plant and a decrease in nodular dry weight can occur. This is due to the C being used for symbiont growth in place of host growth (Fredeen and Terry 1988; Peng et al. 1993).

Despite this C drain imposed by the symbionts, both the rhizobia and the AM fungi generally result in the improved growth of the host plant; however, these growth benefits are most noticeable under low nutrient conditions, when the host plant is most reliant on the respective root symbionts. Numerous studies have highlighted the fact that the dual inoculation of the host plant by AM fungi and nodule bacteria results in enhanced plant growth and symbiont development to a greater extent than singular inoculation (Daft and El-Giahmi 1974; Cluett and Boucher 1983; Kawai and Yamamoto 1986; Pacovsky et al. 1986; Chaturvedi and Singh 1989; Mortimer et al. 2009, 2012). Examples of this synergistic benefit resulting from dual inoculation are seen in the work of Nwoko and Sanginga (1999), who reported a 57 % increase in the percentage AM colonization of host plants, once the host was inoculated with *Bradyrhizobium*; similarly, the dual inoculation led to an increase in nodular weight. Furthermore, studies by Toro et al. (1998) and Jia et al. (2004) have shown that host plants had greater N fixation rates and that dual inoculation led to increased rates of photosynthesis and improved plant productivity.

As a means of compensating for the C being used by the two symbionts, the host plants increase its photosynthetic productivity, which is achieved in a number of different ways. Generally an increase in leaf P is attributed to the presence of AM; this can be used to fuel an increase the rate of photosynthesis; additionally, in a more indirect manner, host plants can increase the specific leaf area and rate of leaf expansion, further boosting photosynthetic productivity (Harris et al. 1985; Fredeen and Terry 1988; Jia et al. 2004). Harris et al. (1985) found 47 % increase in CO_2 fixation of nodulated soybeans and attributed this to the higher leaf P, improved mobilization of starch, and an increase in the specific leaf area of the soybeans. In the same way, Jia et al. (2004) have shown that plants with rhizobia and AM fungi have higher photosynthetic rates per unit leaf area. An alternative mechanism was provided by Fredeen and Terry (1988) who found that host plants had greater photosynthate production due to an increase in the rate of leaf surface expansion and not as a result of an increase in the rate of photosynthesis or photosynthesis on a leaf area basis. From the above studies, it is clear that host plants are able to compensate for the increased demand for photosynthate, resulting from the respective root symbionts, by a number of different mechanisms. This sink stimulation may be dependent upon the developmental stage of the root symbionts and their contributions to nutrient-based growth or uptake efficiencies.

12.5.3 Calculations for Dual Root Symbioses

Carbon and nutrient cost calculations for dual root symbioses in legumes may be calculated as follows:

(a) Specific P absorption rate (SPAR) (mg P g^{-1} root dw d^{-1}) or SNAR in the case of N, is the calculation of the net P absorption rate per unit root dw (Nielson et al. 2001):

$$\text{SPAR} = (M_2 - M_1)/(t_2 - t_1) \times (\log_e R_2 - \log_e R_1)/(R_2 - R_1)$$

where M is the P content per plant and R is the root system (to include nodules, if nodulated) dry weight.

(b) Specific P utilization rate (SPUR) (g dw $mg^{-1}P\ d^{-1}$) or SNUR in the case of N is a measure of the dw gained for the P taken up by the plant (Nielson et al. 2001):

$$\text{SPUR} = (W_2 - W_1)/(t_2 - t_1) \times (\log_e M_2 - \log_e M_1)/(M_2 - M_1)$$

where M is the P content of the plant and W is the plant dry weight. The specific nitrogen utilization rate (SNUR) can be adapted from these equations to include N instead of P, as shown in Mortimer et al. (2008).

The benefits of using SPUR and SPAR for C cost analysis were evident in the comparison of nutritional benefits in terms of uptake and growth efficiency of legume hosts with dual symbiotic roots (Mortimer et al. 2008). In their study on the legume *Phaseolus vulgaris*, Mortimer et al. (2008) found that in the dual symbiosis, the AM colonization proceeded more rapidly than the nodulation when the plants were inoculated at the same time. The AM was the primary belowground sink of host C resources in the dual symbiosis with root nodules, which resulted in the delayed onset of nodular growth. This allowed for AM establishment and the subsequent enhancement of P nutrition, which benefited the later nodular and host development. AM colonization peaked at 17 days after inoculation; the decline in percentage colonization following this stage (24 days) coincided with the increase in nodule mass. It was argued that the increase in nodule development was delayed due to host C being used to support the high levels of AM colonization (thus allowing for increased P acquisition), confirmed by the higher construction costs and growth respiration of these plants. Once AM colonization reached the plateau phase, more C was available for nodule growth (Harris et al. 1985). Coupled with this argument is that nodules require relatively high amounts of P for normal growth and maintenance (Sa and Israel 1991; Al-Niemi et al. 1998; Almeida et al. 2000); thus under low P conditions, the host and nodules would rely on AM-derived P during the AM plateau phase. Therefore, initial AM development up to 17 days took preference over the formation of nodules, so that the plateau phase of AM was reached and the subsequent benefits of P nutrition accrued. This was also noticed in previous studies, where P nutrition was enhanced once the plateau phase of AM development was reached (Mortimer et al. 2005). This AM-induced response was primarily due to the improved P nutrition of AM roots, evidenced by the greater specific P absorption (SPAR) and utilization (SPUR) rates compared to non-mycorrhizal plants. These effects of P on host growth and nutrition are confirmed by previous studies showing improved P nutrition and growth of AM plants (Sanders and Tinker 1971; Smith 1982; Bolan 1991; Orcutt and Nilsen 2000). Confirmation for the preference of AM development over nodular growth (0–17 days) is apparent in the difference between colonization levels at low and high P and the fungal effect on the dual symbiotic belowground sink strength. This sink effect is evidenced by the higher photosynthetic rates and root oxygen consumption rates of the double symbiotic roots at low P than at high P, for the period of 0–17 days. This is consistent with the work of Valentine and Kleinert (2007) who found that mycorrhizal plants had increased respiration resulting from the higher levels of root colonization associated with low P conditions.

Mortimer et al. (2008) also found that there were synergistic effects on nutritional physiology, owing to the dual symbiotic associations. These were most pronounced under low P conditions due to the key role of AM in the tripartite symbiosis. These synergistic effects on nodule development resulted in greater BNF by the nodules, providing the host with more N than their non-AM counterparts. This is consistent with previous studies showing enhanced levels of colonization and nodulation resulting from the synergistic effect of dual inoculation, as well as greater nodular dry weights (Daft and El-Giahmi 1974; Cluett and Boucher 1983; Kawai and Yamamoto 1986; Pacovsky et al. 1986; Chaturvedi and Singh 1989; Nwoko and Sanginga 1999). In addition, Catford et al. (2003) found that the tripartite symbiotic partners may autoregulate the development of the various symbionts. The nutritional benefits of the double symbiosis that resulted in greater host growth under low P are in agreement with the work of Gavito et al. (2000). Gavito et al. (2000) proposed that the improved nutrition of the mycorrhizal legumes led to enhanced N fixation, thereby resulting in the dual symbiotic plants having greater growth.

This cumulative sink effect imposes a considerable drain on host C reserves, as evidenced by the increased oxygen consumption and growth respiration of the dual symbiotic plants in study by Mortimer et al. (2008). The increased demand for host C by the two symbionts resulted in the host plants having higher photosynthetic rates, which concurs with the findings of Jia et al. (2004) for *Vicia faba*. Although Jia et al. (2004) attributed the higher photosynthetic rates to improved N and P nutrition, no evidence was presented for respiratory costs driving belowground sink stimulation of photosynthesis. In this regard, previous studies have found that colonization of host roots by AM led to increased levels of belowground respiration (Peng et al. 1993; Valentine and Kleinert 2007). Therefore, the addition of a further symbiont, such as nodule-producing N_2-fixing bacteria, should lead to an even greater respiratory demand on the host. From the current study, additional evidence for the greater C consumption by the dual symbiosis is the higher construction costs of these plants. This means that the hosts require more C for every gram of tissue produced, thus further increasing the photosynthetic and respiratory C costs. This is consistent with the work done by Peng et al. (1993) who found that mycorrhizal plants had higher construction costs than their non-mycorrhizal counterparts. These findings indicate a more prominent role for the AM in this tripartite symbiosis, because of its role in supplying P more effectively to both host and nodules, as well as contributing to the N economy of the host plant. Physiological and biochemical methods often involve large experimental procedures to determine the carbon costs of the root symbioses. The advent of quantitative tools of gene expression analysis may offer an efficient method to simplify these calculations.

12.6 Perspectives of Molecular Approach to Carbon Costs

The value of a molecular approach to complement the physiological costs assessment is evident in the study by Kiers et al. (2011), where the C allocation of gene expression in an AM symbiotic assemblage was traced as ^{13}C allocation in RNA. Not only is it important to determine the C costs of total gene expression but also to use transcriptional analyses to assess potential C fluxes during the symbiosis.

As described in the previous sections, the impact of nutritional root symbioses on the host's physiology is quite dramatic, not only in the root system but also as consequences to the functioning of aboveground organs. Higher respiratory demand of roots and organic C export imposes greater photosynthetic requirements from the shoots, while increased P (and N) availability to the shoots can lead to greater growth rates. The impact of this positive developmental-nutritional feedback between the participating symbionts goes beyond direct sink of primary C and N pathways in the shoot, such as photosynthesis, TCA cycle, N assimilation, and metabolite transport systems. Notwithstanding, the vast majority of the studies on gene expression of root symbioses concentrate on root tissues.

Using transcriptional analysis to quantify a specific cellular phenomenon is often complex because transcription is only one of many levels of gene expression regulation, including posttranscriptional and posttranslational levels. In metabolic pathways, many enzymes are commonly present in excess, so that variation in gene expression has little impact in the pathway flow, and key enzymes are often regulated by multiple mechanisms (e.g., mRNA degradation by small RNA targeting, translational arrest, protein phosphorylation, feedback allosteric regulation) to guarantee a precise metabolic regulation. Ideal genetic markers representing the carbon costs of the symbiosis from the plant host to the mycorrhizal fungus should present a specific and proportional regulation pattern of gene transcription. This may reflect the level of root infection and metabolic flow of organic C export. Furthermore, there is clearly a requirement for model experiments that correlate plant gene expression and metabolic carbon flow to the mycorrhizal symbiont. In this regard, a quantitative real-time (qPCR) approach comparing symbiotic systems with nonsymbiotic systems in both nutritionally sufficient and limiting conditions could be used in parallel with metabolic analyses (e.g., labeled carbon flow, adenylate energy charge status). This would immensely facilitate the assessment of plant energy costs in terms of reduced C flow to the mycorrhizal fungi. In addition, this would accrue further value if the genes could be assessed in aerial, photosynthetic organs of the plant.

A recent study on the impact of mycorrhization on gene expression in rice leaves revealed that this symbiosis increases pathogen resistance in the shoot by systemically inducing expression of defense-response genes, including diverse transcription factors (notably, AP2/EREBP and bHLH family members), calcium signaling and kinases (MPK6, CBP), and hormone-related genes, such as the jasmonic acid (JA) biosynthesis (Campos-Soriano et al. 2012). As demonstrated in tomato, JA is indeed an important plant hormone for mycorrhization and particularly in relation to carbon partitioning. Prosystemin, a JA precursor, was shown to act as a positive effector of endomycorrhizal colonization in studies using genotypes defective or overexpressing prosystemin (Tejeda-Sartorius et al. 2008). Additionally, other hormones have been shown to be important for mycorrhizal association, such as strigolactone (Yoneyama et al. 2008), abscisic acid (Garrido et al. 2010; Rodriguez et al. 2010), and ethylene (Zsögön et al. 2008). It will be interesting to assess the impact of mycorrhizal symbiosis in the gene expression of shoots of other species and correlate the level of infection with transcriptional variations, in order to better characterize canonical gene expression patterns that are evolutionarily conserved in the plant kingdom.

Good genetic markers of mycorrhization have been described for pre-colonization (expansin EXBL1, specific WRKY transcription factors) and late stages of colonization (DXS-2, chitinase, glutathione S-transferase, β-1,3-glucanase, and an additional WRKY transcription factor) in several plant species (Dermatsev et al. 2010; Gallou et al. 2011). Additionally, the expression of the mycorrhiza-specific phosphate transporter PT4 (Harrison et al. 2002), the H^+-ATPase MTHA1 (Krajinski et al. 2002), and the mycorrhizal high-affinity monosaccharide transporter MST2 (Helber et al. 2011) can be used in quantitative experiments to assess relative mycorrhizal colonization in the root system.

Although the plant model *Arabidopsis thaliana* does not form mycorrhizal associations, the cross talk between the C and N pathways is mostly conserved among plant species, and the results found in *Arabidopsis* might serve to understand the physiological relations of C and N metabolism. A recent study identified two E3 ubiquitin ligases (ATL31 and ATL6) and two transcription factors (MYB51 and WRKY33) involved in both pathogen defense response and C/N metabolism (Maekawa et al. 2012). If future studies in other plants revealed that functional homologs of such genes have the same expression pattern, these genes might become suitable markers of mycorrhizal symbiosis and its impact on C/N physiology.

A comprehensive transcriptional analysis through microarray comparing gene expression profile between mycorrhizal roots and nonsymbiotic roots was performed in the model legume *Medicago truncatula*. Further confirmation by laser microdissection of cortical cells containing arbuscules revealed novel genes expressed specifically in infected cells (Gomez et al. 2009), including several membrane transporters (Benedito et al. 2010). Further studies confirmed and expanded this analysis, demonstrating specific plant and fungal genes during cellular reprogramming of the symbiosis in cortical root cells, including a potential Cu transporter, a H^+-dependent oligopeptide transporter, HAP5 and MYB transcription factors, a protease inhibitor, and a gene of unknown function in infected root cells, while the fungus showed expression changes in several genes, including upregulation of two 18S rRNA genes and downregulation of a RHO-protein GDI gene (Gaude et al. 2012). Importantly, Gomez et al. (2009) found genes specifically related to energy (ADP-ribosylation factor, ARF) and nitrogen metabolism (glutamine synthetase, arginase, ornithine aminotransferase) to be specifically expressed in colonized mycorrhizal root cortical cells, not to mention an ammonium transporter (most likely involved in uptake of ammonium made available by the fungi) and, not surprisingly, several genes involved in lipid metabolism.

The arbuscule formation does indeed require a great deal of lipid synthesis by the plant cells to enable an enormous expansion of the plasma membrane to establish the symbiosome membrane that interfaces the arbuscule that allows chemical exchanges between symbionts. Trépanier et al. (2005) found that palmitic acid synthesis in the symbiotic fungus depends directly on sugar supply from the plant host and can synthesize 16-C fatty acids exclusively in intraradical hyphae. The host also needs a massive lipid biosynthesis, especially in the intraradically colonized cells, and this requirement should be counted as C cost of the symbiosis to the host.

CO_2-enriched atmosphere was shown to increase ectomycorrhizal association in roots of Pinus (Parrent and Vilgalys 2009). The authors used symbiotic plant genes to assess C and N flow (monosaccharide transporter and ammonium transporter, respectively) as well as the expression of fungal 18S rRNA to determine the level of association. Another approach to be considered in understanding carbon costs of mycorrhizal associations is genetics by the using of mutants. For example, three mutants defective in accumulation of starch in root cells of the model legume *Lotus japonicus* were used to assess the effect of root starch accumulation on mycorrhization (Gutjahr et al. 2011). Results pointed out that catabolism of root starch is an alternate route to feeding the symbiotic fungi; however, the authors challenge the hypothesis that plants feed the fungi only during the day, when photosynthetically active, but instead that root starch is used to feed the fungi with hexose at night. This is in agreement with the fact that an α-amylase gene is upregulated in mycorrhizal roots of *Medicago truncatula* (Gomez et al. 2009). It will be interesting to C costs to determine whether or not starch catabolism-related gene expression is regulated in a circadian rhythm pattern in mycorrhizal roots.

Another example of use of genetic tools to characterize carbon metabolism in the mycorrhizal symbiosis was the demonstration of the role of a sucrose synthase gene (MtSucS1) for arbuscule maturation and maintenance in *Medicago truncatula* (Baier et al. 2010). Antisense lines for this gene showed drastic developmental impairment as well as compromised mycorrhization, pointing out the specific role of sucrose synthase activity for this symbiosis.

Additionally, use of CO_2-enriched environments to compare plants under symbiotic and nonsymbiotic conditions using both metabolic and genetic approaches might lead to a more comprehensive understanding of how mycorrhizal fungi utilize plant's resources to grow. The use of mathematical modeling may be a useful resource to integrate genetic and physiological data in biotrophic mutualistic symbiosis.

12.7 Conclusions

The C costs of arbuscular mycorrhizal roots can be assessed using various scales of investigation, ranging from whole-plant physiological to biochemical to molecular biological. Moreover, it may be prudent to integrate these scales of investigation, in order to increase the depth of our understanding of the C costs of the arbuscular mycorrhizal symbiosis. Functional bookkeeping of C costs at various scales is obviously highly complex, and although we might develop approaches to calculate the net costs of this association, one can only appreciate the 400 million years that this symbiosis has been evolving on Earth for the benefit of both plant and fungus.

References

Almeida JPF, Hartwig UA, Frehner M, Nosberger J, Luscher A (2000) Evidence that P deficiency induces N feedback regulation of symbiotic N2 fixation in white clover (*Trifolium repens* L.). J Exp Bot 51:1289–1297

Al-Niemi TS, Kahn ML, McDermott TR (1998) Phosphorus uptake by bean nodules. Plant Soil 198:71–78

Atkins CA, Pate JS, Sanford PJ, Dakora FD, Matthews I (1989) Nitrogen nutrition of nodules in relation to "N-hunger" in cowpea (*Vigna unguiculata* L. Walp). Plant Physiol 90:1644–1649

Azcon R, Gomez M, Tobar R (1992) Effects of nitrogen source on growth, nutrition, photosynthetic rate and nitrogen metabolism of mycorrhizal and phosphorous-fertilized plants of *Lactuca sativa* L. New Phytol 121:227–234

Baier MC, Keck M, Gödde V, Niehaus K, Küster H, Hohnjec N (2010) Knockdown of the symbiotic sucrose synthase MtSucS1 affects arbuscule maturation and maintenance in mycorrhizal roots of *Medicago truncatula*. Plant Physiol 152:1000–1014

Benedito VA, Li H, Dai X, Wandrey M, He J, Kaundal R, Torres-Jerez I, Gomez SK, Harrison MJ, Tang Y, Zhao PX, Udvardi MK (2010) Genomic inventory and transcriptional analysis of *Medicago truncatula* transporters. Plant Physiol 152:1716–1730

Bethlenfalvay GJ, Phillips DA (1977) Effect of light intensity on efficiency of carbon dioxide and nitrogen reduction in *Pisum sativum* L. Plant Physiol 60:868–871

Bethlenfalvay GJ, Brown MS, Pacovsky RS (1982a) Relationships between host and endophyte development in mycorrhizal soybeans. New Phytol 90:537–543

Bethlenfalvay GJ, Pacovsky RS, Brown MS, Fuller G (1982b) Mycotrophic growth and mutualistic development of host plant and fungal endophyte in an endomycorrhizal symbiosis. Plant Soil 68:43–54

Black KG, Mitchell DT, Osborne BA (2000) Effect of mycorrhizal-enhanced leaf phosphate status on carbon partitioning, translocation and photosynthesis in cucumber. Plant Cell Environ 23:797–809

Bolan NS (1991) A critical review on the role of mycorrhizal fungi in the uptake of phosphorous by plants. Plant Soil 134:189–207

Bolan NS, Robson AD, Barrow NJ (1987) Effect of vesicular arbuscular mycorrhiza on the availability of iron phosphates to plants. Plant Soil 99:401–410

Campos-Soriano L, García-Martínez J, Segundo BS (2012) The arbuscular mycorrhizal symbiosis promotes the systemic induction of regulatory defence-related genes in rice leaves and confers resistance to pathogen infection. Mol Plant Pathol 13:579–592

Carling DE, Reihle WG, Brown MF, Johnston DR (1978) Effects of vesicular arbuscular mycorrhizal fungus on nitrate reductase and nitrogenase activities in nodulating and non-nodulating soybeans. Phytopathology 68:1590–1596

Catford JG, Staehelin C, Lerat S, Piché Y, Vierheilig H (2003) Suppression of arbuscular mycorrhizal colonization and nodulation in split-root systems of alfalfa after pre-inoculation and treatment with Nod factors. J Exp Bot 54:1481–1487

Chaturvedi C, Singh R (1989) Response of chickpea (*Cicer arietinum* L.) to inoculation with *Rhizobium* and VA mycorrhiza. Proc Natl Acad Sci Ind B59:443–446

Cluett HC, Boucher DH (1983) Indirect mutualism in the legume- *Rhizobium* mycorrhizal fungus interaction. Oecologia 59:405–408

Constable JVH, Bassirirad H, Lussenhop J, Ayalsew Z (2001) Influence of elevated CO2 and mycorrhizae on nitrogen acquisition: contrasting responses in *Pinus taeda* and *Liquidambar styraciflua*. Tree Physiol 21:83–91

Daft MJ, El-Giahmi AA (1974) Effect of endogone mycorrhiza on plant growth. VII. Influence of infection on the growth and nodulation in French bean (*Phaseolus vulgaris*). New Phytol 73:1139–1147

Dermatsev V, Weingarten-Baror C, Resnick N, Gadkar V, Wininger S, Kolotilin I, Mayzlish-Gati E, Zilberstein A, Koltai H, Kapulnik Y (2010) Microarray analysis and functional tests suggest the involvement of expansins in the early stages of symbiosis of the arbuscular mycorrhizal fungus *Glomus intraradices* on tomato (*Solanum lycopersicum*). Mol Plant Pathol 11:121–135

Drevon JJ, Hartwig UA (1997) Phosphorus deficiency increases the argon- induced decline of nodule nitrogenase activity in soybean and alfalfa. Planta 201:463–469

Fredeen AL, Terry N (1988) Influence of vesicular-arbuscular mycorrhizal infection and soil phosphorous level on growth and carbon metabolism of soybean. Can J Bot 66:2311–2316

Frey B, Schuepp H (1992) Transfer of symbiotically fixed nitrogen from Berseem (*Trifolium alexandrinum* L.) to maize via vesicular-arbuscular mycorrhizal hyphae. New Phytol 122:447–454

Furlan V, Fortin JA (1977) Effects of light intensity on the formation of vesicular arbuscular endomycorrhizas on *Allium cepa* by *Gigaspora calospora*. New Phytol 79:335–340

Gallou A, Declerck S, Cranenbrouck S (2011) Transcriptional regulation of defence genes and involvement of the WRKY transcription factor in arbuscular mycorrhizal potato root colonization. Funct Integr Genomics 12:183–198

Garrido JM, Morcillo RJ, Rodríguez JA, Bote JA (2010) Variations in the mycorrhization characteristics in roots of wild-type and ABA-deficient tomato are accompanied by specific transcriptomic alterations. Mol Plant Microbe Interact 23:651–664

Gaude N, Bortfeld S, Duensing N, Lohse M, Krajinski F (2012) Arbuscule-containing and non-colonized cortical cells of mycorrhizal roots undergo extensive and specific reprogramming during arbuscular mycorrhizal development. Plant J 69:510–528

Gavito ME, Curtis PS, Mikkelson TN, Jakobsen I (2000) Atmospheric CO_2 and mycorrhiza effects on biomass allocation and nutrient uptake of nodulated pea (*Pisum sativum* L.) plants. J Exp Bot 51:1931–1938

Gomez SK, Javot H, Deewatthanawong P, Torres-Jerez I, Tang Y, Blancaflor EB, Udvardi MK, Harrison MJ (2009) *Medicago truncatula* and *Glomus intraradices* gene expression in cortical cells harboring arbuscules in the arbuscular mycorrhizal symbiosis. BMC Plant Biol 9:10

Goss MJ, de Varennes A (2002) Soil disturbance reduces the efficacy of mycorrhizal associations for early soybean growth and N_2 fixation. Soil Biol Biochem 34:1167–1173

Govindarajulu M, Pfeffer PE, Hairu J, Abubaker J, Douds DD, Allen JW, Bucking H, Lammers PJ, Shachar-Hill Y (2005) Nitrogen transfer in the arbuscular mycorrhizal symbiosis. Nature 435:819–823

Gutjahr C, Novero M, Welham T, Wang T, Bonfante P (2011) Root starch accumulation in response to arbuscular mycorrhizal colonization differs among *Lotus japonicus* starch mutants. Planta 234:639–646

Harris D, Paul EA (1987) Carbon requirements of vesicular-arbuscular mycorrhizae. In: Safir GR (ed) Ecophysiology of VA mycorrhizal plants. CRC, Boca Raton, FL, pp 93–105

Harris D, Pacovsky RS, Paul EA (1985) Carbon economy of soybean-*Rhizobium-Glomus* associations. New Phytol 101:427–440

Harrison MJ, Dewbre GR, Liu J (2002) A phosphate transporter from *Medicago truncatula* involved in the acquisition of phosphate released by arbuscular mycorrhizal fungi. Plant Cell 14:2413–2429

Helber N, Wippel K, Sauer N, Schaarschmidt S, Hause B, Requena N (2011) A versatile monosaccharide transporter that operates in the arbuscular mycorrhizal fungus *Glomus* sp. is crucial for the symbiotic relationship with plants. Plant Cell 23:3812–3823

Jakobsen I, Rosendahl L (1990) Carbon flow into soil and external hyphae from roots of mycorrhizal cucumber plants. New Phytol 115:77–83

Jia Y, Gray VM, Straker CJ (2004) The influence of Rhizobium and arbuscular mycorrhizal fungi on Nitrogen and Phosphorous accumulation by *Vicia faba*. Ann Bot Lond 94:251–258

Jones MD, Durall DM, Tinker PB (1991) Fluxes of carbon and phosphorous between symbionts in willow ectomycorrhizas and their changes with time. New Phytol 119:99–106

Kawai Y, Yamamoto Y (1986) Increase in the formation and nitrogen fixation of soybean nodules by vesicular-arbuscular mycorrhiza. Plant Cell Physiol 27:399–405

Kiers ET, Duhamel M, Beesetty Y, Mensah JA, Franken O, Verbruggen E, Fellbaum CR, Kowalchuk GA, Hart MM, Bago A, Palmer TM, West SA, Vandenkoornhuyse P, Jansa J, Bucking H (2011) Reciprocal rewards stabilize cooperation in the mycorrhizal symbiosis. Science 333:880–882

Koch KE, Johnson CR (1984) Photosynthate partitioning in slit root Citrus seedlings with mycorrhizal root systems. Plant Physiol 75:26–30

Koide R (1985) The nature of growth depressions in sunflower caused by vesicular arbuscular mycorrhizal infection. New Phytol 99:449–462

Koide R, Elliott G (1989) Cost, benefit and efficiency of the vesicular-arbuscular mycorrhizal symbiosis. Funct Ecol 3:252–255

Krajinski F, Hause B, Gianinazzi-Pearson V, Franken P (2002) Mtha1, a plasma membrane H + -ATPase gene from *Medicago truncatula*, shows arbuscule-specific induced expression in mycorrhizal tissue. Plant Biol 4:754–761

Kucey RMN, Paul EA (1981) Carbon flow in plant microbial associations. Science 213:473–474

Kucey RMN, Paul EA (1982) Carbon flow, photosynthesis and N_2 fixation in mycorrhizal and nodulated faba beans (*Vicia faba* L.). Soil Biol Biochem 14:407–412

Lodwig E, Poole P (2003) Metabolism of Rhizobium bacteroids. Crit Rev Plant Sci 22:37–78

Luis I, Lim G (1988) Differential response in growth and mycorrhizal colonisation of soybean to inoculation with two isolates of *Glomus clarum* in soils of different P availability. Plant Soil 112:37–43

Maekawa S, Sato T, Asada Y, Yasuda S, Yoshida M, Chiba Y, Yamaguchi J (2012) The Arabidopsis ubiquitin ligases ATL31 and ATL6 control the defense response as well as the carbon/nitrogen response. Plant Mol Biol 79:217–227

Maillet F, Poinsot V, Andre' O, Puech-Pages V, Haouy A, Guenier M, Cromer L, Giraudet D, Formey D, Niebel A, Martinez EA, Driguez H, Becard G, Denarie J (2011) Fungal lipochitoo-ligosaccharide symbiotic signals in arbuscular mycorrhiza. Nature 649:58–63

Marschner H, Dell B (1994) Nutrient uptake in mycorrhizal symbiosis. Plant Soil 59:89–102

Mortimer PE, Archer E, Valentine AJ (2005) Mycorrhizal C costs and nutritional benefits in developing grapevines. Mycorrhiza 15:159–165

Mortimer PE, Pérez-Fernández MA, Valentine AJ (2008) The role of arbuscular mycorrhizal colonization in the carbon and nutrient economy of the tripartite symbiosis with nodulated *Phaseolus vulgaris*. Soil Biol Biochem 40:1019–1027

Mortimer PE, Perez-Fernandez MA, Valentine AJ (2009) NH_4^+ nutrition affects the photosynthetic and respiratory C sinks in the dual symbiosis of a mycorrhizal legume. Soil Biol Biochem 41:2115–2121

Mortimer PE, Perez-Fernandez MA, Valentine AJ (2012) Arbuscular mycorrhiza maintains nodule function during external NH_4^+ supply in *Phaseolus vulgaris* (L.). Mycorrhiza 22:237–245

Motosugi H, Yamamoto Y, Naruo T, Kitabyashi H, Ishi T (2002) Comparison of the growth and leaf mineral concentrations between three grapevine rootstocks and their corresponding tetraploids inoculated with an arbuscular mycorrhizal fungus *Gigaspora margarita*. Vitis 41:21–25

Murphy PM (1986) Effect of light and atmospheric carbon dioxide concentration on nitrogen fixation by herbage legumes. Plant Soil 95:399–409

Nielson KL, Amram E, Lynch JP (2001) The effect of phosphorous availability on the carbon economy of contrasting common bean (*Phaseolus vulgaris* L.) genotypes. J Exp Bot 52:329–339

Nwoko H, Sanginga N (1999) Dependence of promiscuous soybean and herbaceous legumes on arbuscular mycorrhizal fungi and their response to bradyrhizobial inoculation in low P soils. Appl Soil Ecol 13:251–258

Olivera M, Tejera N, Iribarne C, Ocana A, Lluch C (2004) Growth, nitrogen fixation and ammonium assimilation in common bean (*Phaseolus vulgaris*): effect of phosphorous. Physiol Plantarum 121:498–505

Olsson PA, Johansen A (2000) Lipid and fatty acid composition of hyphae and spores of arbuscular mycorrhizal fungi at different growth stages. Mycol Res 104:429–434

Orcutt DM, Nilsen ET (2000) The physiology of plants under stress. Soil and biotic factors. Wiley, New York, NY

Othman WMW, Li TA, Tmannetje L, Wassink GY (1991) Low-level phosphorus supply affecting nodulation, N_2 fixation and growth of Cowpea (*Vigna unguiculata* L. Walp). Plant Soil 135:67–74

Pacovsky RS, Fuller G, Stafford AE, Paul EA (1986) Nutrient and growth interactions in soybeans colonized with *Glomus fasciculatum* and *Rhizobium japonicum*. Plant Soil 92:37–45

Parrent JL, Vilgalys R (2009) Expression of genes involved in symbiotic carbon and nitrogen transport in *Pinus taeda* mycorrhizal roots exposed to CO_2 enrichment and nitrogen fertilization. Mycorrhiza 19:469–479

Pearson JN, Jakobsen I (1993) The relative contribution of hyphae and roots to phosphorus uptake by arbuscular mycorrhizal plants, measured by dual labelling with 32P and 33P. New Phytol 124:489–494

Pearson JN, Schweiger P (1993) *Scutellospora calospora* (Nicol. & Gerd) associated with subterranean clover: dynamics of soluble carbohydrates. New Phytol 124:215–219

Peng S, Eissenstat DM, Graham JH, Williams K, Hodge NC (1993) Growth depression in mycorrhizal citrus at high-phosphorous supply. Plant Physiol 101:1063–1071

Provorov NA, Tikhonovich IA (2003) Genetic resources for improving nitrogen fixation in legume-rhizobia symbioses. Genet Resour Crop Evol 359:907–918

Rodriguez JA, Morcillo RL, Vierheilig H, Ocampo JA, Ludwig-Müller J, Garrido JM (2010) Mycorrhization of the notabilis and sitiens tomato mutants in relation to abscisic acid and ethylene contents. J Plant Physiol 167:606–613

Sa TM, Israel DW (1991) Energy status and function of phosphorous-deficient soybean nodules. Plant Physiol 97:928–935

Sanders FE, Tinker PB (1971) Mechanism of absorption of phosphate from soil by Endogone mycorrhizae. Nature 233:278–279

Schortemeyer J, Atkin OK, McFarlane N, Evans JR (1999) The impact of elevated atmospheric CO2 and nitrate supply on growth, biomass allocation, nitrogen partitioning and N2 fixation of *Acacia melanoxylon*. Aust J Plant Physiol 26:737–747

Selosse MA, Rousset F (2011) The plant-fungal marketplace. Science 333:828–829

Sieverding E (1989) Ecology of VAM fungi in tropical agrosystems. Agr Ecosyst Environ 29:369–390

Smith SE (1980) Mycorrhizas of autotrophic higher plants. Biol Rev 55:475–510

Smith SE (1982) Inflow of phosphate into mycorrhizal and non-mycorrhizal plants of *Trifolium subterraneum* at different levels of soil phosphate. New Phytol 90:293–303

Smith SE, Read DJ (1997) Mycorrhizal symbiosis, 2nd edn. Academic, London

Snellgrove RC, Splittstoesser WE, Stribley DP, Tinker PB (1982) The distribution of carbon and the demand of the fungal symbiont in leek plants with vesicular-arbuscular mycorrhizas. New Phytol 92:75–87

Sussanna JF, Hartwig UA (1996) The effect of elevated CO_2 on symbiotic nitrogen fixation: a link between the carbon and nitrogen cycles in grassland ecosystems. Plant Soil 187:321–332

Tejeda-Sartorius M, Martínez de la Vega O, Délano-Frier JP (2008) Jasmonic acid influences mycorrhizal colonization in tomato plants by modifying the expression of genes involved in carbohydrate partitioning. Physiol Plant 133:339–353

Thompson BD, Robson AD, Abbott LK (1990) Mycorrhizas formed by *Gigaspora calospora* and *Glomus fasciculatum* on subterranean clover in relation to soluble carbohydrate concentrations in roots. New Phytol 114:405–411

Thorneley RNF (1992) Nitrogen fixation-new light on nitrogenase. Nature 360:532–533

Toro M, Azcon R, Barea JM (1998) The use of isotopic dilution techniques to evaluate the interactive effects of Rhizobium genotype, mycorrhizal fungi, phosphate solubilizing rhizobacteria and rock phosphate on nitrogen and phosphorus acquisition by *Medicago sativa*. New Phytol 138:265–273

Toussaint JP, St-Arnaud M, Charest C (2004) Nitrogen transfer and assimilation between the arbuscular mycorrhizal fungus *Glomus intraradices* (Schenck & Smith) and Ri T-DNA roots of *Daucus carota* L. in an in vitro compartmented system. Can J Microbiol 50:251–260

Trépanier M, Bécard G, Moutoglis P, Willemot C, Gagné S, Avis TJ, Rioux JA (2005) Dependence of arbuscular-mycorrhizal fungi on their plant host for palmitic acid synthesis. Appl Environ Microbiol 71:5341–5347

Udvardi MK, Tabata S, Parniske M, Stougaard M (2005) *Lotus japonicus*: legume research in the fast lane. Trends Plant Sci 10:222–228

Vadez V, Beck DP, Lasso JH, Drevon J-J (1997) Utilization of the acetylene reduction assay to screen for tolerance of symbiotic N2 fixation to limiting P nutrition in common bean. Physiol Plantarum 99:227–232

Valentine AJ, Kleinert A (2007) Respiratory responses of arbuscular mycorrhizal roots to short-term alleviation of P deficiency. Mycorrhiza 17:137–143

Vance CP (2002) Root-bacteria interactions: symbiotic nitrogen fixation. In: Waisel Y, Eschel A, Kafkafi U (eds) Plant roots: the hidden half, 3rd edn. Dekker, New York, NY, pp 839–867

Vesjsadova H, Siblikova D, Gryndler M, Simon T, Miksik I (1993) Influence of inoculation with *Bradyrhizobium japonicum* and *Glomus claroideum* on seed yield of soybean under greenhouse and field conditions. J Plant Nutr 16:619–629

Vessey JK, Layzell DB (1987) Regulation of assimilate partitioning in soybean. Initial effects following change in nitrate supply. Plant Physiol 83:341–348

Williams K, Percival F, Merino J, Mooney HA (1987) Estimation of tissue construction cost from heat of combustion and organic nitrogen content. Plant Cell Environ 10:725–734

Yoneyama K, Xie X, Sekimoto H, Takeuchi Y, Ogasawara S, Akiyama K, Hayashi H, Yoneyama K (2008) Strigolactones, host recognition signals for root parasitic plants and arbuscular mycorrhizal fungi, from Fabaceae plants. New Phytol 179:484–494

Zsögön A, Lambais MR, Benedito VA, Figueira AVO, Peres LEP (2008) Reduced arbuscular mycorrhizal colonization in tomato ethylene mutants. Sci Agric 65:259–267

Chapter 13
Arbuscular Mycorrhizal Fungi and Uptake of Nutrients

M. Miransari

13.1 Introduction

There is some kind of soil fungi, arbuscular mycorrhizal (AM) fungi, developing mutual symbiotic with their non-specific host plant. For the initiation of the symbiotic process, the fungal spores must be able to germinate in the presence of the host plant resulting in the production of the extensive network of hyphae (Fig. 13.1). The two symbionts realize their presence by the production of some signal molecules, which can trigger the activation of the related symbiotic genes (Smith and Read 2008). Although the fungal spores are able to grow in the absence of the host plant, the presence of the host plant for the continuation of the fungal development process is necessary. In this symbiosis, the extensive hyphal network is able to significantly increase the uptake of nutrients and water by the host plant. Accordingly, this can very much help the host plant to grow under different conditions including stress (Miransari 2010a, 2011a).

The process of symbiosis between mycorrhizal fungi and the host plant is directed by the communications of signal molecules between the two symbionts and the expression of the related genes. The host plant produces signal molecules, which result in the germination of fungal spores, activation of the symbiotic genes, and the eventual production of the hyphal network in the presence of the host plant. The production of the fungal signals also results in the activation of the host-related genes with some cellular modification, such as the alteration of cellular microtubules. This would allow the fungi to grow inside the plant cells and produce the fungal organelles including arbuscules and vesicles. The arbuscules are

M. Miransari (✉)
Mehrabad Rudehen, Imam Ali Blvd., Mahtab Alley, #55, 3978147395 Tehran, Iran

Abtin Berkeh Limited Co., Imam Blvd., Shariati Blvd. # 107, 3973173831 Rudehen, Tehran, Iran
e-mail: Miransari1@gmail.com

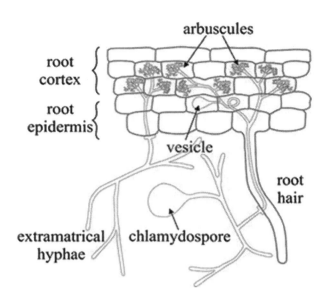

Fig. 13.1 Arbuscular mycorrhizal symbiosis (*source*: http://www.davidmoore.org.uk)

branched like structures and the interfaces for the exchange of nutrients between the fungi and the host plant and the vesicles are the storage organelles for the fungi, which can also be useful under stress (Bonfante and Anca 2009).

Mycorrhizal fungi are able to absorb different macro- and micronutrients; however, the likeliness of phosphorous (P) uptake is higher related to the other nutrients. This is due to the production of some enzymes such as phosphatase by the fungi, which can increase the solubility of insoluble P and hence its absorption by plant (Smith and Read 2008). The uptake of other nutrients such as nitrogen (N), potassium (K), calcium (Ca), magnesium (Mg), iron (Fe), manganese (Mn), zinc (Zn), and copper (Cu) can also be enhanced by mycorrhizal association (Marschner and Dell 1994).

The important point about the efficiency of uptake is the potential gradient of nutrient concentration between the soil and the plant. Under high concentration of nutrients, usually the symbiotic efficiency decreases, which is due to the presence of nutrient receptors in the plant cellular membrane, which are adversely affected. However, under low and medium nutrient concentration, mycorrhizal fungi are able to significantly increase the uptake of nutrients by plant (Smith et al. 2011).

Khanpour Ardestani et al. (2011) examined the effects of different K and Mg concentrations on the root colonization of mycorrhizal maize plants under greenhouse conditions. Using 40 pot cultures (500 ml) with three maize seedlings, they investigated the Mg concentrations of 4.8 (field Mg content), 7.2, and 9.6 meq/l and K concentrations of 0.61 (field K content), 0.92, and 1.23 meq/l on the root colonization. The pots were irrigated for 16 days (every 4 days) with 50 ml of distilled water. They found that at the low and medium level of nutrient concentration, the highest rate of root colonization was resulted by *Glomus* spp. The combined use of K and Mg had also a positive effect on root colonization. Vogel-Mikus et al. (2006) investigated the effects of mycorrhizal fungi on the nutrient and heavy metal uptake of the hyperaccumulator *Thlaspi praecox* Wulfen (Brassicaceae). Mycorrhizal fungi significantly increased and decreased the uptake of nutrients and heavy metals, respectively. They also indicated that high concentrations of nutrients may result in the lower formation of arbuscules (Vogel-Mikus et al. 2006).

With respect to the importance of soil microbes in providing different nutrients for plant fertilization, use of microbial inocula is now very usual in different parts of the world. Microbial inocula include a range of different microbes, which are able to establish symbiotic or nonsymbiotic association with their host plant and fix different nutrients for plant use. The inocula may include mycorrhizal fungi, rhizobium bacteria, and other plant growth-promoting rhizobacteria (PGPR) such as *Azospirillum* sp. and *Bacillus* sp. The important point regarding the use of microbial consortium is the compatibility between the microbes, so that at the time of use, they may not adversely influence their activities and hence plant growth. This means that before using any kind of inocula, they must be tested so that their economical and environmental benefits are proved (Miransari 2011a, b, c, d). Accordingly, some of the latest findings related to the effects of mycorrhizal fungi on the nutrient uptake of plant is analyzed and reviewed.

13.2 Mycorrhizal Fungi and Uptake of Nutrients

The symbiotic association between mycorrhizal fungi and their host plant can greatly contribute to the uptake of nutrients by the host plant. After the establishment of the extensive hyphal network, the fungal hyphae are able to grow into the finest soil pores, where even the finest root hairs are not able to grow and absorb water and nutrients. This can be especially advantageous under stress such as compaction. Plant, soil, and climate properties can affect the uptake of nutrients by the hyphal network. Mycorrhizal fungi are able to absorb almost all nutrients including macro- and micronutrients (Smith and Read 2008).

13.3 Macronutrients

The uptake of macronutrients including nitrogen (N), phosphorous (P), potassium (K), magnesium (Mg), calcium (Ca), and sulfur (S) can be affected by mycorrhizal association. Although P is the nutrient, which is affected most by mycorrhizal fungi, the uptake of other macronutrients may also be influenced by mycorrhizal symbiosis according to the following details (Fig. 13.2).

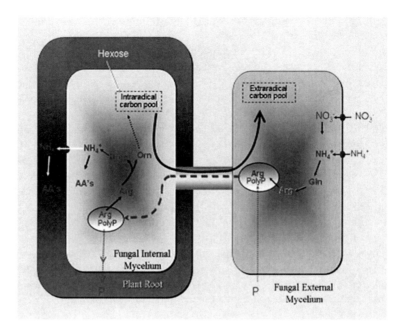

Fig. 13.2 Uptake of N and P by mycorrhizal association (*source*: http://shachar-hill.plantbiology.msu.edu)

13.3.1 Nitrogen (N)

N is a mobile nutrient with significant effects on plant growth. It is the main element in the structure of different plant proteins and chlorophyll molecules. Accordingly, under N-deficient conditions, plant growth is significantly affected. Usually, N is supplied by soil organic matter or the use of chemical fertilization. In addition, using microbial inocula including *Rhizobium* and other PGPRs such as *Azospirillum* sp. is also a suitable way of providing N to the host plant (Miransari 2011d).

Rhizobium is able to establish a symbiotic association with its specific host plant and fix the atmospheric N in the exchange for carbon and nutrients. In the process of symbiotic N fixation, signal molecules are exchanged between the two symbionts. For example, during the process of N symbiotic fixation by soybean (*Glycine max*) and *Bradyrhizobium japonicum*, plant roots exudate a signal molecule called genistein, which is able to activate the bacterial genes. In response, the bacteria produce lipochitooligosaccharides, which can influence the growth of root hairs by bulging or curving the roots, which eventually result in the formation of root nodules. The bacteria can enter the host plant roots by the formation of the infection tread and reside in the nodules and fix N (Long 2001; Miransari et al. 2006; Miransari and Smith 2007, 2008, 2009).

In addition to being advantageous to the host plant, symbiotic N fixation is also of great importance from economical and environmental perspectives. It is because also during the N symbiotic process, a lot of energy must be spent by the host plant, and much less chemical fertilization would be necessary, which are usually subjected to natural processes such as leaching. This is among the most important reasons, which indicate how biofertilization can be beneficial to the plant and environment (Zhang and Smith 1995).

Although there is some indication that mycorrhizal fungi are able to influence plant N uptake, it is yet a matter of debate how mycorrhizal association may influence N dynamic between the soil and plant interface (Miransari 2011b). According to some research work, mycorrhizal fungi are able to provide 50 % of plant N requirement (McFarland et al. 2010). The efficient utilization of inorganic labeled N by mycorrhizal fungi and its transfer to the soil by the fungi can also be another indicator of fungal N uptake (Frey and Schüepp 1992; Subramanian and Charest 1999). Accordingly, mycorrhizal association can increase plant access to mineral N as it was also indicated that the external hyphae of *Glomus intraradices* were able to enhance plant access to mineral N by absorbing NO_3^- and NH_4^+ (Johansen et al. 1993, 1994).

Nitrogen is a more mobile nutrient, relative to P, and hence its uptake by mycorrhizal plant may be of less importance, because N can be supplied to the host plant through mechanisms such as diffusion and mass flow. However, P is not a mobile nutrient in the soil and hence mycorrhizal association can effectively increase its uptake by the development of hyphal network and production of enzymes such as phosphatase (Liu et al. 2007; Atul-Nayyar et al. 2009). The increased water uptake by the fungal hyphae may enhance the uptake of different nutrients including N by the host plant. Nevertheless, different research work has indicated the role of mycorrhizal fungi in enhancing N uptake by the host plant (Tanaka and Yano 2005; Jackson et al. 2008).

The ability of mycorrhizal fungi to utilize mineral N from organic matter and amino acids has been indicated by different researchers (St. John et al. 1983; Hamel 2004). During the process of N mineralization by the soil microbes, mycorrhizal fungi were found on the plant tissues, inside the vascular bundle, using the mineral N produced by the other soil microbes (Aristizábal et al. 2004). Although the ability of mycorrhizal fungi to use organic matter as a source of mineral N has yet to be illustrated (Jin et al. 2005; Talbot et al. 2008), their influence on the activity of other soil microbes mineralizing organic matter may indirectly affect the process of N mineralization (Marschner et al. 2001; Hodge 2003a, b).

The ability of AM fungi *Glomus hoi* to increase the mineralizing process of soil organic matter to acquire N has been illustrated (Hodge et al. 2001). According to Atul-Nayyar et al. (2009), during the process of N mineralization in the presence of mycorrhizal fungi, the C/N ratio of soil organic matter decreased from 17.8 to 13. This illustrates a 228 % increase in the process of N mineralization by mycorrhizal fungi resulting in the higher rate of plant N uptake (Atul-Nayyar et al. 2009).

However, there may be situations, such as N limitation, where the uptake of N by the fungal association may be of great importance to the host plant, for example, in organic patches (Hodge and Fitter 2010). This can also be the case under drought stress, where mycorrhizal fungi are able to increase the uptake of nutrients and water by the host plant (Subramanian and Charest 1999; Hodge and Fitter 2010; Miransari 2011b). Interestingly, recently, Guether et al. (2009) indicated the way by which N is transferred (the activation of N transporters) in the mycorrhizal plant similar to the presence of P transporters in the plant. This can indicate the significance of mycorrhizal association for the uptake of both nutrients by the host plant.

Plant N demand may determine the rate of N mineralization by mycorrhizal fungi, as the high rate of N in the soil can decrease the rate of mycorrhizal hyphae growth into organic patches (Liu et al. 2000; Hodge 2003a, b). There are different factors affecting the rate of N mineralization in the presence of mycorrhizal fungi including (1) the higher the network volume of mycorrhizal fungal hyphae, the higher the rate of N mineralization; (2) the higher production of decomposing (hydrolytic) enzymes such as xyloglucanase, pectinase, and cellulose can increase the mineralization of organic N; and (3) the interaction effects between mycorrhizal fungi and the other soil microbes (Miransari 2011a) and the positive interaction may enhance the fungal ability to mineralize higher amounts of organic N. Mycorrhizal fungi may alter the activity of other microbes by affecting plant growth (signalling pathways), plant-induced resistance, and root exudates (Toljander et al. 2007; Lioussanne et al. 2008). In addition, fungal hyphae can be utilized by the soil microbes as an important source of organic carbon (Schimel and Weintraub 2003).

The effects of mycorrhizal fungi on the process of N uptake are important due to the following (1) the process of N cycling is influenced by the fungi; (2) whether the fungi can control the process of N leaching, especially in humid areas (Miransari and Mackenzie 2011a, b); (3) N uptake by mycorrhizal fungi may provide a part of N necessary for plant use, especially under conditions where N is limited such as arid and semiarid areas (Hodge and Fitter 2010); and (4) the use of fungi as biological fertilization can decrease chemical fertilization (Miransari 2010a, b, c, d).

Using the ^{15}N technique, Ames et al. (1983) indicated that mycorrhizal fungi *Glomus mosseae* increased the uptake and translocation of NH_4^+ by celery. High amounts of N fertilization can adversely affect N uptake by mycorrhizal plant due to its effects on the N carriers and transporters in the plant. Accordingly, the highest effect of mycorrhizal fungi on Plant N uptake is usually at the medium level of fertilization (reviewed by Miransari 2011b) indicating the regulating role of mycorrhizal fungi in plant N uptake from the soil.

The tripartite symbiosis between the host plant, mycorrhizal fungi, and rhizobium can affect N uptake by the host plant. AM fungi can enhance P uptake and to some extent N uptake by the host plant, while rhizobium is able to increase N uptake. There are also interactions between the fungi, bacteria, and the host plant influencing the behavior of the symbionts and the uptake of nutrients. The positive interactions between the symbionts, which can be by producing different biochemicals or the bacterial attachment to the fungal spore, can increase the uptake of different nutrients such as N and P by the host plant (Mortimer et al. 2008).

There are different parameters affecting the tripartite association. For example, according to Wang et al. (2011), plant genotype including root architecture as well as N and P level in the soil may affect the efficiency of the tripartite association including the uptake of nutrients by the host plant. Deep rooting at low level of P (field P content) resulted in the highest mycorrhizal symbiotic efficiency. However, the efficiency of N symbiosis was optimum at the highest rate of P (triple super phosphate, 160 kg/ha). Such results can be used for proper N and P fertilization including biological fertilization. Additionally, the interactions between N and P may also affect root morphology and the processes of mycorrhization and nodulation (Kuang et al. 2005; Miransari et al. 2009a, b).

Hawkins et al. (2000) examined the effects of N uptake by *G. mosseae* from organic and inorganic sources. The hyphae of mycorrhizal wheat were able to absorb organic N (^{15}N-Gly) after 48 h from a sterilized semi-hydroponic culture. The rate of N uptake by mycorrhizal wheat was at 0.2 and 6 % in the form of N-glycine at low and high level of fertilization, respectively. The higher organic N uptake at the high N level was attributed to the more sufficient growth of hyphal network. The fungi were also able to absorb NH_4^+ and NO_3^-. Carrot mycorrhizal roots transformed with Ri-T-DNA absorbed little amounts of organic N from N-glycine and N-glutamate sources. However, according to their results, the uptake of organic N was not significantly different from the control. These results are similar to the results of other researchers (Hodge et al. 2001; Hodge and Fitter 2010).

According to Tian et al. (2010) in addition to the absorption of organic and inorganic N by mycorrhizal fungi, they are able to synthesize arginine in the extra radical hyphae and transfer it to the intra-radical hyphae where N is released and absorbed by the host plant. Accordingly, 11 fungal genes related to the N absorbing pathways were recognized and six of them were sequenced.

13.3.2 Phosphorous (P)

Phosphorous is also another important macronutrient necessary for plant growth. P is required for energy pathways and production in plant as well as for the structure of proteins and production of cellular membranes. However, due to its chemical properties, it is not mobile in the soil and is subjected to precipitation (Schachtman et al. 1998). Accordingly, proper methods of fertilization must be used to provide P for plant growth. In addition to the use of chemical fertilization, use of organic products and biological fertilization is also another useful method to provide P for plant growth. Biological fertilization, which is the use of soil microbes including mycorrhizal fungi and PGPR to supply necessary nutrients for plant growth, is economically and environmentally recommendable. Among the other nutrients, P is the one which is more likely to be supplied by mycorrhizal fungi to its host plant (Smith and Read 2008; Miransari 2011a, d).

The concentration of P, which is usually absorbed in the form of HPO_4^{2-} and $H_2PO_4^-$ by plant roots, is much higher (1,000-fold) in plant cells than the soil; hence plant must use some strategies to be able to absorb P from the soil. Among such strategies are the utilization of energy, the activity of protein transporters from the Pht1 family, and the symbiotic association with mycorrhizal fungi (Marschner 1995; Smith et al. 2011).

There are two different pathways enabling the plant to absorb P from the soil: the direct uptake by plant roots and the indirect uptake by mycorrhizal fungi. Two different P transporters are activated during the uptake of P including the ones which are located in the epidermis and root hairs (direct P uptake) and the fungal hyphal transporters, with a few centimeters distance from the plant roots (indirect uptake). The P absorbed by the fungal hyphae is translocated to the arbuscules and hyphal coils (Smith et al. 2011).

The important point about the uptake of P by mycorrhizal association is the amount and availability of P in the soil. If there are very low amounts of P in the soil, the growth of plant and hence its symbiotic association with mycorrhizal fungi will be adversely affected (Treseder and Allen 2002). They accordingly indicated that the initial amounts of N and P in the soil determine mycorrhizal diversity and abundance. Under high P availability, mycorrhizal fungi may not be able to efficiently colonize the host plant roots (due to decreasing arbuscule development), because under such conditions, the host plant may not be willing to spend energy for the development of symbiotic association. However, under low or medium level of P, the fungi are able to colonize the host plant roots efficiently and significantly and increase P uptake by the host plant (Smith and Read 2008).

Mendoza and Pagani (1997) examined the effects of P fertilization (0–160 µg P g^1 soil) on the growth and morphology of indigenous mycorrhizal fungi colonizing *Lotus tenuis*. *Glomus* spp. was the most dominant fungi colonizing the plant roots. Phosphorous fertilization affected the fungal response, colonization rate, as well as their morphological properties in plant roots. With increasing P level, the infected part of the root length and the corresponding number of arbuscules decreased. Accordingly, the colonizing potential of the fungi and root growth increased with increasing the rate of P fertilization.

The extensive hyphal network is able to grow in even the finest soil micropores (much smaller than the thickness of root hairs) and absorb water and nutrients. This is important for the uptake of P and also for the growth of mycorrhizal plants under stress (Miransari et al. 2009a, b). Hence, mycorrhizal fungi can be advantageous to the host plant under normal and stress conditions. Accordingly, the increased water and nutrient uptake by the fungi under stress is among the most important reasons for the survival of plant under such conditions (Miransari 2010a).

13.3.2.1 Molecular Pathways Affecting P Uptake by Mycorrhizal Fungi

There are some fungal genes such as the Pit ones, which are able to regulate the process of P uptake by the fungi and hence the host plant (Javot et al. 2007). Although it is not yet known how such kind of genes may be activated, it has been indicated that lysophosphatidylcholine is able to activate such genes in tomato and potato (*Solanum tuberosum*) (Drissner et al. 2007). The role of plant hormones has also been elucidated in the process of P uptake due to their effect on the process of root growth and sugar signalling (Rouached et al. 2010). Plant photosynthates are able to activate miR399 in the roots of common bean (*Phaseolus vulgaris*) as a mycorrhizal host plant at the onset of P deficiency (Liu et al. 2010). Accordingly, mycorrhizal symbiosis and the signalling pathways related to P deficiency are partly similar and are also controlled by hormonal and sugar signalling pathways (Smith et al. 2011).

There are different transcription factors activated in non-mycorrhizal and mycorrhizal plants when the plant is subjected to P deficiency (Smith et al. 2011). In non-mycorrhizal plants the transcription factor PHR1 is bound to the P1BS element in the promoters of P-deficiency-related genes and enhances their expression as well as the expression of miR399s. It is likely that miR399s is produced at larger amounts in plant shoots, especially in mycorrhizal plants, indicating the transfer of MYC signalling molecules from the plant roots to shoots. Such kind of transfer is under the influence of PHP1. Under P-deficient conditions high miR399s levels decrease the activity of enzyme activated by PHO2 and hence the related responses including the enhanced expression of PiTs (phosphate transporter genes). Eventually, PHO2 may control plant responses under P-deficient conditions and hence decreases PiTs expression (Branscheid et al. 2010).

13.3.3 Potassium, Magnesium, Calcium, and Sulfur

Mycorrhizal fungi are able to increase the uptake of different nutrients under different conditions including stress. The following may indicate how mycorrhizal association is able to increase plant nutrient uptake. In association with their host plant, mycorrhizal fungi develop an extensive network of hyphae, which is able to increase plant nutrient uptake significantly. This is due to the volume of mycorrhizal hyphae and their very fine thickness, which allows the fungal hyphae to grow into the even finest soil pores producing different enzymes such as phosphatases. In addition, the enhanced uptake of water by mycorrhizal network can substantially increase the rate of nutrient uptake by the host plant. The production of enzymes by the fungal hyphae can increase the solubility of nutrients and hence their subsequent uptake by the host plant. Nutrient uptake by the host plant is also increased indirectly due to enhanced plant growth with a larger root medium, which significantly increases the uptake of nutrients (Smith and Read 2008).

Different parameters such as soil, climate, and plant species may affect the uptake of nutrients by non-mycorrhizal and mycorrhizal plants. Accordingly, under optimum conditions and at low or medium level of fertility, the uptake of nutrients by the mycorrhizal plant may be at the highest. Some plant species are able to develop more efficient symbiotic association with mycorrhizal fungi. However, the production of some root exudates avoids some plant species from developing symbiotic association with mycorrhizal fungi (Smith and Read 2008; Miransari 2011c). Although mycorrhizal fungi are able to increase plant nutrient uptake under stress, high levels of stress may adversely influence the uptake of nutrients by the fungi (Miransari et al. 2007, 2008; Miransari 2010a).

Cimen et al. (2010) investigated the effects of mycorrhizal fungi *G. intraradices* on the uptake of different nutrients including K, Mg, and Ca by tomato (*Lycopersicon esculentum* L.) in a split-plot experiment under field conditions. The fungi increased the uptake of K, Mg, and Ca by tomato. Miransari et al. (2009a, b) found similar results by investigating the effects of different species of mycorrhizal fungi including *Glomus etunicatum* and *G. mosseae* on the nutrient uptake of corn and wheat in a compacted soil. They indicated that mycorrhizal fungi were able to alleviate the unfavorite effects of compaction on corn and wheat growth by enhancing the uptake of different nutrients such as K, Mg, and Ca.

The research work related to the use of mycorrhizal fungi under stress has also shown that the fungi are able to alleviate the unfavorable effects of stress under other stresses such as salinity and drought reviewed by Augé (2001). The fungi are able to alleviate the stress by adjusting the uptake and hence the ratio of different nutrients in plant including K, Mg, Ca, Na, and Cl (Boomsma and Vyn 2008; Daei et al. 2009).

Sulfur is absorbed in the form of sulfate by the activity of plant sulfate transporters. Although there is not much findings regarding the enhancing effects of mycorrhizal association on sulfur uptake by the host plant, a few research work has indicated that the fungi may positively influence plant sulfur uptake. For example, Casieri et al. (2012) analyzed the expression of genes, which can affect the activation of *Medicago* sulfate transporters (MtSULTRs), by *G. intraradices* at different sulfur levels. Mycorrhizal fungi significantly influenced plant growth and sulfur uptake. Eight MtSULTRs were recognized, some of which were expressed in plant leaf and root at different sulfur concentrations indicating the role of mycorrhizal fungi on the uptake of sulfur by the host plant. The effects of mycorrhizal colonization on the uptake of sulfur by the host plant can be through the higher production of root exudates, the increased activity of other soil microbes such as *Thiobacillus*, the extensive hyphal network, and the production of different enzymes, which may acidify the rhizosphere and hence increase sulfur availability to the host plant.

13.4 Micronutrients

Among the most important micronutrients are Zn, Fe, and Cu. The chemical properties of micronutrients and hence their mobility indicate their availability to the plant. There are different parameters such as the activity of soil microbes, which can significantly influence nutrient availability in the soil. Such activities include (1) their interaction by other microbes, (2) production of plant hormones, (3) production of different enzymes such as ACC deaminase, and (4) their effects on the roots and hence the production of exudates (Jalili et al. 2009; Miransari 2011a). Different research work has indicated that mycorrhizal fungi are able to increase the uptake of different micronutrients in association with their host plant. This can be related to the mobilization of micronutrients by the fungi (Liu et al. 2000). This can be due to (1) the production of different enzymes by the fungi, (2) the positive interactions of the fungi with the other soil microbes, (3) the effects of the fungi on the plant rhizosphere, and (4) the effects of the fungi on the morphology (root growth) and physiology of the host plant affecting the root exudates.

Miransari et al. (2009a, b) evaluated the effects of soil compaction on the uptake of nutrients including macro- and micronutrients by mycorrhizal corn and mycorrhizal wheat. Among the most important reasons to alleviate the stress by the fungi was enhanced nutrient uptake of different nutrients including N, P, K, Fe, Zn, Mn, and Cu. Interestingly, mycorrhizal corn plants were more sensitive to the stress relative to the mycorrhizal wheat plants, which may be due to different root architecture.

Zaefarian et al. (2011) investigated the effects of different fungal species including *G. mosseae*, *G. etunicatum*, and *G. intraradices* (as single treatments) and the combined treatment of *G. mosseae*, *Gigaspora hartiga*, and *Glomus fasciculatum* on the uptake of N, P, K, Fe, Zn, and Cu. The fungal species were able to increase the uptake of all nutrients (with the exception of Fe) by alfalfa (*Medicago sativa*), and *G. mosseae* was the most efficient one.

There is some kind of interaction between P and Zn, which are significantly absorbed by mycorrhizal association. High amounts of P decrease the uptake of Zn and vice versa. Hence, under high P fertilization, the uptake of Zn decreases by plant. Watts-Williams and Cavagnaro (2012) examined the combined effects of different P and Zn levels on the activity and formation of mycorrhizal fungi. They used a tomato mutant (non-mycorrhizal) and a wild-type plant (mycorrhizal). Five concentrations of Zn ranging from deficient to toxic and two P levels were tested in the experiment. The addition of P and Zn significantly affected plant growth and the formation of mycorrhizal fungi. The highest mycorrhizal benefits were resulted at low P and Zn levels. The effects of Zn on the activity and formation of mycorrhizal fungi were significantly dependent on the amounts of P in the soil indicating their interactive effects. Hence, to exactly indicate the interactive effects between P and Zn in plants by mycorrhizal symbiosis, different concentrations of such elements from deficient to toxic must be tested.

13.5 Mycorrhizal Nutrient Uptake and Alleviation of Soil Stress

Under drought stress the alteration of plant enzymatic activity can help the plant to tolerate the stress (Sajedi et al. 2010, 2011). Mycorrhizal association can enhance such kind of ability under the stress conditions. Interestingly, the enhanced activity of enzymes such as nitrate reductase, glutamate synthetase, and glutamate synthase indicates that the absorbed N by mycorrhizal association can be directly incorporated into the organic structures. The production of these enzymes is also increased under drought stress (Cliquet and Stewart 1993; Subramanian and Charest 1999). Hence, by increasing the production of such enzymes and the uptake of N, mycorrhizal association can improve plant growth under stress. This is because mycorrhizal fungi are able to affect plant physiology in a way so that the host plant can handle the stress (Miransari et al. 2008).

Different researchers have investigated the effects of mycorrhizal fungi on the alleviation of different stresses (Miransari 2010a). For example Daei et al. (2009) evaluated the effects of different mycorrhizal species on the growth and nutrient uptake of different wheat genotypes under field saline conditions. The soil and irrigation water salinity were equal to 7.41 and 13.87 dS/m, respectively. The mycorrhizal species including *G. mosseae*, *G. etunicatum*, and *G. intraradices* and different wheat genotypes were tested in the experiment. Mycorrhizal symbiosis alleviated the adverse effects of salinity through significantly increasing plant growth and nutrient uptake including P, K, and Zn and by decreasing the absorption of Na^+ and Cl^-. The higher uptake of K by mycorrhizal symbiosis under salinity is an important parameter alleviating the unfavorable effects of salinity through adjusting the uptake of Na^+ and Cl^- by the host plant. Researchers also indicated that mycorrhizal association increased plant growth under salinity by enhancing the uptake of P, K, Zn, Cu, and Fe (Al-Karaki et al. 2001; Al-Karaki 2006).

Mardukhi et al. (2011) investigated the *G. intraradices* effects of mycorrhizal association on the nutrient uptake of wheat genotypes in a saline soil under greenhouse and field conditions. The single and combined use of *G. mosseae* and *G. etunicatum* were tried. Under greenhouse conditions the single and coinoculation of mycorrhizal fungi significantly increased the uptake of different nutrients including N, K, P, Ca, Mg, Mn, Cu, Fe, and Zn by the host plant. However, under field conditions this was only the case for N, Ca, and Mn. Interestingly, and similar to the results of Evelin et al. (2012) that evaluated the effects of *G. intraradices* on the nutrient uptake of fenugreek (*Trigonella foenum-graecum*) plants under different salinity levels (0, 50, 100, and 200 mM NaCl), the fungal treatment was also able to alleviate the stress by inhibiting the extra uptake of Na^+ rather than Cl^-.

In addition to the analyses of plant tissues for the uptake of nutrients, leaf senescence (permeability of membrane and concentration of chlorophyll) and lipid peroxidation were also determined. Relative to control treatment, mycorrhizal inoculation increased plant growth under salinity by decreasing leaf senescence and lipid peroxidation. The uptake of NPK and root nodulation was adversely affected by salinity stress. However, the fungal treatment was able to alleviate the stress. As previously mentioned, decreased Na^+ uptake by the host plant was among the alleviating effects of fungi under the stress. The fungal treatment was also able to enhance plant growth under the stress by imparting the allocation of Na^+ to plant shoot and hence decreasing the shoot:root ratio of Na^+ (Mardukhi et al. 2011).

In addition higher concentrations of K^+ and Ca^{2+} are resulted from mycorrhizal inoculation under salinity stress. The ratios of $K^+:Na^+$, $Ca^{2+}:Na^+$, and $Ca^{2+}:Mg^{2+}$ were also modified by the fungal treatment under salinity. Micronutrient concentration (Mn^{2+}, Zn^{2+}, Cu, and Fe) was increased in plant tissues by the fungi relative to the control treatment. Accordingly, mycorrhizal fungi alleviated the stress of salinity by increasing the integrity of cellular membrane (deceased electrolyte leakage) and increasing the uptake of nutrients under salinity (Evelin et al. 2012). Such results indicate the importance of mycorrhizal association to alleviate the stress by increasing the uptake of nutrients such as K and Mg. These nutrients are able to adjust plant water behavior by modifying cell water potential and nutrient entrance into the cell through the cellular membrane (Marschner 1995; Miransari 2012a, b).

There are other types of soil stresses such as compaction, which may adversely affect the uptake of nutrients by plant. Under such kind of stress due to the decreased root growth and oxygen amount in the soil, the absorption of nutrients by the host plant decreases. For example, as a result of denitrification in the soil, N is emitted from the soil. The cluster (pancake) growth of roots in a compacted soil also decreases the uptake of water and nutrients (Miransari et al. 2007, 2008). It has been indicated that mycorrhizal fungi are able to alleviate such kind of stress by increasing the uptake of water and nutrients. In a compacted soil, the ratio of fine pores increases relative to the large pores. This can decrease the growth of non-mycorrhizal plant related to the mycorrhizal plant (due to the growth of the extensive hyphal network) under such conditions (Miransari et al. 2009a, b). The role of plant hormones under stress is also of great significance. For example, under nutrient stress, which can also be resulted in a compacted soil (like P deficiency), some root morphological alterations such as the higher production of dense root hairs and hence higher root surface are induced by plant hormones including auxin, ethylene, and cytokinin (Wittenmayer and Merbach 2005). According to the abovementioned details, the important role of mycorrhizal fungi on the alleviation of stress by increasing the uptake of water and nutrients is illustrated.

13.6 Conclusion

Mycorrhizal fungi are able to absorb almost all nutrients at different rate. However, the fungi are able to absorb P with the highest efficiency, which is due to production of enzymes such as phosphatase. Although it has been indicated that the fungi are able to absorb mineral and organic N from the soil, it is yet a matter of debate how the fungi may affect N uptake by its host plant. This may be due to the chemical differences that exist among such elements. The fungi are also able to absorb K, Mg, Ca, and S. Nutrient uptake by the fungi is also important under stress conditions, where the fungi can significantly increase plant nutrient uptake. For example, under salinity stress the increased uptake of nutrients such as K, Mg, and Ca by the fungi can help the plant to tolerate the stress. The uptake of micronutrients is also affected by the fungi. Although the fungi are usually able to increase micronutrient uptake, there may be situations where mycorrhizal association decreases the uptake of micronutrients. Elucidating the molecular pathways related to the uptake of nutrients in the fungi and the host plant may result in the development of fungal and plant species, which are more effective in the absorption of nutrients under different conditions including stress. The role of mycorrhizal fungi in the uptake and recycling of nutrients is of ecological and environmental significance.

References

Al-Karaki GN (2006) Nursery inoculation of tomato with arbuscular mycorrhizal fungi and subsequent performance under irrigation with saline water. Sci Hortic 109:1–7

Al-Karaki GN, Hammad R, Rusan M (2001) Response of two tomato cultivars differing in salt tolerance to inoculation with mycorrhizal fungi under salt stress. Mycorrhiza 11:43–47

Ames RN, Reid CPP, Porter KL, Cambardella C (1983) Hyphal uptake and transport of nitrogen from two N-labelled sources by *Glomus mosseae*, a vesicular-arbuscular mycorrhizal fungus. New Phytol 95:381–396

Aristizábal C, Rivera EL, Janos DP (2004) Arbuscular mycorrhizal fungi colonize decomposing leaves of *Myrica parvifolia, M. pubescens* and *Paepalanthus* sp. Mycorrhiza 14:221–228

Atul-Nayyar A, Hamel C, Hanson K, Germida J (2009) The arbuscular mycorrhizal symbiosis links N mineralization to plant demand. Mycorrhiza 19:239–246

Augé RM (2001) Water relations, drought and vesicular–arbuscular mycorrhizal symbiosis. Mycorrhiza 11:3–42

Bonfante P, Anca LA (2009) Plants, mycorrhizal fungi, and bacteria: a network of interactions. Annu Rev Microbiol 63:363–383

Boomsma C, Vyn T (2008) Maize drought tolerance: potential improvements through arbuscular mycorrhizal symbiosis? Field Crops Res 108:14–31

Branscheid A, Sieh D, Pant BD, May P, Devers EA, Elkrog A, Schauser L, Scheible WR, Krajinski F (2010) Expression pattern suggests a role of MiR399 in the regulation of the cellular response to local Pi increase during arbuscular mycorrhizal symbiosis. Mol Plant Microbe Interact 23:915–926

Casieri L, Gallardo K, Wipf D (2012) Transcriptional response of *Medicago truncatula* sulphate transporters to arbuscular mycorrhizal symbiosis with and without sulphur stress. Planta 235:1431–1447

Cimen I, Pirinc V, Doran I, Turgay B (2010) Effect of soil solarization and arbuscular mycorrhizal fungus (*Glomus intraradices*) on yield and blossom-end rot of tomato. Int J Agric Biol 12:551–555

Cliquet JB, Stewart GR (1993) Ammonia assimilation in maize infected with the VAM fungus *Glomus fasciculatum*. Plant Physiol 101:65–871

Daei G, Ardakani M, Rejali F, Teimuri S, Miransari M (2009) Alleviation of salinity stress on wheat yield, yield components, and nutrient uptake using arbuscular mycorrhizal fungi under field conditions. J Plant Physiol 166:617–625

Drissner D, Kunze G, Callewaert N, Gehrig P, Tamasloukht M, Boller T, Felix G, Amrhein N, Bucher M (2007) Lyso-phosphatidylcholine is a signal in the arbuscular mycorrhizal symbiosis. Science 318:265–268

Evelin H, Giri B, Kapoor R (2012) Contribution of *Glomus intraradices* inoculation to nutrient and mitigation of ionic imbalance in NaCl-stressed *Trigonella foenum-graecum*. Mycorrhiza 22:203–217

Frey B, Schüepp H (1992) Transfer of symbiotically Wxed nitrogen from berseem (*Trifolium alexandrium* L.) to maize via vesicular arbuscular mycorrhizal hyphae. New Phytol 122:447–454

Guether M, Neuhauser B, Balestrini R, Dynowski M, Ludewig U, Bonfante P (2009) A mycorrhizal-specific ammonium transporter from *Lotus japonicus* acquires nitrogen released by arbuscular mycorrhizal fungi. Plant Physiol 150:73–83

Hamel C (2004) Impact of arbuscular mycorrhizal fungi on N and P cycling in the root zone. Can J Soil Sci 84:383–395

Hawkins HJ, Johansen A, George E (2000) Uptake and transport of organic and inorganic nitrogen by arbuscular mycorrhizal fungi. Plant Soil 226:275–285

Hodge A (2003a) N capture by *Plantagolanceolata* and *Brassicanapus* from organic material: the influence of spatial dispersion, plant competition and an arbuscular mycorrhizal fungus. J Exp Bot 57:401–411

Hodge A (2003b) Plant nitrogen capture from organic matter as affected by spatial dispersion, interspecific competition and mycorrhizal colonization. New Phytol 157:303–314

Hodge A, Fitter AH (2010) Substantial nitrogen acquisition by arbuscular mycorrhizal fungi from organic material has implications for N cycling. Proc Natl Acad Sci USA 107:13754–13759

Hodge A, Campbell CD, Fitter AH (2001) An arbuscular mycorrhizal fungus accelerates decomposition and acquires nitrogen directly from organic material. Nature 413:297–299

Jackson LE, Burger M, Cavagnaro TR (2008) Nitrogen transformations and ecosystem services. Annu Rev Plant Biol 59:341–363

Jalili F, Khavazi K, Pazira E, Nejati A, Asadi Rahmani H, Rasuli Sadaghiani H, Miransari M (2009) Isolation and characterization of ACC deaminase producing fluorescent pseudomonads, to alleviate salinity stress on canola (*Brassica napus* L.) growth. J Plant Physiol 166:667–674

Javot H, Pumplin N, Harrison MJ (2007) Phosphate in the arbuscular mycorrhizal symbiosis: transport properties and regulatory roles. Plant Cell Environ 30:310–322

Jin H, PfeVer PE, Douds DD, Piotrowski E, Lammers PJ, Shachar-Hill Y (2005) The uptake, metabolism, transport and transfer of nitrogen in an arbuscular mycorrhizal symbiosis. New Phytol 168:687–696

Johansen A, Jakobsen I, Jensen ES (1993) Hyphal transport by a vesicular-arbuscular mycorrhizal fungus of N applied to the soil as ammonium or nitrate. Biol Fertil Soils 16:66–70

Johansen A, Jakobsen I, Jensen ES (1994) Hyphal N transport by a vesicular-arbuscular mycorrhizal fungus associated with cucumber grown at three nitrogen levels. Plant Soil 160:1–9

Khanpour Ardestani N, Zare-Maivan H, Ghanati F (2011) Effect of different concentrations of potassium and magnesium on mycorrhizal colonization of maize in pot culture. Afr J Biotechnol 10:16548–16550

Kuang R, Liao H, Yan X, Dong Y (2005) Phosphorus and nitrogen interactions in weld-grown soybean as related to genetic attributes of root morphological and nodular traits. J Integr Plant Biol 47:549–559

Lioussanne L, Beauregard MS, Hamel C, Jolicoeur M, St-Arnaud M (2008) Interactions between arbuscular mycorrhiza and soil microorganisms. In: Khasa D, Piché Y, Coughlan A (eds) Advances in mycorrhizal science and technology. NRC, Ottawa

Liu A, Hamel C, Hamilton R, Smith D (2000) Mycorrhizae formation and nutrient uptake of new corn (*Zea mays* L.) hybrids with extreme canopy and leaf architecture as influenced by soil N and P levels. Plant Soil 221:157–166

Liu A, Plenchette C, Hamel C (2007) Soil nutrient and water providers: how arbuscular mycorrhizal mycelia support plant performance in a resource-limited world. In: Hamel C, Plenchette C (eds) Mycorrhizae in crop production. Haworth, Binghamton, NY, pp 38–66

Liu JQ, Allan DL, Vance CP (2010) Systemic signaling and local sensing of phosphate in common bean: cross-talk between photosynthate and microRNA399. Mol Plant 3:428–437

Long S (2001) Gene and signals in the rhizobium-legume symbiosis. Plant Physiol 125:69–72

Mardukhi B, Rejali F, Daei G, Ardakani MR, Malakouti MJ, Miransari M (2011) Arbuscular mycorrhizas enhance nutrient uptake in different wheat genotypes at high salinity levels under field and greenhouse conditions. C R Biol 334:564–571

Marschner H (1995) Mineral nutrition of higher plants, 2nd edn. Academic, London

Marschner H, Dell B (1994) Nutrient uptake in mycorrhizal symbiosis. Plant Soil 159:89–102

Marschner P, Crowley D, Lieberei R (2001) Arbuscular mycorrhizal infection changes the bacterial 16 S rDNA community composition in the rhizosphere of maize. Mycorrhiza 11:297–302

McFarland JW, Ruess RW, Kielland K, Pregitzer K, Hendrick R, Allen M (2010) Cross-ecosystem comparisons of in situ plant uptake of amino acid-N and NH_4. Ecosystems 13:177–193

Mendoza RE, Pagani EA (1997) Influence of phosphorus nutrition on mycorrhizal growth response and morphology of mycorrhizae in *Lotus tenuis*. J Plant Nutr 20:625–639

Miransari M (2010a) Contribution of arbuscular mycorrhizal symbiosis to plant growth under different types of soil stresses. Review article. Plant Biol 12:563–569

Miransari M (2010b) Arbuscular mycorrhiza and soil microbes. In: Thangadurai D, Busso CA, Hijri M (eds) Mycorrhizal biotechnology. Science Publishers, New York, NY, p 226

Miransari M (2010c) Biological fertilization. In: Méndez-Vilas A (ed) Current research, technology and education topics in applied microbiology and microbial biotechnology. Formatex Microbiology book series—2010 edition. Spain

Miransari M (2010d) Mycorrhizal fungi and ecosystem efficiency. In: Fulton SM (ed) Mycorrhizal fungi: soil, agriculture and environmental implications. Nova Publishers, Hauppauge, NY. ISBN 978-1-61122-659-1

Miransari M (2011a) Interactions between arbuscular mycorrhizal fungi and soil bacteria. Review article. Appl Microbiol Biotechnol 89:917–930

Miransari M (2011b) Arbuscular mycorrhizal fungi and nitrogen uptake. Review article. Arch Microbiol 193:77–81

Miransari M (2011c) Hyperaccumulators, arbuscular mycorrhizal fungi and stress of heavy metals. Biotechnol Adv 29:645–653

Miransari M (2011d) Soil microbes and plant fertilization. Review article. Appl Microbiol Biotechnol 92:875–885

Miransari M (2012a) Soil microbes and environmental health. Nova Publishers, Hauppauge, NY. ISBN 978-1-61209-647-6

Miransari M (2012b) Soil nutrients. Nova Publishers, Hauppauge, NY. ISBN 978-1-61324-785-3

Miransari M, Mackenzie AF (2011a) Development of a soil N-test for fertilizer requirements for corn (*Zea mays* L.) production in Quebec. Commun Soil Sci Plant Anal 42:50–65

Miransari M, Mackenzie AF (2011b) Development of a soil N test for fertilizer requirements for wheat. J Plant Nutr 34:762–777

Miransari M, Smith DL (2007) Overcoming the stressful effects of salinity and acidity on soybean [*Glycinemax* (L.) Merr.] nodulation and yields using signal molecule genistein under field conditions. J Plant Nutr 30:1967–1992

Miransari M, Smith DL (2008) Using signal molecule genistein to alleviate the stress of suboptimal root zone temperature on soybean-*Bradyrhizobium* symbiosis under different soil textures. J Plant Interact 3:287–295

Miransari M, Smith DL (2009) Alleviating salt stress on soybean (*Glycine max* (L.) Merr.) – *Bradyrhizobium japonicum* symbiosis, using signal molecule genistein. Eur J Soil Biol 45:146–152

Miransari M, Balakrishnan P, Smith DL, Mackenzie AF, Bahrami HA, Malakouti MJ, Rejali F (2006) Overcoming the stressful effect of low pH on soybean root hair curling using lipochitooligosaccharides. Commun Soil Sci Plant Anal 37:1103–1110

Miransari M, Bahrami HA, Rejali F, Malakouti MJ, Torabi H (2007) Using arbuscular mycorrhiza to reduce the stressful effects of soil compaction on corn (*Zea mays* L.) growth. Soil Biol Biochem 39:2014–2026

Miransari M, Bahrami HA, Rejali F, Malakouti MJ (2008) Using arbuscular mycorrhiza to reduce the stressful effects of soil compaction on wheat (*Triticum aestivum* L.) growth. Soil Biol Biochem 40:1197–1206

Miransari M, Rejali F, Bahrami HA, Malakouti MJ (2009a) Effects of soil compaction and arbuscular mycorrhiza on corn (*Zea mays* L.) nutrient uptake. Soil Till Res 103:282–290

Miransari M, Rejali F, Bahrami HA, Malakouti MJ (2009b) Effects of arbuscular mycorrhiza, soil sterilization, and soil compaction on wheat (*Triticum aestivum* L.) nutrients uptake. Soil Till Res 104:48–55

Mortimer P, Prez-Fernandez M, Valentine A (2008) The role of arbuscular mycorrhizal colonization in the carbon and nutrient economy of the tripartite symbiosis with nodulated *Phaseolus vulgaris*. Soil Biol Biochem 40:1019–1027

Rouached H, Arpat AB, Poirier Y (2010) Regulation of phosphate starvation responses in plants: signaling players and cross-talks. Mol Plant 3:288–299

Sajedi NA, Ardakani MR, Rejali F, Mohabbati F, Miransari M (2010) Yield and yield components of hybrid corn (*Zea mays* L.) as affected by mycorrhizal symbiosis and zinc sulfate under drought stress. Physiol Mol Biol Plants 16:343–351

Sajedi NA, Ardakani MR, Madani H, Naderi A, Miransari M (2011) The effects of selenium and other micronutrients on the antioxidant activities and yield of corn (*Zeamays* L.) under drought stress. Physiol Mol Biol Plants 17:215–222

Schachtman DP, Reid RJ, Ayling SM (1998) Phosphorus uptake by plants: from soil to cell. Plant Physiol 116:447–453

Schimel JP, Weintraub MN (2003) The implications of exoenzyme activity on microbial carbon and nitrogen limitation in soil: a theoretical model. Soil Biol Biochem 35:549–563

Smith SE, Read DJ (2008) Mycorrhizal symbiosis, 3rd edn. Academic, New York, NY

Smith SE, Jakobsen I, Grønlund M, Smith FA (2011) Roles of arbuscular mycorrhizas in plant phosphorus nutrition: interactions between pathways of phosphorus uptake in arbuscular mycorrhizal roots have important implications for understanding and manipulating plant phosphorus acquisition. Plant Physiol 156:1050–1057

St. John TV, Coleman DC, Reid CPP (1983) Association of vesicular–arbuscular mycorrhizal hyphae with soil organic particles. Ecology 64:957–959

Subramanian K, Charest C (1999) Acquisition of N by external hyphae of an arbuscular mycorrhizal fungus and its impact on physiological responses in maize under drought-stressed and well-watered conditions. Mycorrhiza 9:69–75

Talbot JM, Allison SD, Treseder KK (2008) Decomposers in disguise: mycorrhizal fungi as regulators of soil C dynamics in ecosystems under global change. Funct Ecol 22:955–963

Tanaka Y, Yano K (2005) Nitrogen delivery to maize via mycorrhizal hyphae depends on the form of N supplied. Plant Cell Environ 28:1247–1254

Tian C, Kasiborski B, Koul R, Lammers PJ, Bucking H, Shachar-Hill Y (2010) Regulation of the nitrogen transfer pathway in the arbuscular mycorrhizal symbiosis: gene characterization and the coordination of expression with nitrogen Xux. Plant Physiol 153:1175–1187

Toljander JF, Lindahl BD, Paul LR, Elfstrand M, Finlay RD (2007) Influence of arbuscular mycorrhizal mycelial exudates on soil bacterial growth and community structure. FEMS Microbiol Ecol 61:295–304

Treseder K, Allen M (2002) Direct nitrogen and phosphorus limitation of arbuscular mycorrhizal fungi: a model and field test. New Phytol 155:507–515

Vogel-Mikus K, Pongrac P, Kump P, Neceman M, Regvar M (2006) Colonization of a Zn, Cd and Pb hyperaccumulator *Thlaspi praecox* Wulfen with indigenous arbuscular mycorrhizal fungi mixture induces changes in heavy metal and nutrient uptake. Environ Pollut 139:362–371

Wang X, Pan Q, Chen F, Yan X, Liao H (2011) Effects of co-inoculation with arbuscular mycorrhizal fungi and rhizobia on soybean growth as related to root architecture and availability of N and P. Mycorrhiza 21:173–181

Watts-Williams S, Cavagnaro T (2012) Arbuscular mycorrhizas modify tomato responses to soil zinc and phosphorus addition. Biol Fertil Soils 48:285–294

Wittenmayer L, Merbach W (2005) Plant responses to drought and phosphorus deficiency: contribution of phytohormones in root-related processes. J Plant Nutr Soil Sci 168:531–540

Zaefarian F, Rezvani M, Rejali F, Ardakani MR, Noormohammadi G (2011) Effect of heavy metals and arbuscular mycorrhizal fungal on growth and nutrients (N, P, K, Zn, Cu and Fe) accumulation of alfalfa (*Medicago sativa* L.). Am Eurasian J Agric Environ Sci 11:346–352

Zhang F, Smith DL (1995) Preincubation of *Bradyrhizobium japonicum* with genistein accelerates nodule development of soybean at suboptimal root zone temperature. Plant Physiol 108:961–968

Chapter 14
Arbuscular Mycorrhizal Fungi and the Tolerance of Plants to Drought and Salinity

Mónica Calvo-Polanco, Beatriz Sánchez-Romera, and Ricardo Aroca

14.1 Introduction

Plant species need to overcome a wide variety of biotic and abiotic stresses that affect their growth, production, and development. Salinity and drought are within the main factors that will trigger crop development and yield. The basic physiology of the different stresses overlaps with each other and are interconnected, affecting plants mainly by disturbing their water balance and/or inducing photoinhibition and photooxidative stress. Plants will try to reduce their water loss by closing the stomata (Yang et al. 2005; Schachtman and Goodger 2008) and adjusting their root hydraulic conductance (Maurel et al. 2008; Aroca et al. 2012).

The presence of microorganisms within the soil, specially a main group known as arbuscular mycorrhizal (AM) fungi, is a key factor in the adaptation of plants to the different ecosystems. AM fungi have been widely studied for their capacity to form symbiosis with plants in nature (Smith and Read 2008). AM fungi form intraradical and extraradical structures when associate with plants. Intraradical structures include intercellular hyphae, vesicles, and the formation of hyphal branches within the cortical cells called arbuscules. These arbuscules increase the contact area between the plant and the fungus and are considered to be the primary area of exchange. AM fungi also form a hyphal network of extraradical hyphae that allows extending the absorption area of the plant beyond the nutrient-depleted areas around the root zone (George 2000; Smith and Read 2008). From these symbioses, the fungi obtain carbon compounds from the plant and a niche to complete their life cycle, while plants get access to nutrients and water resources (Harrison 2005). The presence of these fungi also helps plants to enhance their tolerance to several biotic (Pozo and Azcón-Aguilar 2007) and abiotic stresses (Miransari 2010; Dodd and

M. Calvo-Polanco (✉) • B. Sánchez-Romera • R. Aroca
Department of Soil Microbiology and Symbiotic Systems, Estación Experimental del Zaidín, Consejo Superior de Investigaciones Científicas (CSIC), C/Profesor Albareda 1, 18008 Granada, Spain
e-mail: monica.calvo@eez.csic.es; beatriz.sanchez@eez.csic.es; ricardo.aroca@eez.csic.es

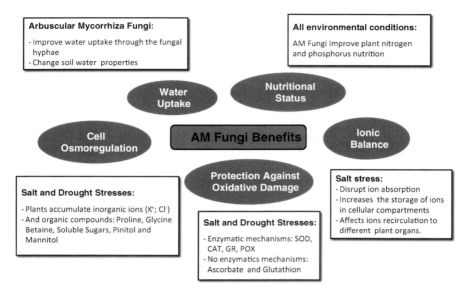

Fig. 14.1 Main benefits that arbuscular mycorrhizal symbioses contribute to the host plants under stress conditions. *SOD* superoxide dismutase, *CAT* catalase, *GR* glutathione reductase, *POX* peroxidase

Ruíz-Lozano 2012) and plays a major role in the plant tolerance to water deficit with the regulation of their root hydraulic properties (Aroca et al. 2007, 2008; Ruiz-Lozano and Aroca 2010) (Fig. 14.1).

In the present chapter, we will revise the physiological, biochemical, and molecular plant responses to salt and drought stress. We will also revise how the presence of AM fungi may affect tolerance of plants to these stresses and the different mechanisms involved.

14.2 Salt and Drought Stresses: The Role of AM Fungi

Drought and salt stresses are considered between the most widespread worldwide problems limiting plant growth, productivity, and survival (Kramer and Boyer 1997; Wang et al. 2004; Porcel et al. 2012). Both drought and salinity generate low water potentials within the soil that lead to a primary osmotic stress in the plants. Furthermore, the presence of high salt concentrations within the soil will also cause an ionic and toxic effect in the plant tissues. Plants would have to adapt to low water potential soils by controlling leaf transpiration rates and water uptake (Martínez-Ballesta et al. 2003; Boursiac et al. 2005). The reduction of leaf transpiration and root hydraulic conductivity will be one of the first symptoms of plants not being able to maintain their water balance (Aroca et al. 2008; Gao et al. 2010). The initial decrease of root hydraulic conductivity under drought/salt conditions could be a mechanism to avoid water flow from the roots to the soil while soil water potential is decreasing. Both drought and salt may also affect membrane selective permeability, inhibit protein synthesis, and will alter reactive oxygen species (ROS) production, causing oxidative damage to the plants (Ding et al. 2010).

The presence of AM fungi within the plant roots can determine the success of plants in colonizing salt- and/or drought-affected areas. There is a variety of AM fungi that have been found in different saline and dry ecosystems with a diverse capacity to tolerate/avoid the negative effects of salt (Juniper and Abbott 2006; Sheng et al. 2008; Evelin et al. 2012; Estrada et al. 2013). When AM fungi are in symbiosis with plant roots, they have been generally reported to have a positive effect in the plants, although this will depend on the host plant species as well as in the AM fungi involved (Marulanda et al. 2003; Wu et al. 2007). In general, AM fungi have been shown to increase plants resistance to soil salinity (Zuccarini and Okurowska 2008; Hajiboland et al. 2010; Wu et al. 2010) and drought (Azcón et al. 1996; Porcel and Ruiz-Lozano 2004), while improving their productivity (Navarro et al. 2011; Abbaspour et al. 2012), nutrient status (Farzaneh et al. 2011; Lee et al. 2012), and alleviate their water stress (Sheng et al. 2008; Bárzana et al. 2012).

14.2.1 Cell Osmoregulation

Drought and high salt levels will affect the water available for plants by reducing the osmotic potential of the soil solution. Plants accumulate metabolites to balance their cytoplasm water potential and to prevent water moving from the roots to the soil (Maurel et al. 2008; Ruiz-Lozano et al. 2012). Plants will active accumulate inorganic ions (K^+ and Cl^-) and uncharged organic compounds such as proline, glycine betaine, soluble sugars, pinitol, and mannitol (Hoekstra et al. 2001; Flowers and Colmer 2008). Proline and sugars accumulation within the plant cells has been recognized as a response to several environmental stresses (Samaras et al. 1995). Proline accumulates within the cell cytoplasm without interfering with the metabolism and regulating its osmotic potential. Proline can also work as a sink for energy to regulate redox potentials, as a hydroxyl radical scavenger, and as a solute that protects macromolecules against denaturation (Kishor et al. 1995). Proline regulation has been studied by measuring the expression of the gene P5CS which catalyzes the limiting step in the synthesis of proline (Aral and Kamoun 1997) and that is induced by drought, salinity, and ABA. Porcel et al. (2004) found that P5CS gene increased their expression under drought conditions more in non-AM plants than in AM ones. Thus, the use of this gene as stress marker can be useful in studies involving AM symbiosis and osmotic stresses (Aroca et al. 2008; Jahromi et al. 2008).

The presence of AM fungi under drought and high salt levels may modify the production and accumulation of different osmoregulators within the plant, mainly sugars and amino acids (proline) (Fig. 14.1). Plant proline content has been found to be higher in plants inoculated with AM fungi compared with non-inoculated plants under salt (Sharifi et al. 2007; Talaat and Shawky 2011; Abdel-Fattah et al. 2012) and drought (Trotel-Aziz et al. 2000; Porcel et al. 2004). However, according to the plant tissue studied and the intensity of the stress, we can also find reductions on proline content under salt (Rabie and Almadini 2005; Jahromi et al. 2008; Sheng et al. 2011) or drought (Porcel et al. 2004) in AM fungi-inoculated plants. These discrepancies are caused as to the degree of the stress developed between AM and non-AM plant, to the point that, if AM plants do not experienced any stress, they will not accumulate any proline (Porcel and Ruiz-Lozano 2004).

Sugar accumulation within the cells is also believed to play a role in the alleviation of drought (Porcel and Ruiz-Lozano 2004) and salt (Talaat and Shawky 2011). Sugars are directly involved in the synthesis of other compounds and in the production of energy as well as on membrane stabilization (Hoekstra et al. 2001) and regulation of genes expression (Koch 1996) and are well known as signal molecules (Sheen et al. 1999; Smeekens 2000). Similar to proline, there were positive correlations between the increase on root sugar content and the degree of mycorrhization under salt (Feng et al. 2002) and drought (Porcel and Ruiz-Lozano 2004), although the opposite has also been reported (Sheng et al. 2011). This higher accumulation of soluble sugars in mycorrhizal plant tissue, especially in roots, could make mycorrhizal plants more resistant to osmotic stresses. However, it has been suggested whether the sugar accumulation in the roots of AM plants is due to the sink effect of the mycorrhizal fungus demanding sugars from the shoot (Augé 2001; Porcel et al. 2004).

The use of plant mutants with different needs of proline and sugar content under salt and drought stress may help understand the role of AM fungi enhancing drought and salt tolerance. This knowledge should be always combined with the study of the genes involved in plant responses to drought and salt and with the role of signal molecules.

14.2.2 Ion Balance and Compartmentalization

Salt-tolerant plants usually have a greater ability to maintain low Na^+ levels in the cytoplasm (Hajibagheri et al. 1989). Na^+ can easily enter plant cells via nonselective K^+ channels. K^+ plays a key role in stomatal movement, activating a range of enzymes and protein synthesis. Under high salt levels, Na^+ will compete with K^+ for binding sites (Tester and Davenport 2003) and the $K^+:Na^+$ ratio will reflect the level of salt stress of an organism. A low $K^+:Na^+$ ratio will disrupt the ionic balance in the cytoplasm as well as several metabolic pathways (Giri et al. 2007). The increase of the $K^+:Na^+$ ratio is regulated through the compartmentalization of Na^+ within the plant tissues and the regulation of Na^+ uptake through the Na^+/H^+ antiporter SOS1 (Olías et al. 2009), the Na^+ influx HKT transporters family, and the tonoplast Na^+/H^+ antiporter family NHX (Ruiz-Lozano et al. 2012). Arbuscular mycorrhizal fungi exclude Na^+ during its transfer to the plants or by discriminating its uptake from the soil (Hammer et al. 2011), maintaining a fine balance of $K^+:Na^+$ and $Ca^{2+}:Na^+$ ratios (Fig. 14.1). Several studies have shown the importance of AM fungi in maintaining higher rate of $K^+:Na^+$ under salt stress (Giri et al. 2007; Porras-Soriano et al. 2009; Shokri and Maadi 2009; Evelin et al. 2012). The presence of AM fungi can reverse the effect of salt on K^+ by enhancing K^+ absorption (Alguacil et al. 2003; Giri et al. 2007; Sharifi et al. 2007; Zuccarini and Okurowska 2008). This higher $K^+:Na^+$ will allow plants to maintain a fine balance in their nutrient status, preventing the disruption of metabolic processes and the absorption and translocation of Na^+ to the shoot tissues. There is not, to our knowledge, any information available on AM fungal K^+ transporters or how the presence of AM fungi may affect K^+ plant transporters; however, Corratgé et al. (2007) described K^+ transporters on ectomycorrhizal fungi *Hebeloma cylindrosporum*. How these transporters may be modified by the presence of AM fungi could be critical for the resistance of plants to the transport of Na^+. The description of these transporters may also play a key role in the understanding of K^+ and Na^+ transport in the fungi–plant symbiosis and open a new aspect in the molecular study of AM fungi.

Sodium chloride is the main soil-present salt in nature. At high levels, it will cause toxic effects in plants through the accumulation of Na^+ within the plant tissues. However, several studies on plants under NaCl stress have suggested that Cl^- could be more toxic than Na^+ and the main factor contributing to membrane integrity loss (Bernstein 1975; Calvo-Polanco et al. 2009; Tavakkoli et al. 2010). Chloride ion is an essential micronutrient that regulates enzyme activities and photosynthesis, helps maintain membrane potential, and is involved in cell turgor and cytoplasmatic pH regulation (Xu et al. 2000). Many plant species are not able to effectively regulate Cl^- entry into the shoot, and they accumulate Cl^- in greater amounts than Na^+ (Jacoby 1994; Tavakkoli et al. 2010). Chloride ions reach the xylem by the symplastic pathway by ion channels or carriers as an active process (Tyerman and Skerrett 1999). Chloride ion levels within the plant tissues can be alleviated with the presence of AM fungi (Copeman et al. 1996; Zuccarini and Okurowska 2008), as Cl^- ions can be compartmentalized into the fungal vacuolar membranes. However, AM fungi have been also shown to increase the Cl^- concentration in citrus seedlings (Graham and Syvertsen 1989). Although plant Cl^- channels have been well characterized (White and Broadley 2001), their role in water uptake and nutrient imbalances has been less studied. Also, little is known about the transport and compartmentalization of Cl^- across hyphal membranes in mycorrhizal fungi and the effect that these may have in fungal resistance.

Cytoplasmatic Ca^{2+} is known as a second messenger in plant cells. High salt levels inhibit uptake and transport of Ca^{2+}, lowering the $Ca^{2+}:Na^+$ ratio. This ratio is an important measure of salinity stress since Na^+ replaces Ca^{2+} in the plasma membrane and cell wall when Na^+ levels are high, reducing cell turgor and hydraulic conductivity and disturbing Ca^{2+} signaling (Läuchli and Luttge 2002). In the presence of AM fungi, $Ca^{2+}:Na^+$ ratios have been shown to increase (Garg and Manchanda 2009; Evelin et al. 2012), although no changes in Ca^{2+} tissue concentrations have also been observed (Giri and Mukerji 2004). The involvement of AM fungi in these processes is not clear, although it has also been suggested that AM fungi may not be able to fine regulate the uptake of certain nutrients, including Ca^{2+} (Evelin et al. 2012).

14.2.3 Plant Water Balance and Aquaporins

Stomatal closure is among the initial symptoms under water deficit in the root zone. The maintenance of a fine plant water balance between water loss and water uptake would require the adjustment of the hydraulic conductivity within the plant tissues. The main barrier for water uptake from the soil to the vascular tissues is the radial transport between the root epidermis and the xylem (Doussan et al. 1998). According to the composite model, radial transport will involve the flow of water along three major pathways: apoplastic, symplastic, and transcellular, where symplastic and transcellular are considered as cell to cell (Steudle and Peterson 1998). Apoplastic pathway is preferred when hydrostatic forces are created by the transpiration stream. When stomata are closed, there will be an osmotic gradient created between the soil solution and the xylem sap, and water will be transported mainly through the cell-to-cell pathway that will involve water transport across membranes, plasmodesmata, and the symplast. The regulation of water across membranes, and hence the regulation of root hydraulic conductivity, has been mainly attributed to the transmembrane aquaporins. Aquaporins are membrane intrinsic proteins present in all living organisms, including fungi (Agre et al. 1993), that facilitate the transport of water and another small uncharged molecules across cell membranes following a gradient (Maurel 2007; Maurel et al. 2008). In plants, they can be divided into five subgroups according to their amino acid sequence similarity: PIPs, plasma membrane intrinsic proteins; TIPs, tonoplast intrinsic proteins; NIPs, nodulin-26-like intrinsic proteins; SIPs, small and basic intrinsic proteins; and the uncharacterized intrinsic proteins (XIPs) (Chaumont et al. 2000; Sakurai et al. 2005; Park et al. 2010).

Arbuscular mycorrhizal fungi have been shown to transfer water from the soil to the root of the host plants (Marulanda et al. 2003; Ruth et al. 2011) and to regulate root hydraulic properties through the regulation of plant aquaporins (Ruiz-Lozano and Aroca 2010) (Fig. 14.1). The role of AM fungi in water transport and aquaporin function and regulation is poorly understood (Lehto and Zwiazek 2011). Fungal mycelia contain their own aquaporins, although little is known about their contribution to the water transport of mycorrhizal plants (Aroca et al. 2009; Navarro-Ródenas et al. 2012; Li et al. 2013). Arbuscular mycorrhizal fungi will alter the aquaporin expression of the plants they are colonizing by increasing (Ouziad et al. 2006; Aroca et al. 2007), decreasing (Porcel et al. 2006; Jahromi et al. 2008), or with no effect (Aroca et al. 2007; Jahromi et al. 2008) on them. These results support the idea that each aquaporin has its specific function under each environmental stress condition (Aroca et al. 2007; Jang et al. 2007) and that each plant will respond differently to each colonizing fungi.

Aroca et al. (2009) described the first putative aquaporin gen from the AM fungus *Rhizophagus irregularis* (former *G. intraradices*) (*GintAQP1*), and more recently another two have been discovered (Tisserant et al. 2012; Li et al. 2013). The expression of *GintAQP1* was shown to be coordinated with the expression of plant aquaporins under different stresses including salt and drought. In *P. vulgaris* inoculated plants, *GintAQP1* did not change its expression under salt stress, but there was an increase in expression on the four *P. vulgaris* PIP genes analyzed. Under drought conditions, expression of *GintAQP1* increased, with the corresponding downregulation of two of the four *P. vulgaris* PIP genes (Aroca et al. 2009). However, expression of *GintAQP1* was downregulated in maize plants inoculated with *R. irregularis* at the time there was an upregulation in all the PIP genes analyzed (El-Mesbahi et al. 2012). Li et al. (2013) described two functional aquaporin genes from AM fungi *Gint*AQPF1 and *Gint*AQPF2 under drought tolerance. These genes were highly expressed in fungal structures, and their regulation was upregulated under drought conditions. The results of the different studies have been interpreted as a compensatory mechanism between host plant aquaporins and the AM fungal aquaporins. It has been also demonstrated (Aroca et al. 2009) that *GintAQP1* expression changed in nonstressed parts of the mycelium when another section was stressed by NaCl, indicating a possible communication between different parts of the stressed and nonstressed mycelia. It is also plausible that the AM fungi may affect the gating factors that are involved in aquaporin regulation, including phosphorylation and dephosphorylation (Johansson et al. 1998), cytoplasmic pH (Tournaire-Roux et al. 2003; Sutka et al. 2005), and divalent cations (Gerbeau et al. 2002; Alleva et al. 2006). Phosphorylation and dephosphorylation events enable to switch a channel on or off. Channel control by phosphorylation was suggested as a main mechanism in aquaporin control and hence in water transport regulation. It has been also demonstrated that soil pH can affect root hydraulic properties in plants (Tang et al. 1993; Kamaluddin and Zwiazek 2004). Furthermore, it has been shown that cell cytosolic pH plays a key role in the regulation of plant water transport capacity and that aquaporins may account for the cytosolic pH levels (Tournaire-Roux et al. 2003; Törnroth-Horsefield et al. 2006) by a pH-sensitive gating mechanism involving the protonation of a histidine (His) residue (Zelenina et al. 2003) located within a highly conserved PIP region (Fischer and Kaldenhoff 2008). There have been some studies showing that acidic pH decreased both root water flux and aquaporin-mediated flux (Alleva et al. 2006; Tournaire-Roux et al. 2003). Acidic conditions result in a conformational change with the closure of the pore on the cytosolic side (Törnroth-Horsefield et al. 2006). Some studies have also reported optimal aquaporin activity at pH 4 and minimal activity at pH 7 (Németh-Cahalan and Hall 2000).

Although there are no studies of arbuscular mycorrhizal fungi directly affecting the cytosolic pH of the host plants and hence aquaporins functioning, it is widely known that AM fungi are able to regulate the pH of the root and the root zone after the symbiosis establishment (Bago and Azcón-Aguilar 1997) via extrusion of protons and organic acids. Protons (H^+) may have an important role on the fungal growth and host signal perception through the activity of the P-type plasma membrane H^+-ATPase (Ramos et al. 2008), which seems to be the major responsible by the H^+ efflux across plasma membrane. AM fungal cells also energize their plasma membrane using P-type H^+ pumps quite similar to the plant. It would be possible that the effect of pH within the root hypha would affect the plant responses to stresses or even to act as a signal for the increase or decrease of the aquaporin expression.

In conclusion, aquaporins, both in plants and fungi, play a key role in the plant resistance and tolerance to different stresses. The responses to these stresses seem to be plant/fungi dependent and with a strong dependence on several other responses and physiological parameters. A more compressive examination, possible with the molecular technology available, will help determine the whole response of plant and fungal aquaporins and how they affect each other under different circumstances. The study of fungal aquaporins and how they influence plant aquaporins as symbionts would help to explain the water status of the plant and help understand their survival to certain environments.

14.2.4 Antioxidant Mechanisms

During salt and drought stresses, plants will induce the production of harmful reactive oxygen species (ROS), including singlet oxygen ($^1O_2$), hydrogen peroxide (H_2O_2), and hydroxyl radicals (OH·). These radicals are harmful at high concentrations causing oxidative damage to biomolecules, denaturation of proteins, and DNA mutations (Bowler et al. 1991). The mechanisms of prevention and repair of the damage caused by ROS will include the production of nonenzymatic molecules called ROS scavengers (ascorbate, glutathione, α-tocopherol, flavonoids, anthocyanins, and carotenoids) and the production of ROS-scavenging antioxidative enzymes [revised in Miller et al. (2010) and in Scheibe and Beck (2011)]. Ascorbate and glutathione are considered within the most important antioxidative metabolites under stress conditions. Ascorbate is an antioxidant that serves as an electron donor to several metabolic reactions and has a special role on the photosynthesis protection (Shao et al. 2008). Glutathione is a tripeptide abundant in the cell in its reduced form. The ratio to its oxidized form (GSSG—glutathione disulfide) maintains redox equilibrium during H_2O_2 degradation (Shao et al. 2008). Glutathione plays a key role in the regeneration of reduced ascorbate in the ascorbate–glutathione cycle (Halliwell and Foyer 1976).

Increasing antioxidant capacity has been shown to improve salt and drought tolerance in several plant species (Türkan and Demiral 2009; Abbaspour et al. 2012). The presence of AM fungi will, as well, enhance the activities of several antioxidant enzymes as superoxide dismutase (SOD), catalase (CAT), glutathione reductase (GR), and peroxidase (POX) and the accumulation of ascorbate and glutathione under salt (Garg and Manchanda 2009; Wu et al. 2010; Talaat and Shawky 2011) and drought (Wu and Zou 2009; Ruiz-Sanchez et al. 2010;

Abbaspour et al. 2012) (Fig. 14.1). Although the general response in AM fungi is in general positive, each of the enzymes involved in the different processes is nutrient dependent, and their activity will be dependent on the AM fungi genotype, the mycorrhizal capacity to acquire nutrients, specially nitrogen and phosphorus (Alguacil et al. 2003; Wu et al. 2007; Evelin et al. 2009), and the water status of the plant (Porcel and Ruiz-Lozano 2004).

The mechanisms that control the antioxidant capacity of AM plants are still not fully understood, but there have been some very interesting studies in AM fungi uncovering the genes that encode different proteins involved in the cellular defense against oxidative stress. There are several examples of genes encoding SOD (Lanfranco et al. 2005; González-Guerrero et al. 2010), glutaredoxins (GRXs) (Benabdellah et al. 2009a), pyridoxine 5′-phosphate synthase (PDX) (Benabdellah et al. 2009b), and a metallothionein (MT)-encoding gene (González-Guerrero et al. 2007). The responses to these genes to salt and drought are still not studied. The information available at the fungal side together with the current knowledge of the fungal/plant symbiosis is a very promising first step to elucidate the mechanisms controlling antioxidant defenses that may lead to improve the resistance of plants to salt and drought.

14.2.5   The Role of Phytohormones

Plant hormones are signal molecules that regulate the growth of plants and are important pieces in the mechanisms of plant tolerance under different stresses (Shaterian et al. 2005; Wolters and Jürgens 2009). Hormonal regulation is achieved through a complex regulatory network that connects the different hormonal pathways enabling each hormone to assist or antagonize the others (Peleg and Blumwald 2011). Auxin, brassinosteroid, cytokinin (CK), and gibberellins (GA) are major developmental growth regulators, while abscisic acid (ABA), ethylene (ET), jasmonic acid (JA), salicylic acid (SA), and strigolactones (SL) are often implicated in stress responses. In general, ABA and JA frequently show similar biological effects (Wang et al. 2001; Aroca et al. 2013); CK (Peleg and Blumwald 2011) and ET (Wilkinson and Davies 2010) are antagonist of ABA, and SA is an inhibitor of ET (Leslie and Romani 1988).

ABA accumulates rapidly in response to stress-regulating interacting signaling pathways involved in plant responses to salt (Keskin et al. 2010; Wan 2010; Raghavendra et al. 2010) and drought (Aroca et al. 2008; Parent et al. 2009; Ruiz-Lozano et al. 2009), through its effect on stomatal closure (Zhang et al. 2009; Wilkinson and Davies 2010), root hydraulic conductivity (revised in Aroca et al. 2012), PIP aquaporin expression and abundance (Jang et al. 2004; Aroca et al. 2006; Mahdieh and Mostejaran 2009), aquaporin phosphorylation/dephosphorylation (Kline et al. 2010), as well as by induction of genes involved in water stress tolerance (Zhang et al. 2006; Ruiz-Lozano et al. 2006; Aroca et al. 2008, 2013). ABA has been also found to be necessary for arbuscular development in the AM symbiosis (Herrera-Medina et al. 2007; Martín-Rodríguez et al. 2011). Once the fungus is established within the host roots, it will regulate ABA content when plants are exposed to drought (Estrada-Luna and Davies 2003; Aroca et al. 2008) or salt stress (Jahromi et al. 2008; Aroca et al. 2013) with lower contents of ABA in AM plants. This effect will allow a faster recovery of leaf transpiration and water balance of AM plants (Aroca et al. 2008), which has been seen directly correlated with the accumulation pattern of PIP aquaporins (Ruiz-Lozano et al. 2009). The presence of AM fungi will also affect the induction of the gene LsNCED2 in response to salinity (Aroca et al. 2013) and drought (Aroca et al. 2008), regulating plant ABA contents.

Other plant hormones that have been reported to affect several plant responses to stress are ET, SA, SL, and JA. It is known that low levels of ET and SA will promote root AM colonization (Herrera-Medina et al. 2003; Riedel et al. 2008) and that there is a relationship between the multipurpose signaling molecules of ethylene and the generation of ROS, but the interaction mechanism remains unclear (Wang et al. 2002). Recently, Aroca et al. (2013) have reported that AM symbiosis alleviates salt stress by altering the hormonal profiles in correlation with SL production and ABA content. However, the information available is scarce, especially when liking the effects of AM fungi in ET, SL, and JA production and their effect on water relations and aquaporin expression and functioning.

14.3 Conclusion

We have revised some aspects of plant responses to drought and salt with a special interest in those affecting plant water balance. The effects of AM fungi in plant roots under these stresses are still not well understood although it is known that their presence will alter plant responses at the molecular, physiological, and biochemistry levels. The different responses found under salt and drought are usually related with the plant/host affinity and the different environmental circumstances of the study that would lead to differences on the rates of colonization at the root level.

We have concentrated our efforts in the study of cell osmoregulation, water balance control through aquaporins, hormones, ROS species, and ion balance. All the subjects revised are interconnected and will affect each other especially at the plant water relations level. Water balance in the fungi–plant system will be dependent to an osmotic and membrane-aquaporin adjustment, together with a fine balance with hormones and ROS species and the control of Na^+ and Cl^- in the case of salt stress. How AM fungi affect these parameters under salt and drought is still uncertain due to the contradictory of the results as the different responses found seem to be plant/fungi dependent. The molecular techniques can help determine the whole response of plant/fungi symbiosis at different levels, although a special effort should be put into the study of the fungal side and how they affect plants at the different circumstances. The study of plant and fungal aquaporins, the response of genes to salt and drought, the determination of the genes involved in ROS scavenging, how hormones infer in the whole process, and the different transporters involved in the Na^+ and Cl^- transport is a thrilling field. The use of both plant and fungal mutants will be a precious tool to understand basic processes and a wide field for future studies.

In general, combined studies where different fields as molecular biology, biochemistry, and physiology can meet together are still needed. These studies should be combined with a comprehensive analysis of the transfer of this knowledge to natural ecology, taking always in account that AM fungi have a notorious role in ecosystems and are a key factor for plant survival under a changing environment where plants are going to be exposed to extreme circumstances on the oncoming years.

References

Abbaspour H, Saeidi-Sar S, Abdel-Wahhab MA (2012) Improving drought tolerance of *Pistacia vera* L. Seedlings by arbuscular mycorrhiza under greenhouse conditions. J Plant Physiol 169:704–709

Abdel-Fattah GM, Wasea A, Asrar A (2012) Arbuscular mycorrhizal fungal application to improve growth and tolerance of wheat (*Triticum aestivum* L.) plants grown in saline soil. Acta Physiol Plant 34:267–277

Agre P, Sasaki S, Chrispeels MJ (1993) Aquaporins: a family of water channel proteins. Am J Physiol Renal Physiol 265:F461

Alguacil MM, Hernández JA, Caravaca F, Portillo B, Roldán A (2003) Antioxidant enzyme activities in shoots from three mycorrhizal shrub species afforested in a degraded semi-arid soil. Physiol Plant 118:562–570

Alleva K, Niemiets CM, Sutka M, Maurel C, Parasi M, Tyerman SD, Amodeo G (2006) Plasma membrane of *Beta vulgaris* storage root shows high water channel activity by cytoplasmic pH and a dual range of calcium concentrations. J Exp Bot 57:609–621

Aral B, Kamoun P (1997) The proline biosynthesis in living organisms. Amino Acids 13:189–217

Aroca R, Ferrante A, Vernieri P, Chrispeels MJ (2006) Drought, abscisic acid and transpiration rate effects on the regulation of PIP aquaporin gene expression and abundance in *Phaseolus vulgaris* plants. Ann Bot 98:1301–1310

Aroca R, Porcel R, Ruiz-Lozano JM (2007) How does arbuscular mycorrhizal symbiosis regulate root hydraulic properties and plasma membrane aquaporin in *Phaseolus vulgaris* under drought, cold or salinity stresses? New Phytol 173:808–816

Aroca R, Vernieri P, Ruiz-Lozano JM (2008) Mycorrhizal and nonmycorrhizal *Lactuca sativa* plants exhibit contrasting responses to exogenous ABA during drought stress and recovery. J Exp Bot 59:2029–2041

Aroca R, Bago A, Sutka M, Paz JA, Cano C, Amodeo G, Ruiz-Lozano JM (2009) Expression analysis of the first arbuscular mycorrhizal fungi aquaporin described reveals concerted gene expression between salt-stressed and nonstressed mycelium. Mol Plant Microbe Interact 22:1169–1178

Aroca R, Porcel R, Ruiz-Lozano JM (2012) Regulation of root water uptake under abiotic stress conditions. J Exp Bot 63:43–57

Aroca R, Ruiz-Lozano JM, Zamarreño AM, Paz JA, García-Mina JM, Pozo MJ, López-Ráez JA (2013) Arbuscular mycorrhizal symbiosis influences strigolactone production under salinity and alleviates salt stress in lettuce plants. J Plant Physiol 170:47–55

Augé RM (2001) Water relations, drought and vesicular–arbuscular mycorrhizal symbiosis. Mycorrhiza 11:3–42

Azcón R, Gómez M, Tobar RM (1996) Physiological and nutritional responses by *Lacta sativa* L. to nitrogen sources and mycorrhizal fungi under drought conditions. Biol Fertil Soils 22:156–161

Bago B, Azcón-Aguilar C (1997) Changes in the rhizospheric pH induced by arbuscular mycorrhiza formation in onion (*Allium cepa* L.). Z Pflanzenernaehr Bodenk 160:333–339

Bárzana G, Aroca R, Paz JA, Chaumont F, Martinez-Ballesta M, Carvajal M, Ruiz-Lozano JM (2012) Arbuscular mycorrhizal symbiosis increases relative apoplastic water flow in roots of the host plant under both well-watered and drought stress conditions. Ann Bot 109:1009–1017

Benabdellah K, Merlos MA, Azcón-Aguilar C, Ferrol N (2009a) *GintGRX1*, the first characterized glomeromycotan glutaredoxin, is a multifunctional enzyme that responds to oxidative stress. Fungal Genet Biol 46:94–103

Benabdellah K, Azcón-Aguilar C, Valderas A, Speziga D, Fitzpatrick TB, Ferrol N (2009b) *GintPDX1* encodes a protein involved in vitamin B6 biosynthesis that is up-regulated by oxidative stress in the arbuscular mycorrhizal fungus *Glomus intraradices*. New Phytol 184:682–693

Bernstein L (1975) Effects of salinity and sodicity on plant growth. Annu Rev Phytopathol 13:295–312

Boursiac Y, Chen S, Luu DT, Sorieul M, Dries N, Maurel C (2005) Early effects of salinity on water transport in Arabidopsis roots. Molecular and cellular features of aquaporin expression. Plant Physiol 139:790–805

Bowler C, Slooten L, Vandenbranden S, De Rycke R, Botterman J, Sybesma C, Van Montagu M, Inze D (1991) Manganese superoxide dismutase can reduce cellular damage mediated by oxygen radicals in transgenic plants. EMBO J 10:1723–1732

Calvo-Polanco M, Zwiazek JJ, Jones MD, MacKinnon MD (2009) Effects of NaCl on responses of ectomycorrhizal black spruce (*Picea mariana*), white spruce (*Picea glauca*) and jack pine (*Pinus banksiana*) to fluoride. Physiol Plantarum 135:51–61

Chaumont F, Barrieu F, Jung R, Chrispeels MJ (2000) Plasma membrane intrinsic proteins from maize cluster in two sequence subgroups with differential aquaporin activity. Plant Physiol 122:1025–1034

Copeman RH, Martin CA, Stutz JC (1996) Tomato growth in response to salinity and mycorrhizal fungi from saline or nonsaline soils. Hortic Sci 31:341–344

Corratgé C, Zimmermann S, Lambilliotte RRL, Plassard C, Marmeisse R, Thibaud JB, Lacombe B, Sentenac H (2007) Molecular and functional characterization of a Na^+-K^+ transporter from the Trk family in the ectomycorrhizal fungus *Hebeloma cylindrosporum*. J Biol Chem 282:26057–26066

Ding M, Hou P, Shen X et al (2010) Salt-induced expression of genes related to $Na(+)/K(+)$ and ROS homeostasis in leaves of salt resistant and salt-sensitive poplar species. Plant Mol Biol 73:251–269

Dodd IC, Ruíz-Lozano JM (2012) Microbial enhancement of crop resource use efficiency. Curr Opin Biotechnol 23:236–242

Doussan C, Vercambre G, Page L (1998) Modelling of the hydraulic architecture of root systems: an integrated approach to water absorption—distribution of axial and radial conductances in maize. Ann Bot 81:225–232

El-Mesbahi MN, Azcón R, Ruiz-Lozano JM, Aroca R (2012) Plant potassium content modifies the effects of arbuscular mycorrhizal symbiosis on root hydraulic properties in maize plants. Mycorrhiza 22:555–564

Estrada B, Barea JM, Aroca R, Ruiz-Lozano JM (2013) A native *Glomus intraradices* strain from a Mediterranean saline area exhibits salt tolerance and enhanced symbiotic efficiency with maize plants under salt stress conditions. Plant Soil. doi:10.1007/s11104-012-1409-y

Estrada-Luna AA, Davies FT (2003) Arbuscular mycorrhizal fungi influence water relations, gas Exchange, abscisic acid and growth of micropropagated chile ancho pepper (*Capsicum annuum*) plantlets during acclimatization and post-acclimatization. J Plant Physiol 160:1073–1083

Evelin H, Kapoor R, Giri B (2009) Arbuscular mycorrhizal fungi in alleviation of salt stress: a review. Ann Bot 104:1263–1280

Evelin H, Giri B, Kapoor R (2012) Contribution of *Glomus intraradices* inoculation to nutrient acquisition and mitigation of ionic imbalance in NaCl-stressed *Trigonella foenum-graecum*. Mycorrhiza 22:203–217

Farzaneh M, Vierheiling H, Lössl A, Kaul HP (2011) Arbuscular mycorrhiza enhances nutrient uptake in chickpea. Plant Soil Environ 57:465–470

Feng G, Zhang FS, Li XL, Tian CY, Tang C, Rengel Z (2002) Improved tolerance of maize plants to salt stress by arbuscular mycorrhiza is related to higher accumulation of soluble sugars in roots. Mycorrhiza 12:185–190

Fischer M, Kaldenhoff R (2008) On the pH regulation of plant aquaporins. J Biol Chem 283:33889–33892

Flowers TJ, Colmer TD (2008) Salinity tolerance in halophytes. New Phytol 179:945–963

Gao YX, Li Y, Yang XX, Li HJ, Shen QR, Guo SW (2010) Ammonium nutrition increases water absorption in rice seedlings (*Oryza sativa* L.) under water stress. Plant Soil 331:193–201

Garg N, Manchanda G (2009) Role of arbuscular mycorrhizae in the alleviation of ionic, osmotic and oxidative stresses induced by salinity in *Cajanus cajan* (L.) Millsp. (pigeon pea). J Agron Crop Sci 195:110–123

George E (2000) Nutrient uptake. In: Douds KY, Jr DD (eds) Arbuscular mycorrhizas: physiology and function. Kluwer Academic, Dordrecht

Gerbeau P, Amodeo G, Henzler T, Santoni V, Ripoche P, Maurel C (2002) The water permeability of Arabidopsis plasma membrane is regulated by divalent cations and pH. Plant J 30:71–81

Giri B, Mukerji KG (2004) Mycorrhizal inoculant alleviates salt stress in *Sesbania aegyptiaca* and *Sesbania grandiflora* under field conditions: evidence for reduced sodium and improved magnesium uptake. Mycorrhiza 14:307–312

Giri B, Kapoor R, Mukerji KG (2007) Improved tolerance of *Acacia nilotica* to salt stress by arbuscular mycorrhiza, *Glomus fasciculatum* may be partly related to elevated K/Na ratios in root and shoot tissues. Microb Ecol 54:753–760

González-Guerrero M, Cano C, Azcón-Aguilar C, Ferrol N (2007) GintMT1 encodes a functional metallothionein in *Glomus intraradices* that responds to oxidative stress. Mycorrhiza 17:327–335

González-Guerrero M, Oger E, Benabdellah K, Azcón-Aguilar C, Lanfranco L, Ferrol N (2010) Characterization of a CuZn superoxide dismutase gene in the arbuscular mycorrhizal fungus. *Glomus intraradices*. Curr Genet 56:265–274

Graham JH, Syvertsen JP (1989) Vesicular-arbuscular mycorrhizas increase chloride concentration in citrus seedlings. New Phytol 113:29–36

Hajibagheri MA, Yeo AR, Flowers TJ, Collins JC (1989) Salinity resistance in *Zea mays*-fluxes of potassium, sodium and chloride, cytoplasmic concentrations and microsomal membrane lipids. Plant Cell Environ 12:753–757

Hajiboland R, Aliasgharzadeh N, Laiegh SF, Poschenrieder C (2010) Colonization with arbuscular mycorrhizal fungi improves salinity tolerance of tomato (*Solanum lycopersicum* L.) plants. Plant Soil 331:313–327

Halliwell B, Foyer CH (1976) Ascorbic acid, metal ions and the superoxide radical. Biochem J 155:697–700

Hammer EC, Nasr H, Pallon J, Olsson PA, Wallander H (2011) Elemental composition of arbuscular mycorrhizal fungi at high salinity. Mycorrhiza 21:117–129

Harrison MJ (2005) Signaling in the arbuscular mycorrhizal symbiosis. Annu Rev Microbiol 59:19–42

Herrera-Medina MJ, Gagnon H, Piché Y, Ocampo JA, García-Garrido JM, Vierheilig H (2003) Root colonization by arbuscular mycorrhizal fungi is affected by the salicylic acid content of the plant. Plant Sci 164:993–998

Herrera-Medina MJ, Steinkellner S, Vierheilig H, Ocampo Bote JA, García Garrido JM (2007) Abscisic acid determines arbuscule development and functionality in the tomato arbuscular mycorrhiza. New Phytol 175:554–564

Hoekstra FA, Golovina EA, Buitink J (2001) Mechanisms of plant desiccation tolerance. Trends Plant Sci 6:431–438

Jacoby B (1994) Mechanisms involved in salt tolerance by plants. In: Pessarakli M (ed) Handbook of plant and crop stress. Dekker, New York, NY

Jahromi F, Aroca R, Porcel R, Ruiz-Lozano JM (2008) Influence of salinity on the in vitro development of *Glomus intraradices* and on the in vivo physiological and molecular responses of mycorrhizal lettuce plants. Microb Ecol 55:45–53

Jang JY, Kim DG, Kim YO, Kim JS, Kang H (2004) An expression analysis of a gene family encoding plasma membrane aquaporins in response to abiotic stresses in *Arabidopsis thaliana*. Plant Mol Biol 54:713–725

Jang JY, Lee SH, Rhee JY, Chung GC, Ahn SJ, Kang H (2007) Transgenic *Arabidopsis* and tobacco plants overexpressing an aquaporin respond differently to various abiotic stresses. Plant Mol Biol 64:621–632

Johansson I, Karlsson M, Shukla VK, Chrispeels MJ, Larsson C, Kjellbom P (1998) Water transport activity of the plasma membrane aquaporin PM28A is regulated by phosphorylation. Plant Cell 10:451–459

Juniper S, Abbott LK (2006) Soil salinity delays germination and limits growth of hyphae from propagules of arbuscular mycorrhizal fungi. Mycorrhiza 16:371–379

Kamaluddin M, Zwiazek JJ (2004) Effects of root medium pH on water transport in paper birch (*Betula papyrifera*) seedlings in relation to root temperature and abscisic acid treatments. Tree Physiol 24:1173–1180

Keskin BC, Sarikaya AT, Yuksel B, Memon AR (2010) Abscisic acid regulated gene expression in bread wheat. Aust J Crop Sci 4:617–625

Kishor PB, Hong Z, Miao GH, Hu CA, Verma DPS (1995) Overexpression of Δ^1-pyrroline-5-carboxylate synthetase increases proline production and confers osmotolerance in transgenic plants. Plant Physiol 108:1387–1394

Kline KG, Barrett-Wilt GA, Sussman MR (2010) *In planta* changes in protein phosphorylation induced by the plant hormone abscisic acid. Proc Natl Acad Sci USA 107:15986–15991

Koch K (1996) Carbohydrate-modulated gene expression in plants. Annu Rev Plant Physiol Plant Mol Biol 47:509–540

Kramer PJ, Boyer JS (1997) Water relations of plants and soils. Academic, San Diego, CA

Lanfranco L, Novero M, Bonfante P (2005) The mycorrhizal fungus *Gigaspora margarita* possesses a CuZn superoxide dismutase that is up-regulated during symbiosis with legume hosts. Plant Physiol 137:1319–1330

Läuchli A, Luttge U (eds) (2002) Salinity: environment-plants-molecules. Kluwer Academic, Dordrecht

Lee BR, Muneer S, Avice JC, Jung WJ, Kim TH (2012) Mycorrhizal colonisation and P-supplement effects on N uptake and N assimilation in perennial ryegrass under well-watered and drought-stressed conditions. Mycorrhiza 22:525–534

Lehto T, Zwiazek JJ (2011) Ectomycorrhizas and water relations of trees: a review. Mycorrhiza 21:71–90

Leslie CA, Romani RJ (1988) Inhibition of ethylene biosynthesis by salicylic acid. Plant Physiol 88:833–837

Li T, Hu YJ, Hao ZP, Li H, Wang YS, Chen BD (2013) First cloning and characterization of two functional aquaporin genes from an arbuscular mycorrhizal fungus *Glomus intraradices*. New Phytol 197:617–630

Mahdieh M, Mostejaran A (2009) Abscisic acid regulates root hydraulic conductance via aquaporin expression in *Nicotiana tabacum*. J Plant Physiol 166:1993–2003

Martínez-Ballesta MC, Martínez V, Carvajal M (2003) Aquaporin functionality in relation to H$^+$-ATPase activity in root cells of *Capsicum annuum* grown under salinity. Physiol Plantarum 117:413–420

Martín-Rodríguez JA, León-Morcillo R, Vierheilig H, Ocampo JA, Ludwig-Müller J, García-Garrido JM (2011) Ethylene-dependent/ethylene-independent ABA regulation of tomato plants colonized by arbuscular mycorrhiza fungi. New Phytol 190:193–205

Marulanda A, Azcón R, Ruíz-Lozano JM (2003) Contribution of six arbuscular mycorrhizal fungal isolates to water uptake by *Lactuca sativa* L. plants under drought stress. Physiol Plantarum 119:526–533

Maurel C (2007) Plant aquaporins: novel functions and regulation properties. FEBS Lett 581:2227–2236

Maurel C, Verdoucq L, Luu D-T, Santoni V (2008) Plant aquaporins: membrane channels with multiple integrated functions. Annu Rev Plant Biol 59:595–624

Miller G, Suzuki N, Ciftci-Yilmaz S, Mittler R (2010) Reactive oxygen species homeostasis and signalling during drought and salinity stress. Plant Cell Environ 33:453–467

Miransari M (2010) Contribution of arbuscular mycorrhizal symbiosis to plant growth under different types of soil stress. Plant Biol 1:563–569

Navarro GA, Del P, Bañón AS, Morte A, Sánchez-Blanco MJ (2011) Effects of nursery pre-conditioning through mycorrhizal inoculation and drought in *Arbutus unedo* L. plants. Mycorrhiza 21:53–64

Navarro-Ródenas A, Ruíz-Lozano JM, Kaldenhoff R, Morte A (2012) The aquaporin TcAQP1 of the desert truffle *Terfezia claveryi* is a membrane pore for water and CO$_2$ transport. Mol Plant Microbe Interact 25:259–266

Németh-Cahalan KL, Hall JE (2000) pH and calcium regulate the water permeability of aquaporin 0. J Biol Chem 275:6777–6782

Olías R, Eljakaoui Z, Pardo JM, Belver A (2009) The Na$^+$/H$^+$ exchanger SOS1 controls extrusion and distribution of Na$^+$ in tomato plants under salinity conditions. Plant Signal Behav 4:973–976

Ouziad F, Wilde P, Schmelzer E, Hildebrandt U, Bothe H (2006) Analysis of expression of aquaporins and Na$^+$/H$^+$ transporters in tomato colonized by arbuscular mycorrhizal fungi and affected by salt stress. Environ Exp Bot 57:177–186

Parent B, Hachez C, Redondo E, Simonneau T, Chaumont F, Tardieu F (2009) Drought and abscisic acid effects on aquaporin content translate into changes in hydraulic conductivity and leaf growth rate: a trans-scale approach. Plant Physiol 149:2000–2012

Park W, Scheffler E, Bauer PJ, Campbell BT (2010) Identification of the family of aquaporin genes and their expression in upland cotton (*Gossypium hirsutum* L.). BMC Plant Biol 10:142

Peleg Z, Blumwald E (2011) Hormone balance and abiotic stress tolerance in crop plants. Curr Opin Plant Biol 14:290–295

Porcel R, Ruiz-Lozano JM (2004) Arbuscular mycorrhizal influence on leaf water potential, solute accumulation, and oxidative stress in soybean plants subjected to drought stress. J Exp Bot 55:1743–1750

Porcel R, Azcón R, Ruiz-Lozano JM (2004) Evaluation of the role of genes encoding for Δ^1-pyrroline-5-carboxylate synthetase (P5CS) during drought stress in arbuscular mycorrhizal *Glycine max* and *Lactuca sativa* plants. Physiol Mol Plant Pathol 65:211–221

Porcel R, Aroca R, Azcón R, Ruiz-Lozano JM (2006) PIP aquaporin gene expression in arbuscular mycorrhizal *Glycine max* and *Lactuca sativa* plants in relation to drought stress tolerance. Plant Mol Biol 2006:389–404

Porcel R, Aroca R, Ruiz-Lozano JM (2012) Salinity stress alleviation using arbuscular mycorrhizal fungi. Agron Sustain Dev 32:181–200

Porras-Soriano A, Soriano-Martin ML, Porras-Piedra A, Azcon R (2009) Arbuscular mycorrhizal fungi increased growth, nutrient uptake and tolerance to salinity in olive trees under nursery conditions. J Plant Physiol 166:1350–1359

Pozo MJ, Azcón-Aguilar C (2007) Unravelling mycorrhiza-induced resistance. Curr Opin Plant Biol 10:393–398

Rabie GH, Almadini AM (2005) Role of bioinoculants in development of salt-tolerance of *Vicia faba* plants under salinity stress. Afr J Biotechnol 4:210–222

Raghavendra AS, Gonugunta VK, Christmann A, Grill E (2010) ABA perception and signalling. Trends Plant Sci 15:395–401

Ramos AC, Façanha AR, Feijó JA (2008) A proton (H^+) flux signature of the presymbiotic development of the arbuscular mycorrhizal fungi. New Phytol 178:177–188

Riedel T, Groten K, Baldwin IT (2008) Symbiosis between *Nicotiana attenuate* and *Glomus intraradices*: ethylene plays a role, jasmonic acid does not. Plant Cell Environ 31:1203–1213

Ruiz-Lozano JM, Aroca R (2010) Modulation of aquaporin genes by the arbuscular mycorrhizal symbiosis in relation to osmotic stress tolerance. In: Sechback J, Grube M (eds) Symbiosis and stress. Springer, Berlin

Ruiz-Lozano JM, Porcel R, Aroca R (2006) Does the enhanced tolerance of arbuscular mycorrhizal plants to water deficit involve modulation of drought-induced plant genes? New Phytol 171:693–698

Ruiz-Lozano JM, Alguacil MM, Bárzana G, Vernieri P, Aroca R (2009) Exogenous ABA accentuates the differences in root hydraulic properties between mycorrhizal and non mycorrhizal maize plants through regulation of PIP aquaporins. Plant Mol Biol 70:565–579

Ruiz-Lozano JM, Porcel R, Azcón C, Aroca R (2012) Regulation by arbuscular mycorrhizae of the integrated physiological response to salinity in plants: new challenges in physiological and molecular studies. J Exp Bot 63:4033–4044

Ruiz-Sanchez M, Aroca R, Munoz Y, Polon R, Ruiz-Lozano JM (2010) The arbuscular mycorrhizal symbiosis enhances the photosynthetic efficiency and the antioxidative response of rice plants subjected to drought stress. J Plant Physiol 167:862–869

Ruth B, Khalvati M, Schmidhalter U (2011) Quantification of mycorrhizal water uptake via high-resolution on-line water content sensors. Plant Soil 342:459–468

Sakurai J, Ishikawa F, Yamaguchi T, Uemura M, Maeshima M (2005) Identification of 33 rice aquaporin genes and analysis of their expression and function. Plant Cell Physiol 46:1568–1577

Samaras Y, Bressan RA, Csonka LN, Garcia-Rios M, Paino D'Urzo M, Rhodes D (1995) Proline accumulation during water deficit. In: Smirnoff N (ed) Environment and plant metabolism. Flexibility and acclimation. Bios Scientific, Oxford

Schachtman DP, Goodger JQD (2008) Chemical root to shoot signaling under drought. Trends Plant Sci 13:281–287

Scheibe R, Beck E (2011) Drought, desiccation, and oxidative stress. In: Lüttge U, Bech E, Bartels D (eds) Plant desiccation tolerance. Ecological studies, vol 215. Springer, Berlin

Shao HB, Chu LY, Shao MA, Jaleel CA, Mi HM (2008) Higher plant antioxidants and redox signaling under environmental stresses. C R Biol 331:433–441

Sharifi M, Ghorbanli M, Ebrahimzadeh H (2007) Improved growth of salinity-stressed soybean after inoculation with salt pre-treated mycorrhizal fungi. J Plant Physiol 164:1144–1151

Shaterian J, Waterer D, De Jong H, Tanino KK (2005) Differential stress responses to NaCl salt application in early- and late maturing diploid potato (*Solanum* sp.) clones. Environ Exp Bot 54:202–212

Sheen J, Zhou L, Jang JC (1999) Sugars as signaling molecules. Curr Opin Plant Biol 2:410–418

Sheng M, Tang M, Chen H, Yang BW, Zhang FF, Huang YH (2008) Influence of arbuscular mycorrhizae on photosynthesis and water status of maize plants under salt stress. Mycorrhiza 18:287–296

Sheng M, Tang M, Zhang F, Huang Y (2011) Influence of arbuscular mycorrhiza on organic solutes in maize leaves under salt stress. Mycorrhiza 21:423–430

Shokri S, Maadi B (2009) Effect of arbuscular mycorrhizal fungus on the mineral nutrition and yield of *Trifolium alexandrinum* plants under salinity stress. J Agron 8:79–83

Smeekens S (2000) Sugar-induced signal transduction in plants. Annu Rev Plant Biol 51:49–81

Smith SE, Read DJ (eds) (2008) Mycorrhizal symbiosis. Academic, San Diego, CA

Steudle E, Peterson CA (1998) How does water get through roots? J Exp Bot 49:775–788

Sutka M, Alleva K, Parisi M, Amodeo G (2005) Tonoplast vesicles of *Beta vulgaris* storage root show functional aquaporins regulated by protons. Biol Cell 97:837–846

Talaat NB, Shawky BT (2011) Influence of arbuscular mycorrhizae on yield, nutrients, organic solutes, and antioxidant enzymes of two wheat cultivars under salt stress. J Plant Nutr Soil Sci 174:283–291

Tang C, Cobley BT, Mokhtara S, Wilson CE, Greenway H (1993) High pH in the nutrient solution impairs water uptake in *Lupinus angustifolius* L. Plant Soil 155–156:517–519

Tavakkoli E, Rengasamy P, McDonald GK (2010) High concentrations of Na^+ and Cl^- ions in soil solution have simultaneous detrimental effects on growth of faba bean under salinity stress. J Exp Bot 61:4449–4459

Tester M, Davenport R (2003) Na^+ tolerance and Na^+ transport in higher plants. Ann Bot 91:503–527

Tisserant E, Kohler A, Dozolme-Seddas P et al (2012) The transcriptome of the arbuscular mycorrhizal fungus *Glomus intraradices* (DAOM 197198) reveals functional tradeoffs in an obligate symbiont. New Phytol 193:755–769

Törnroth-Horsefield S, Wang Y, Hedfalk K, Johanson U, Karlsson M, Tajkhorshid E, Neutze R, Kjellbom P (2006) Structural mechanism of plant aquaporin gating. Nature 439:688–694

Tournaire-Roux C, Sutka M, Javot H, Gout E, Gerbeau P, Luu DT, Bligny R, Maurel C (2003) Cytosolic pH regulates root water transport during anoxic stress through gating of aquaporins. Nature 425:393–397

Trotel-Aziz P, Niogret MF, Larher F (2000) Proline level is partly under the control of abscisic acid in canola leaf discs recovery from hyper-osmotic stress. Physiol Plantarum 110:376–383

Türkan I, Demiral T (2009) Recent developments in understanding salinity tolerance. Environ Exp Bot 67:2–9

Tyerman SD, Skerrett IM (1999) Root ion channels and salinity. Sci Hortic 78:175–235

Wan X (2010) Osmotic effects of NaCl on cell hydraulic conductivity of corn roots. Acta Biochim Biophys Sin 42:351–357

Wang Y, Mopper S, Hasentein KH (2001) Effects of salinity on endogenous ABA, IAA, JA and SA in *Iris hexagona*. J Chem Ecol 27:327–342

Wang KLC, Li H, Ecker JR (2002) Ethylene biosynthesis and signaling networks. Plant Cell 14:S131–S151

Wang FY, Liu RJ, Lin XG, Zhou JM (2004) Arbuscular mycorrhizal status of wild plants in saline-alkaline soils of the Yellow River Delta. Mycorrhiza 14:133–137

White PJ, Broadley MR (2001) Chloride in soils and its uptake and movement within the plant: a review. Ann Bot 88:967–988

Wilkinson S, Davies WJ (2010) Drought, ozone, ABA and ethylene: new insights from cell to plant to community. Plant Cell Environ 33:510–525

Wolters H, Jürgens G (2009) Survival of the flexible: hormonal growth control and adaptation in plant development. Nat Rev Genet 10:305–317

Wu QS, Zou YN (2009) Mycorrhiza has a direct effect on reactive oxygen metabolism of drought-stressed citrus. Plant Soil Environ 55:436–442

Wu QS, Zou YN, Xia RX, Wang MY (2007) Five *Glomus* species affect water relations of *Citrus tangerine* during drought stress. Bot Stud 48:147–154

Wu QS, Zou YN, Liu W, Ye XF, Zai HF, Zhao LJ (2010) Alleviation of salt stress in citrus seedlings inoculated with mycorrhiza: changes in leaf antioxidant defense systems. Plant Soil Environ 56:470–475

Xu G, Magen H, Tarchitzky J, Kafkaki U (2000) Advances in chloride nutrition. Adv Agr 68:96–150

Yang HM, Zhang JH, Zhang XY (2005) Regulation mechanisms of stomatal oscillation. J Integr Plant Biol 47:1159–1172

Zelenina M, Bondar AA, Zelenin S, Aperia A (2003) Nickel and extracellular acidification inhibit the water permeability of human aquaporin-3 in lung epithelial cells. J Biol Chem 278:30037–30043

Zhang J, Jia W, Yang J, Ismail AM (2006) Role of ABA in integrating plant responses to drought and salt stresses. Field Crop Res 97:111–119

Zhang Y, Zhu H, Zhang Q, Li M, Yan M, Wang R, Wang L, Welti R, Zhang W, Wang X (2009) Phospholipase Dα1 and phosphatidic acid regulate NADPH oxidase activity and production of reactive oxygen species in ABA-mediated stomatal closure in *Arabidopsis*. Plant Cell 21:2357–2377

Zuccarini P, Okurowska P (2008) Effects of mycorrhizal colonization and fertilization on growth and photosynthesis of sweet basil under salt stress. J Plant Nutr 31:497–513

Chapter 15
Root Allies: Arbuscular Mycorrhizal Fungi Help Plants to Cope with Biotic Stresses

María J. Pozo, Sabine C. Jung, Ainhoa Martínez-Medina, Juan A. López-Ráez, Concepción Azcón-Aguilar, and José-Miguel Barea

15.1 Introduction

Microbial populations living at the root–soil interfaces, the rhizosphere, are immersed in a framework of interactions known to influence plant growth and health and soil quality, key issues for agroecosystem sustainability (Barea and Azcón-Aguilar 2012). Rhizosphere microbes, mainly bacteria and fungi, have different trophic/living habits and establish a variety of saprophytic or symbiotic relationships with the plant, either detrimental or beneficial. They are able to affect plant growth and to influence plant responses to biotic and abiotic stresses (Barea et al. 2005). Some of the mutualistic microorganisms are known to enter the root system of their hosts adopting an endophytic lifestyle, which benefit both the plant and the microbial endophyte. These endophytic microorganisms include (1) plant mutualistic symbionts, such as N_2-fixing rhizobial and actinorhizal bacteria and mycorrhizal fungi; (2) root endophytic fungi; and (3) certain plant growth-promoting rhizobacteria (Barea et al. 2005). The growth of beneficial endophytic microbes inside plant roots requires mutual recognition and substantial coordination of plant and microbial responses. Defense mechanisms are coordinated by the plant immune system which allows the plant to distinguish non-self alien organisms by recognizing structurally conserved microbe-associated molecules, collectively termed *m*icrobe-*a*ssociated *m*olecular *p*atterns (MAMPs). Then different signal molecules including phytohormones such as salicylic acid (SA) and jasmonic acid (JA) orchestrate the plant defense response (Pieterse et al. 2009).

M.J. Pozo (✉) • S.C. Jung • A. Martínez-Medina • J.A. López-Ráez • C. Azcón-Aguilar • J.-M. Barea
Department of Soil Microbiology and Symbiotic Systems, Estación Experimental del Zaidín (CSIC), Professor Albareda 1, 18008 Granada, Spain
e-mail: mariajose.pozo@eez.csic.es; mjpozo@eez.csic.es; sabine.jung@eez.csic.es; ammedina@cebas.csic.es; juan.lopezraez@eez.csic.es; concepcion.azcon@eez.csic.es; jmbarea@eez.csic.es

Like pathogens, beneficial organisms are confronted with the innate immune system of the roots and colonization success essentially depends on the evolution of strategies for immune evasion. The modulation of plant defense responses by microbial symbionts aids establishing the delicate balance between the two partners and may result in an enhanced defensive capacity of the plant (Pozo and Azcón-Aguilar 2007; Zamioudis and Pieterse 2012). Moreover, this activated defense is often also expressed in aboveground plant tissues, thereby providing the plant with an induced systemic resistance effective against a broad spectrum of plant pathogens (Pozo and Azcón-Aguilar 2007; van Wees et al. 2008).

Well-known examples of beneficial endophytic soil fungi are the mycorrhizal fungi, common components of the soil microbial biomass which establish mutualistic symbioses with most terrestrial plants, including economically important crop species (Smith and Read 2008). Mycorrhizal symbioses can be found in almost all ecosystems worldwide to improve plant growth and health through key ecological processes (Azcón-Aguilar et al. 2009). Most of the major plant families form arbuscular mycorrhizal (AM) associations, the most common mycorrhizal type (Smith and Read 2008). The AM symbiosis increases plant growth; enhances available soil nutrient uptake; increases plant resistance to drought; protects plants against a wide range of belowground attackers such as soilborne fungal and bacterial pathogens, nematodes, or root-chewing insects; and, as reported during the last decade, induces resistance against shoot pathogens (Whipps 2004; Pozo and Azcón-Aguilar 2007; Gianinazzi et al. 2010; Jeffries and Barea 2012; Jung et al. 2012). Therefore, the AM fungi, a major group of plant mutualistic endophytic microorganisms, can be considered as "root allies" which support plants to cope with biotic stresses. Because of their role in enhancing plant resistance to potential deleterious organisms, including microbial pathogens, nematodes, phytopathogenic insects, and parasitic plants (Jung et al. 2012), these microorganisms will be the target endophytic symbiont of this chapter. Accordingly, we will first summarize some key concepts related to (1) the biology and ecology of AM fungi, (2) the biology and functions of the AM symbiosis, and (3) how the AM symbiosis can be managed, as a low-input biotechnology, to help sustainable environmentally friendly agro-technological practices. Then, and as the main information core of this article, we will summarize and discuss the impact of AM on plant resistance to biotic stresses and the underlying mechanisms. Special emphasis will be devoted to the role of plant defense mechanisms, as experimental evidences support that mycorrhizal interactions can effectively stimulate the host's defensive capacity. This can lead to a primed state of the plant, thereby boosting defense mechanisms triggered upon attack by potential enemies, not only locally in the roots but also in aboveground tissues.

15.2 Nature and Functions of Arbuscular Mycorrhizal Fungi: The AM Symbiosis

The AM fungi establish the so-called AM symbiosis with most vascular plants where both partners exchange nutrients and obtain important benefits. The soilborne AM fungi colonize the root cortex biotrophically and then develop an external mycelium, a bridge connecting the root with the surrounding soil microhabitats (Smith and Read 2008). In this section, the most significant aspects of the available information on the biology and ecology of AM fungi and the biology, functions, and management of the AM symbiosis are summarized.

15.2.1 Biology and Ecology of AM Fungi

The AM fungi are ubiquitous soilborne microbial fungi, whose origin and divergence dates back more than 450 million years (Redecker et al. 2000; Bonfante and Genre 2008; Schüßler and Walker 2011). Molecular analyses and fossil records indicate that AM associations evolved as a symbiosis, facilitating the adaptation of plants to the terrestrial environment and suggesting that AM fungi played a crucial role in land colonization by plants (Schüßler and Walker 2011). Actually, the primitive roots developed in association with AM fungi and coevolved with them to build up the mycorrhizal root systems of extant vascular plants (Brundrett 2002). As a consequence of this coevolution, the AM relationship became an integral component of plant ecology in both natural and agricultural ecosystems (Brundrett 2002).

Because of their very peculiar evolutionary history, underground lifestyle, and genetic makeup, AM fungi are endowed with unusual biological traits (Parniske 2008; Bonfante and Genre 2008; Barea and Azcón-Aguilar 2012). AM fungi develop a typically aseptate and coenocytic mycelial network containing hundreds of nuclei sharing the same cytoplasm and produce very large multinucleate spores having abundant storage lipids and resistant thick walls containing chitin (Smith and Read 2008). Among other important characteristics, AM fungi are asexual, unculturable, and obligatorily biotrophic microbes (Schüßler and Walker 2011). The character of obligate symbionts, unabling them to complete their life cycle without colonizing a host plant, has hampered the study of the biology and the biotechnological applications of AM fungi (Bago and Cano 2005).

The AM fungi belong to the phylum *Glomeromycota* (Schüßler et al. 2001; Schüßler and Walker 2011). Different polymorphic DNA-sequence variants are distributed among different nuclei in the same coenocytic hypha or spore (Rosendahl 2008; Sanders and Croll 2010; Schüßler and Walker 2011). Thus, from a taxonomic point of view, there are difficulties for defining clear species concepts of individuals and also boundaries within populations. However, molecular phylogenetic and evolutionary analyses are substantially contributing to our knowledge on AM fungi speciation. Diversity studies based on molecular approaches allowed us to ascertain that individual fungal strains exhibit little host specificity, while a single plant can be colonized by many different AM fungal species within the same root. A certain degree of host preference (functional compatibility) was evidenced to occur and this has been shown to play an important role in regulating the diversity, stability, and productivity of agroecosystems (Jeffries and Barea 2012).

Undoubtedly, the more relevant biological characteristic of AM fungi is their capacity to form AM associations with members of all phyla of land plants, whatever their taxonomic position, life form, or geographical distribution (Smith and Read 2008).

15.2.2 Biology and Functioning of AM Symbiosis

The information generated during the last years on the cellular and molecular events taking place during AM establishment and the ecophysiological and molecular components of AM functioning have recently been reviewed (Bonfante and Genre 2008; Parniske 2008; Smith and Read 2008; Gianinazzi-Pearson et al. 2009; Barea and Azcón-Aguilar 2012).

AM fungi can colonize plant roots from three main types of soil-based propagules: spores, fragments of mycorrhizal roots, and extraradical hyphae, all of them producing a more or less well-developed mycelial network expanding in the soil. When a hypha from an asymbiotic, soil-based AM mycelium approaches a host root, an exchange of signaling molecules between both symbionts takes place (Parniske 2008; Gianinazzi-Pearson et al. 2009; Genre and Bonfante 2010). This molecular dialogue activates specific signaling pathways affecting fungal development and plant gene expression. Several plant regulatory genes are involved in reprogramming processes from a direct cell-to-cell contact on the root surface to the intracellular accommodation of the fungal symbiont. When finally a hypha contacts the plant root, it adheres to epidermal cells forming a characteristic fungal structure called appressorium (or hyphopodium). This event marks the initiation of the symbiotic phase which ends with the production of the characteristic treelike structures, termed "arbuscules," that the fungus develops within the root cortical cells. The arbuscules, which give name to the symbiosis, are the structures where most of the nutrient exchange between the fungus and the plant is thought to occur (Smith and Read 2008).

Following root colonization, AM fungi form extensive mycelial networks outside the root—the extraradical mycelium (ERM)—where spores are developed completing their life cycle. The ERM results in a tridimensional structure specialized in the acquisition of mineral nutrients from soil, particularly those whose ionic forms have poor mobility or are present in low concentration in the soil solution, as it is the case with phosphate and ammonia (Barea and Azcón-Aguilar 2012). Through the activities of the interlinked and extensive soil ERM,

AM fungi affect the distribution and movement of nutrients within the soil ecosystem (Richardson et al. 2009). The major flux is the transfer of carbon from plant to fungus (and thereby to the soil) and the reciprocal movement of phosphate and ammonium from fungus to plant. In addition to the uptake of nutrients, the AM symbiosis improves plant performance through increased protection against environmental stresses, whether they be biotic (e.g., pathogen and herbivore attack) or abiotic (e.g., drought, salinity, heavy metals toxicity, or presence of organic pollutants), and also enhances soil structure through the formation of hydro-stable aggregates necessary for good soil tilth (Barea and Azcón-Aguilar 2012). The ecological, physiological, and molecular basis of AM functioning and their implications in enabling the plant to cope more effectively with natural or anthropogenic environmental stress, either biotic or abiotic, have been the subject of diverse experimental and review studies during the last decade (Barea and Azcón-Aguilar 2012).

15.2.3 Managing the AM Fungi and Symbiosis

Nowadays, it is well accepted that AM associations, which helped plants to thrive in hostile environments such as those prior to their origin and during their evolution, continue helping plants to develop in stressed environments (Barea et al. 2011). This is fundamental because adverse conditions, particularly exacerbated in the current scenario of global climate change, generate a great array of stress situations affecting the stability of both natural and agricultural ecosystems. Plants must be able to cope with these stresses. Consequently, adaptive strategies should be developed in order to increase their resilience to overcome negative impacts. AM establishment can be considered one of these adaptive strategies, through their ability to increase host tolerance to environmental constraints (Barea et al. 2011; Jeffries and Barea 2012).

In nonagricultural situations, as plant diversity has been related to AM diversity, maintenance of a sustainable mixed plant population depends on the maintenance of a diverse AM fungi population and vice versa (Maherali and Klironomos 2007). Thus, it is relevant to recognize that the activity and diversity of mycorrhizal fungi are key elements linking biodiversity and ecosystem functioning (Read 1998).

On the other hand, the increasing demand for low-input agriculture has resulted in greater interest in the manipulation and use of beneficial soil microorganisms. Thus, management of native populations of AM fungi is recognized as a sustainable strategy in agriculture because it can reduce the use of chemicals and energy in agriculture leading to a more sustainable production, while minimizing environmental degradation (Jeffries and Barea 2012). These biological interventions are becoming more attractive as the use of chemicals for fumigation and disease control is progressively discouraged and fertilizers have become more and more expensive (Atkinson 2009).

Because of the importance of the AM symbiosis in sustainable agriculture or restoration of ecosystems management, the development of techniques for AM inoculant production and inoculation has become a focal point of research. The difficulty in culturing obligate symbionts such as AM fungi in the absence of their host plant is a major obstacle for massive inoculum productions (Baar 2008). Despite these problems, the beneficial effects of AM fungi on plant growth have led to their development as bioinoculants for forestry, agriculture, and horticulture (Ijdo et al. 2011), and several companies worldwide are producing AM inoculum products which are now commercially available (Gianinazzi and Vosátka 2004; Vosátka et al. 2008; Ijdo et al. 2011).

15.3 Disease Control by AM Fungi

As stated, AM symbiosis implies important changes in the plant physiology. As a consequence, the association may impact the plant interaction with other organisms. Many studies have shown the protective effect of colonization by AM fungi against infections by microbial pathogens and other deleterious organisms in different plant systems (reviewed in Whipps 2004; Jung et al. 2012). This bioprotection has been termed mycorrhiza-induced resistance (MIR) (Pozo and Azcón-Aguilar 2007). In this section, we will summarize the main conclusions from recent studies about MIR, schematized in Fig. 15.1.

15.3.1 Protection Against Soilborne Pathogens

AM symbioses are able to reduce the damage caused by many soilborne pathogens on a variety of crop species. Although there are some examples of protection against pathogenic bacteria, most reports focus on harmful fungi and oomycetes, including major pathogens from the genera *Fusarium*, *Rhizoctonia*, *Verticillium*, *Phytophthora*, *and Pythium* (Whipps 2004; Jung et al. 2012). A reduction of the detrimental effects by parasitic nematodes has also been reported in AM plants (Pinochet et al. 1996; De La Peña et al. 2006; Li et al. 2006; Elsen et al. 2008; Vos et al. 2011; Hao et al. 2012). There are, however, relatively few studies on the impact of AM fungi on root-feeding insects, and these mostly focus on members of the genus *Otiorhynchus*, or weevils (Koricheva et al. 2009). The larvae of these insects are rhizophagous, whereas the adults feed on the foliage of the same plant. A clear protective effect of AM fungi on these pests has been reported (Gange 1996, 2001; Koricheva et al. 2009).

Several mechanisms operate simultaneously in the alleviation of the damage caused by soilborne pathogens. For example, AM fungi have been proven to directly compete with soilborne pathogens and nematodes for space and nutrients (Azcón-Aguilar and Barea 1996; Cordier et al. 1998; Filion et al. 1999; Norman and

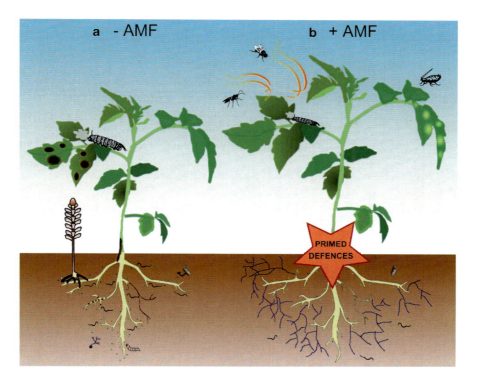

Fig. 15.1 (a) Non-mycorrhizal plant (−AMF). Absence of root colonization by AMF leads to stronger development of symptoms in response to necrotrophic pathogens and more damage upon feeding by chewing insects in roots and shoots when compared to mycorrhizal plants. Release of strigolactones (SLs) as part of the root exudates induces branching of AMF hyphae to promote mycorrhization but also induces germination of Orobanchaceae seeds which then parasitize the host plants' root system. (**b**) Mycorrhizal plant (+AMF). Growth promotion is often observed due to improved acquisition of mineral nutrients through the AM fungal hyphal network (represented in *blue*). Changes in the root exudate patterns repel nematodes and induce changes in the soil microbial community, possibly attracting antagonists of pathogens, and a reduced release of SLs minimizes the risk of infection by root parasitic plants. Priming of plant defenses leads to a general reduction of the incidence and/or damage caused by soilborne pathogens, nematodes, and chewing insects. In aboveground plant parts, viral and fungal biotrophs, as well as phloem-feeding insects, perform better on mycorrhizal plants. In contrast, the primed jasmonate-regulated plant defense mechanisms restrict the development of necrotrophic pathogens and the performance of phytophagous insects. Indirect defenses, such as the release of volatiles, are boosted and parasitoids are efficiently attracted

Hooker 2000; De La Peña et al. 2006) and to alter root morphology that may change the dynamics of pathogen infection (Schellenbaum et al. 1991; Norman et al. 1996). Colonization by AM fungi can lead to alterations in the quality and quantity of root exudates (Sood 2003; Pivato et al. 2008). These changes impact the microbial community of the mycorrhizosphere and, among other effects, may lead to a shift in its composition favoring certain components of the microbiota with the capacity to antagonize possible root pathogens (Barea et al. 2005; Badri and Vivanco 2009). Altered root exudation may also directly impact microbial pathogens and nematodes (see Jung et al. 2012 for references).

The use of experimental split-root systems has confirmed that the protection by mycorrhiza is manifested in non-colonized areas of the root system (Cordier et al. 1998; Pozo et al. 2002; Zhu and Yao 2004; Khaosaad et al. 2007; Hao et al. 2012). These experiments allowing physical separation of the AM fungi and the aggressor have highlighted the involvement of plant-mediated responses in the enhanced resistance, pointing out a major role for plant defense mechanisms. The plant defense mechanisms involved in root protection have been analyzed (Pozo et al. 2002; Jung et al. 2012) and will be discussed in Sect. 15.4.

15.3.2 Systemic Protection Against Leaf Pathogens

Recent findings have shown the ability of certain beneficial soil fungi for controlling shoot pathogens by eliciting a plant-mediated resistance response. The pathogens' lifestyles have been shown to determine the outcome of the interaction with regard to the biocontrol of shoot diseases by the AM symbiosis (Pozo and Azcón-Aguilar 2007; Jung et al. 2012). Early studies reported a higher susceptibility of AM plants to viruses, and biotrophic pathogens appear to develop better on AM plants, although an increased tolerance was observed in terms of plant productivity (Gernns et al. 2001; Whipps 2004). Regarding to pathogens with a hemibiotrophic lifestyle, the effect of the symbiosis varies from no effect to a reduction of the disease (Lee et al. 2005; Chandanie et al. 2006). In contrast, several studies evidence the positive effect of AM symbiosis on plant resistance against necrotrophic shoot pathogens (Fritz et al. 2006; de la Noval et al. 2007; Møller et al. 2009; Pozo et al. 2010; Campos-Soriano et al. 2012).

Two main plant-mediated mechanisms may be operative in the protective effect of beneficial soil fungi against leaf pathogens. One is the improvement of the nutritional status of the host plants and/or alterations of the source–sink relation that may help a plant overcome a herbivore attack or quickly recover after the attacker has been fought off. The other mechanism involves the activation of the plant defense response by the beneficial microorganisms (Jung et al. 2012). Because of the special interest of this modulation of the plant immune responses and its implications in plant resistance to biotic stresses, the topic is discussed in detail in Sect. 15.4.

15.3.3 Effect on Herbivorous Insects

Insects may be deleterious to plants by directly damaging them through herbivory or acting as vectors for pathogens such as viruses and phytoplasmas. However, they can also have positive effects on plant health acting as natural enemies of pests or as pollinators (Jung et al. 2012). The outcome of the AM-plant–herbivore interaction depends on many factors, such as the AM fungus, host plant, insect species involved, and prevailing environmental factors (Gange 2007; Pineda et al. 2010). Several reviews have tried to compile the published studies dealing with these multitrophic interactions, most of them from an ecological point of view (Gehring and Bennett 2009; Hartley and Gange 2009; Jung et al. 2012). The AM symbiosis can actually influence insect herbivore performance, but the magnitude and direction of the effect depend mainly upon the feeding mode and lifestyle of the insect (Hartley and Gange 2009; Koricheva et al. 2009). The improvement of plant nutrition by the symbiosis may have opposite effects: On the one hand, by improving nutrient and water uptake, mycorrhizas can facilitate the regrowth of tissues after herbivory by promoting plant tolerance through compensation of biomass losses. On the other hand, as nutrition improves, plants may become more nutritive or attractive to insects (Kula et al. 2005; Bennett and Bever 2007; Hoffmann et al. 2011). Furthermore, the induction of defense mechanisms operating in resistance against microbial pathogens may also impact herbivorous insects (Jung et al. 2012). The final impact on insect performance will depend on the interplay between a positive effect derived from the enhanced plant growth and a negative effect derived from the induced resistance in the plant and is depending on the type of the attacking insect. Generalist insects, able to feed on diverse plants and sensible to the plant defense mechanisms, are usually negatively affected by the presence of AM fungi (Gange and West 1994; Fontana et al. 2009). However, specialist insects, which feed from one or only a small number of host species and show a high degree of adaptation to their hosts' defense responses, usually perform better on AM plants, probably because of the improved nutritional quality of the host (Gehring and Bennett 2009; Hartley and Gange 2009). Some even prefer mycorrhizal plants for oviposition to improve growth and development of their offspring (Cosme et al. 2011).

The degree of protection also depends on the feeding guild of the attacking herbivore. Phloem-sucking insects produce minimal damage to the plant while feeding and thereby avoid detection by the host's immune system (Walling 2008). Thus, it is unlikely that potentiation of plant defense mechanisms in AM plants may have a significant impact on them. Moreover, they may profit from its higher nutritional value. In fact, higher incidence of phloem-sucking insects in AM plants has been reported (Gange et al. 1999; Goverde et al. 2000). In contrast, leaf chewers and miners are usually negatively affected by AM fungi (Gange and West 1994; Vicari et al. 2002). These insects feed on the leaf tissue and cause massive damage which activates defenses that depend on the plant hormone JA (Howe and Jander 2008). Remarkably, as discussed later on, JA seems to be a key signal in MIR.

15.3.4 Impact on Parasitic Plants

Plants of the genera *Striga* and *Orobanche* can parasitize a number of important crop plants. They attach to the host roots and acquire nutrients and water from them, constituting one of the most damaging agricultural pests (Bouwmeester et al. 2003). Several studies reported that the attachment and the emergence of *Striga* are reduced in AM plants (see López-Ráez et al. 2011b for references). It is known that seeds of these weeds germinate upon perception of strigolactones (SLs), a group of carotenoid-derived signaling molecules that are exuded by the roots of the host plant. These signals are produced by the plant under conditions of phosphate starvation and promote AM hyphal branching and thereby facilitate mycorrhiza establishment (Akiyama et al. 2005; Bouwmeester et al. 2007). Root parasitic plants have intercepted this recruitment system and utilize the signal for the detection of an appropriate host plant. Remarkably, mycorrhization downregulates the level of SLs once a well-established mycorrhiza is achieved, thus reducing the germination rate of weed seeds (Lendzemo et al. 2007; López-Ráez et al. 2011a). This reduction seems to be the underlying reason for the decrease in the incidence and damage of root parasitic plants on mycorrhizal plants (López-Ráez et al. 2012).

In addition to the root parasitic plants, it has been suggested that certain AM fungi may suppress growth of other aggressive agricultural weeds which cause crop yield losses every year (Rinaudo et al. 2010). These authors reported that the presence of AM fungi reduced total weed biomass, because most aggressive weeds are non-mycorrhizal, while the crop plant benefitted from AM symbiosis via enhanced phosphorus nutrition. Overall, these observations indicate that the use or stimulation of AM fungi in agroecosystems may suppress some aggressive weeds and suggest a possible applicability of the AM symbiosis in weed control.

15.4 Overview of Mechanisms Underlying the Impact of AM Fungi on Plant Protection Against Pathogen and Pests

As indicated above, AM fungi may alleviate biotic stresses through a combination of different mechanisms ranging from direct interactions as competition with the agressor to indirect, plant-mediated effects.

Direct effects include competition for carbon, nitrogen, and other growth factors and competition for niches or specific infection sites. Direct competition has been suggested as mechanism by which AM fungi can reduce the abundance of pathogenic fungi in roots (Filion et al. 2003). Presumably, pathogenic and AM fungi exploit common resources within the root, including infection/colonization sites, space, and photosynthates (reviewed in Whipps 2004). Negative correlations between the abundance of AM fungal structures and pathogenic microorganisms have been found in roots and soil (Filion et al. 2003; St Arnaud and Elsen 2005).

Full exclusion of the pathogenic oomycete *Phytophthora* from arbusculated cells was also evidenced (Cordier et al. 1998).

Root colonization by AM fungi also induces changes in the root system architecture, in morphology, and in root exudates (Schellenbaum et al. 1991; Norman et al. 1996; Pivato et al. 2008). These changes may alter the dynamics of infection by the pathogen or impact on the microbial community of the mycorrhizosphere favoring components of the microbiota with the capacity to antagonize root pathogens (Barea et al. 2005; Badri and Vivanco 2009). Changes in root exudation can directly impact microbial pathogens and nematodes (Norman and Hooker 2000; Vos et al. 2011). More recent findings indicate that a primary mechanism of pathogen control occurs through the ability of AM fungi to reprogram plant gene expression (Liu et al. 2007; López-Ráez et al. 2010a; Campos-Soriano et al. 2012). As consequence, alterations in the primary and secondary metabolism of the plant do occur, many of these changes being related to plant defense (Hause et al. 2007; Schliemann et al. 2008; López-Ráez et al. 2010b).

Actually, as other biotrophs, AM fungi are able to trigger plant defense responses at initial stages (Paszkowski 2006). Thus, for a successful colonization, the fungus has to cope with these reactions and actively modulate plant defense responses. This modulation may result in pre-conditioning of the tissues for efficient activation of plant defenses upon a challenger attack, a phenomenon that is called priming (Pozo and Azcón-Aguilar 2007). Priming set the plant is an "alert" state in which defenses are not actively expressed but in which the response to an attack occurs faster and/or stronger compared to plants not previously exposed to the priming stimulus, efficiently increasing plant resistance. Thus, priming confers important plant fitness benefits (Conrath et al. 2006; Walters and Heil 2007) thereby defense priming by AM has a great ecological relevance (Jung et al. 2012).

15.4.1 Modulation of the Host Plant's Immune System by AM Fungi and Induced Systemic Resistance

Both mutualistic and pathogenic biotroph fungi are initially recognized as alien organisms and the plant reacts with the activation of an immune response. As stated before, the AM fungus has to deal with the plant's immune system, contend with the defense mechanisms, and overcome them for a successful colonization of the host (Kloppholz et al. 2011; Zamioudis and Pieterse 2012). Once established, the plant has to regulate the level of AM fungal proliferation within the roots to prevent excessive colonization and carbon drainage, thus maintaining the interaction at an equilibrium to limit the colonization by the mutualistic symbionts. Actually, plants possess a feedback system to prevent excessive colonization over a critical threshold, a phenomenon termed autoregulation of the symbiosis (Vierheilig et al. 2008). In summary, from pre-symbiotic stages and throughout a well-established AM association, plant defense mechanisms are tightly regulated to control the symbiosis. As a side effect, this regulation of plant defenses in the root may directly impact root pathogens.

During the early stages of the interaction, the plant reacts to the presence of AM fungi activating some defense-related responses that are subsequently suppressed (García-Garrido and Ocampo 2002). A quick but transient increase of endogenous salicylic acid (SA) occurs in the roots with a concurrent accumulation of defensive compounds, such as reactive oxygen species, specific isoforms of hydrolytic enzymes, and the activation of the phenylpropanoid pathway (reviewed in Jung et al. 2012). These initial reactions are temporally and spatially limited compared to the reaction during plant–pathogen interactions, suggesting a role in the establishment or control of the symbiosis (Dumas-Gaudot et al. 1996; García-Garrido and Ocampo 2002). To promote successful colonization, AM fungi likely have to evade and manipulate the host innate immune system. Indeed, recent studies support that AM fungi can actively suppress plant defense reactions by secreting effector proteins that interfere with the host's immune system (Kloppholz et al. 2011). However, even in later stages of the interaction, the levels of SA and other defense-related phytohormones, such as jasmonic acid (JA), abscisic acid (ABA), and ethylene (ET), may be altered in mycorrhizal roots. These changes may contribute to control the extension of fungal colonization and the functionality of the symbiosis on mutualistic terms (Hause et al. 2007; López-Ráez et al. 2010a). Indeed, regulation of JA has been reported to have a central role in the correct functioning of the AM symbiosis (Hause and Schaarschmidt 2009). As regulation of defense signaling molecules occurs, the plant immunity system is altered in AM plants, and this may play a major role in MIR. Often, this induced resistance is also expressed in aboveground plant tissues, giving rise to a systemic response that is typically effective against a broad spectrum of plant pathogens and even herbivores. The dependence of successful mycorrhization on the control of JA and SA signaling would explain the range of protection conferred by this symbiosis (Pozo and Azcón-Aguilar 2007; Jung et al. 2012). Mycorrhizal plants are more resistant to necrotrophs and chewing insects, which are targeted by JA-dependent defense responses, while frequently being more susceptible to biotrophs, as these are targeted by SA-regulated defenses. This pattern correlates with an activation of JA-dependent defenses and repression of SA-dependent ones in a well-established mycorrhiza. The antagonistic interaction between SA and JA signaling is a conserved mechanism for plant defense regulation (Thaler et al. 2012).

15.5 Priming for Enhanced Defense by AM Fungi

The induction of resistance after plant root colonization by AM fungi does not necessarily require direct activation of defense mechanisms but can result from a sensitization of the tissue upon appropriate stimulation to express basal defense mechanisms more efficiently after subsequent pathogen attack (Pozo and Azcón-Aguilar 2007; Jung et al. 2012). This priming of the plant's innate immune system is common upon interaction with beneficial microorganisms and has important fitness benefits compared to direct activation of defenses (Van Hulten et al. 2006; Van Wees et al. 2008; Conrath 2009). Several mechanisms have been proposed to mediate the induction of the primed state as a moderate accumulation of defense-related regulatory molecules, such as transcription factors or MAP kinases and chromatin modifications (Pozo et al. 2008; Beckers et al. 2009; Van der Ent et al. 2009; Pastor et al. 2012). For example, rhizobacteria-induced systemic resistance in *Arabidopsis* is related to priming of JA-dependent responses through the accumulation of MYC2, a transcription factor with a key role in the regulation of JA responses (Pozo et al. 2008).

Examples of primed defense responses in AM plants were first observed in root tissues. Mycorrhizal-transformed carrot roots displayed stronger defense reactions at sites challenged by *Fusarium* (Benhamou et al. 1994). In tomato, AM colonization systemically protected roots against *Phytophthora parasitica* infection. Only mycorrhizal plants formed papilla-like structures around the sites of pathogen infection through deposition of non-esterified pectins and callose, preventing the pathogen from spreading further, and they accumulated significantly more PR-1a and basic β-1,3 glucanases than non-mycorrhizal plants upon *Phytophthora* attack (Cordier et al. 1998; Pozo et al. 1999, 2002). Similarly, mycorrhizal potatoes showed amplified accumulation of the phytoalexins rishitin and solavetivone upon *Rhizoctonia* infection, whereas AM fungi alone did not affect the levels of these compounds (Yao et al. 2003). Primed accumulation of phenolic compounds in AM date palm trees has been related to protection against *F. oxysporum* (Jaiti et al. 2007), and priming has also been involved in AM induction of resistance against nematodes (Li et al. 2006; Hao et al. 2012).

However, the primed response is not restricted to the root system as priming of defenses has also been shown in shoots of AM plants (Pozo et al. 2010). Actually, the AM symbiosis induced systemic resistance in tomato plants against the necrotrophic foliar pathogen *Botrytis cinerea*. While the amount of pathogen in leaves of mycorrhizal plants was significantly lower, the expression of some jasmonate-regulated, defense-related genes was higher in those plants (Pozo et al. 2010; Jung et al. unpublished). A primed response of JA-dependent defenses was confirmed by transcript profiling of leaves after exogenous application of JA, since JA-responsive genes were induced earlier and to a higher extend in AM plants (Pozo et al. 2009). The use of tomato mutants impaired in JA signaling confirmed that JA is required for AM-induced resistance against *Botrytis* (Jung et al. 2012), corroborating that MIR is similar to the well-studied rhizobacteria-induced systemic resistance (ISR) in *Arabidopsis* and requires a functional JA signaling pathway for the efficient induction of resistance (Pieterse et al. 1998).

15.6 Conclusions

Arbuscular mycorrhizal fungi are key elements in natural and man-made ecosystems. The establishment of the AM symbiosis with plant roots significantly alters the host plant physiology and has far-reaching consequences for the plant and its biotic interactions. Generally, mycorrhizal symbioses enhance the plant's ability to cope with biotic stresses. Even though the individual outcome always depends on the AMF-plant attacker combination, protective effects against deleterious organisms ranging from microbial pathogens to herbivorous insects and parasitic plants have been widely described.

Experimental evidences confirm that this protection is based not only on improved nutrition or local changes within the roots and the rhizosphere, but that plant defense mechanisms play a key role. Mycorrhizal colonization can prime plant immunity by boosting the plant ability to respond to an attack. In this process, jasmonate signaling appears as a central element. Unveiling the principles behind a successful symbiosis and the functional interplay between plant and fungus is of great interest. Particularly, the identification of defense regulatory elements coordinating mycorrhizal development and mycorrhiza-induced resistance is a major challenge for research. This identification will pave the way to the development of biotechnological strategies for improving mycorrhiza establishment and the use of AMF in the integrated management of pests and diseases.

References

Akiyama K, Matsuzaki KI, Hayashi H (2005) Plant sesquiterpenes induce hyphal branching in arbuscular mycorrhizal fungi. Nature 435:824–827

Atkinson D (2009) Soil microbial resources and agricultural policies. In: Azcón-Aguilar C, Barea JM, Gianinazzi S, Gianinazzi-Pearson V (eds) Mycorrhizas functional processes and ecological impact. Springer, Berlin, pp 33–45

Azcón-Aguilar C, Barea JM (1996) Arbuscular mycorrhizas and biological control of soil-borne plant pathogens—an overview of the mechanisms involved. Mycorrhiza 6:457–464

Azcón-Aguilar C, Barea JM, Gianinazzi S, Gianinazzi-Pearson V (2009) Mycorrhizas functional processes and ecological impact. Springer, Berlin

Baar J (2008) From production to application of arbuscular mycorrhizal fungi in agricultural systems: requirements and needs. In: Varma A (ed) Mycorrhiza: state of the art, genetics and molecular biology, eco-function, biotechnology, eco-physiology, structure and systematics. Springer, Berlin, pp 361–373

Badri DV, Vivanco JM (2009) Regulation and function of root exudates. Plant Cell Environ 32:666–681

Bago B, Cano C (2005) Breaking myths on arbuscular mycorrhizas in vitro biology. In: Declerck S, Strullu FG, Fortin JA (eds) In vitro culture of mycorrhizas, vol 4, Soil biology. Springer, Berlin, pp 111–138

Barea JM, Azcón-Aguilar C (2012) Evolution, biology and ecological effects of arbuscular mycorrhizas. In: Camisão AF, Pedroso CC (eds) Symbiosis: evolution, biology and ecological effects. Nova Publishers, Hauppauge, NY, pp 1–34

Barea JM, Pozo MJ, Azcon R, Azcon-Aguilar C (2005) Microbial co-operation in the rhizosphere. J Exp Bot 56:1761–1778
Barea JM, Palenzuela J, Cornejo P, Sánchez-Castro I, Navarro-Fernández C, Lopéz-García A, Estrada B et al (2011) Ecological and functional roles of mycorrhizas in semi-arid ecosystems of southeast Spain. J Arid Environ 75:1292–1301
Beckers GJM, Jaskiewicz M, Liu Y, Underwood WR, He SY, Zhang S, Conrath U (2009) Mitogen-activated protein kinases 3 and 6 are required for full priming of stress responses in *Arabidopsis thaliana*. Plant Cell 21:944–953
Benhamou N, Fortin JA, Hamel C, St Arnaud M, Shatilla A (1994) Resistance responses of mycorrhizal Ri T-DNA-transformed carrot roots to infection by *Fusarium oxysporum* f. sp. *chrysanthemi*. Phytopathology 84:958–968
Bennett AE, Bever JD (2007) Mycorrhizal species differentially alter plant growth and response to herbivory. Ecology 88:210–218
Bonfante P, Genre A (2008) Plants and arbuscular mycorrhizal fungi: an evolutionary-developmental perspective. Trends Plant Sci 13:492–498
Bouwmeester HJ, Matusova R, Zhongkui S, Beale MH (2003) Secondary metabolite signalling in host–parasitic plant interactions. Curr Opin Plant Biol 6:358–364
Bouwmeester HJ, Roux C, López-Ráez JA, Bécard G (2007) Rhizosphere communication of plants, parasitic plants and AM fungi. Trends Plant Sci 12:224–230
Brundrett MC (2002) Coevolution of roots and mycorrhizas of land plants. New Phytol 154:275–304
Campos-Soriano L, García-Martínez J, Segundo BS (2012) The arbuscular mycorrhizal symbiosis promotes the systemic induction of regulatory defence-related genes in rice leaves and confers resistance to pathogen infection. Mol Plant Pathol 13:579–592
Chandanie W, Kubota M, Hyakumachi M (2006) Interactions between plant growth promoting fungi and arbuscular mycorrhizal fungus *Glomus mosseae* and induction of systemic resistance to anthracnose disease in cucumber. Plant Soil 286:209–217
Conrath U (2009) Priming of induced plant defense responses. In: Loon LCV (ed) Advances in botanical research. Academic, Burlington, MA, pp 361–395
Conrath U, Beckers GJM, Flors V, García-Agustín P, Jakab G, Mauch F, Newman MA, Pieterse CMJ, Poinssot B, Pozo MJ, Pugin A, Schaffrath U, Ton J, Wendehenne D, Zimmerli L, Mauch-Mani B (2006) Priming: getting ready for battle. Mol Plant Microbe Interact 19:1062–1071
Cordier C, Pozo MJ, Barea JM, Gianinazzi S, Gianinazzi-Pearson V (1998) Cell defense responses associated with localized and systemic resistance to *Phytophthora parasitica* induced in tomato by an arbuscular mycorrhizal fungus. Mol Plant Microbe Interact 11:1017–1028
Cosme M, Stout MJ, Wurst S (2011) Effect of arbuscular mycorrhizal fungi (*Glomus intraradices*) on the oviposition of rice water weevil (*Lissorhoptrus oryzophilus*). Mycorrhiza 21:651–658
De la Noval B, Pérez E, Martínez B, León O, Martínez-Gallardo N, Délano-Frier J (2007) Exogenous systemin has a contrasting effect on disease resistance in mycorrhizal tomato (*Solanum lycopersicum*) plants infected with necrotrophic or hemibiotrophic pathogens. Mycorrhiza 17:449–460
De La Peña E, Echeverría SR, Van Der Putten WH, Freitas H, Moens M (2006) Mechanism of control of root-feeding nematodes by mycorrhizal fungi in the dune grass *Ammophila arenaria*. New Phytol 169:829–840
Dumas-Gaudot E, Slezack S, Dassi B, Pozo M, Gianinazzi-Pearson V, Gianinazzi S (1996) Plant hydrolytic enzymes (chitinases and β-1,3-glucanases) in root reactions to pathogenic and symbiotic microorganisms. Plant Soil 185:211–221
Elsen A, Gervacio D, Swennen R, De Waele D (2008) AMF-induced biocontrol against plant parasitic nematodes in *Musa* sp.: a systemic effect. Mycorrhiza 18:251–256
Filion M, St Arnaud M, Fortin JA (1999) Direct interaction between the arbuscular mycorrhizal fungus *Glomus intraradices* and different rhizosphere microorganisms. New Phytol 141:525–533

Filion M, St Arnaud M, Jabaji-Hare SH (2003) Quantification of *Fusarium solani* f. sp. *phaseoli* in mycorrhizal bean plants and surrounding mycorrhizosphere soil using Real-Time Polymerase Chain Reaction and direct isolations on selective media. Phytopathology 93:229–235

Fontana A, Reichelt M, Hempel S, Gershenzon J, Unsicker S (2009) The effects of arbuscular mycorrhizal fungi on direct and indirect defense metabolites of *Plantago lanceolata* L. J Chem Ecol 35:833–843

Fritz M, Jakobsen I, Lyngkjær MF, Thordal-Christensen H, Pons-Kühnemann J (2006) Arbuscular mycorrhiza reduces susceptibility of tomato to *Alternaria solani*. Mycorrhiza 16:413–419

Gange AC (1996) Reduction in vine weevil larval growth by mycorrhizal fungi. Mitt Biol Bund Forst 316:56–60

Gange AC (2001) Species-specific responses of a root- and shoot-feeding insect to arbuscular mycorrhizal colonization of its host plant. New Phytol 150:611–618

Gange AC (2007) Insect–mycorrhizal interactions: patterns, processes, and consequences. In: Ohgushi T, Craig TP, Price PW (eds) Ecological communities: plant mediation in indirect interaction webs. Cambridge University Press, New York, NY, pp 124–144

Gange AC, West HM (1994) Interactions between arbuscular mycorrhizal fungi and foliar-feeding insects in *Plantago lanceolata* L. New Phytol 128:79–87

Gange AC, Bower E, Brown VK (1999) Positive effects of an arbuscular mycorrhizal fungus on aphid life history traits. Oecologia 120:123–131

García-Garrido JM, Ocampo JA (2002) Regulation of the plant defence response in arbuscular mycorrhizal symbiosis. J Exp Bot 53:1377–1386

Gehring C, Bennett A (2009) Mycorrhizal fungal-plant-insect interactions: the importance of a community approach. Environ Entomol 38:93–102

Genre A, Bonfante P (2010) The making of symbiotic cells in arbuscular mycorrhizal roots. In: Kapulnick Y, Douds DD (eds) Arbuscular mycorrhizas: physiology and function. Springer, Dordrecht, pp 57–71

Gernns H, Von Alten H, Poehling HM (2001) Arbuscular mycorrhiza increased the activity of a biotrophic leaf pathogen—is a compensation possible? Mycorrhiza 11:237–243

Gianinazzi S, Vosátka M (2004) Inoculum of arbuscular mycorrhizal fungi for production systems: science meets business. Can J Bot 82:1264–1271

Gianinazzi S, Gollotte A, Binet MN, van Tuinen D, Redecker D, Wipf D (2010) Agroecology: the key role of arbuscular mycorrhizas in ecosystem services. Mycorrhiza 20:519–530

Gianinazzi-Pearson V, Tollot M, Seddas PMA (2009) Dissection of genetic cell programmes driving early arbuscular mycorrhiza interactions. In: Azcón-Aguilar C, Barea JM, Gianinazzi S, Gianinazzi-Pearson V (eds) Mycorrhizas functional processes and ecological impact. Springer, Berlin, pp 33–45

Goverde M, van der Heijden MGA, Wiemken A, Sanders IR, Erhardt A (2000) Arbuscular mycorrhizal fungi influence life history traits of a lepidopteran herbivore. Oecologia 125:362–369

Hao Z, Fayolle L, van Tuinen D, Chatagnier O, Li X, Gianinazzi S, Gianinazzi-Pearson V (2012) Local and systemic mycorrhiza-induced protection against the ectoparasitic nematode *Xiphinema index* involves priming of defence gene responses in grapevine. J Exp Bot 63:3657–3672

Hartley SE, Gange AC (2009) Impacts of plant symbiotic fungi on insect herbivores: mutualism in a multitrophic context. Annu Rev Entomol 54:323–342

Hause B, Schaarschmidt S (2009) The role of jasmonates in mutualistic symbioses between plants and soil-born microorganisms. Phytochemistry 70:1589–1599

Hause B, Mrosk C, Isayenkov S, Strack D (2007) Jasmonates in arbuscular mycorrhizal interactions. Phytochemistry 68:101–110

Hoffmann D, Vierheilig H, Schausberger P (2011) Arbuscular mycorrhiza enhances preference of ovipositing predatory mites for direct prey-related cues. Physiol Entomol 36:90–95

Howe GA, Jander G (2008) Plant immunity to insect herbivores. Annu Rev Plant Biol 59:41–66

Ijdo M, Cranenbrouck S, Declerck S (2011) Methods for large-scale production of am fungi: past, present, and future. Mycorrhiza 21:1–16

Jaiti F, Meddich A, El Hadrami I (2007) Effectiveness of arbuscular mycorrhizal fungi in the protection of date palm (*Phoenix dactylifera* L.) against bayoud disease. Physiol Mol Plant Pathol 71:166–173

Jeffries P, Barea JM (2012) Arbuscular mycorrhiza—a key component of sustainable plant-soil ecosystems. In: Hock B (ed) The Mycota, vol IX. Fungal Associations. Springer-Verlag, Berlin, Heidelberg, pp 51–75. ISBN: 978-3-642-30825-3

Jung SC, Martinez-Medina A, Lopez-Raez JA, Pozo MJ (2012) Mycorrhiza-induced resistance and priming of plant defenses. J Chem Ecol 38:651–664

Khaosaad T, García-Garrido JM, Steinkellner S, Vierheilig H (2007) Take-all disease is systemically reduced in roots of mycorrhizal barley plants. Soil Biol Biochem 39:727–734

Kloppholz S, Kuhn H, Requena N (2011) A secreted fungal effector of *Glomus intraradices* promotes symbiotic biotrophy. Curr Biol 21:1204–1209

Koricheva J, Gange AC, Jones T (2009) Effects of mycorrhizal fungi on insect herbivores: a meta-analysis. Ecology 90:2088–2097

Kula AAR, Hartnett DC, Wilson GWT (2005) Effects of mycorrhizal symbiosis on tallgrass prairie plant–herbivore interactions. Ecol Lett 8:61–69

Lee CS, Lee YJ, Jeun YC (2005) Observations of infection structures on the leaves of cucumber plants pre-treated with arbuscular mycorrhiza *Glomus intraradices* after challenge inoculation with *Colletotrichum orbiculare*. Plant Pathol J 21:237–243

Lendzemo VW, Kuyper TW, Matusova R, Bouwmeester HJ, van Ast A (2007) Colonization by arbuscular mycorrhizal fungi of *Sorghum* leads to reduced germination and subsequent attachment and emergence of *Striga hermonthica*. Plant Signal Behav 2:58–62

Li HY, Yang GD, Shu HR, Yang YT, Ye BX, Nishida I, Zheng CC (2006) Colonization by the arbuscular mycorrhizal fungus *Glomus versiforme* induces a defense response against the root-knot nematode *Meloidogyne incognita* in the grapevine (*Vitis amurensis* Rupr.), which includes transcriptional activation of the class III chitinase gene VCH3. Plant Cell Physiol 47:154–163

Liu J, Maldonado-Mendoza I, Lopez-Meyer M, Cheung F, Town CD, Harrison MJ (2007) Arbuscular mycorrhizal symbiosis is accompanied by local and systemic alterations in gene expression and an increase in disease resistance in the shoots. Plant J 50:529–544

López-Ráez JA, Verhage A, Fernández I, García JM, Azcón-Aguilar C, Flors V, Pozo MJ (2010a) Hormonal and transcriptional profiles highlight common and differential host responses to arbuscular mycorrhizal fungi and the regulation of the oxylipin pathway. J Exp Bot 61:2589–2601

López-Ráez JA, Flors V, García JM, Pozo MJ (2010b) AM symbiosis alters phenolic acid content in tomato roots. Plant Signal Behav 5:1138–1140

López-Ráez JA, Pozo MJ, García-Garrido JM (2011a) Strigolactones: a cry for help in the rhizosphere. Botany 89:513–522

López-Ráez JA, Charnikhova T, Fernández I, Bouwmeester H, Pozo MJ (2011b) Arbuscular mycorrhizal symbiosis decreases strigolactone production in tomato. J Plant Physiol 168:294–297

López-Ráez JA, Bouwmeester H, Pozo MJ (2012) Communication in the rhizosphere, a target for pest management. In: Lichtfouse E (ed) Agroecology and strategies for climate change. Springer Science+Business Media B.V, Dordrecht, pp 109–133

Maherali H, Klironomos JN (2007) Influence of phylogeny on fungal community assembly and ecosystem functioning. Science 316:1746–1748

Møller K, Kristensen K, Yohalem D, Larsen J (2009) Biological management of gray mold in pot roses by co-inoculation of the biocontrol agent *Ulocladium atrum* and the mycorrhizal fungus *Glomus mosseae*. Biol Control 49:120–125

Norman JR, Hooker JE (2000) Sporulation of *Phytophthora fragariae* shows greater stimulation by exudates of non-mycorrhizal than by mycorrhizal strawberry roots. Mycol Res 104:1069–1073

Norman J, Atkinson D, Hooker J (1996) Arbuscular mycorrhizal fungal-induced alteration to root architecture in strawberry and induced resistance to the root pathogen *Phytophthora fragariae*. Plant Soil 185:191–198

Parniske M (2008) Arbuscular mycorrhiza: the mother of plant root endosymbioses. Nat Rev Microbiol 6:763–775

Pastor V, Luna E, Mauch-Mani B, Ton J, Flors V (2012) Primed plants do not forget. Environ Exp Bot. http://dx.doi.org/10.1016/j.envexpbot.2012.02.013

Paszkowski U (2006) Mutualism and parasitism: the yin and yang of plant symbioses. Curr Opin Plant Biol 9:364–370

Pieterse CMJ, Van Wees SCM, Van Pelt JA, Knoester M, Laan R, Gerrits H, Weisbeek PJ, Van Loon LC (1998) A novel signaling pathway controlling induced systemic resistance in *Arabidopsis*. Plant Cell 10:1571–1580

Pieterse CM, Leon-Reyes A, Van der Ent S, Van Wees SCM (2009) Networking by small-molecule hormones in plant immunity. Nat Chem Biol 5:308–316

Pineda A, Zheng S-J, van Loon JJA, Pieterse CMJ, Dicke M (2010) Helping plants to deal with insects: the role of beneficial soil-borne microbes. Trends Plant Sci 15:507–514

Pinochet J, Calvet C, Camprubí A, Fernández C (1996) Interactions between migratory endoparasitic nematodes and arbuscular mycorrhizal fungi in perennial crops: a review. Plant Soil 185:183–190

Pivato B, Gamalero E, Lemanceau P, Berta G (2008) Colonization of adventitious roots of *Medicago truncatula* by *Pseudomonas fluorescens* C7R12 as affected by arbuscular mycorrhiza. FEMS Microbiol Lett 289:173–180

Pozo MJ, Azcón-Aguilar C (2007) Unraveling mycorrhiza-induced resistance. Curr Opin Plant Biol 10:393–398

Pozo MJ, Azcón-Aguilar C, Dumas-Gaudot E, Barea JM (1999) β-1,3-glucanase activities in tomato roots inoculated with arbuscular mycorrhizal fungi and/or *Phytophthora parasitica* and their possible involvement in bioprotection. Plant Sci 141:149–157

Pozo MJ, Cordier C, Dumas-Gaudot E, Gianinazzi S, Barea JM, Azcón-Aguilar C (2002) Localized versus systemic effect of arbuscular mycorrhizal fungi on defence responses to *Phytophthora* infection in tomato plants. J Exp Bot 53:525–534

Pozo MJ, Van Der Ent S, Van Loon LC, Pieterse CMJ (2008) Transcription factor MYC2 is involved in priming for enhanced defense during rhizobacteria-induced systemic resistance in *Arabidopsis thaliana*. New Phytol 180:511–523

Pozo MJ, Verhage A, García-Andrade J, García JM, Azcón-Aguilar C (2009) Priming plant defence against pathogens by arbuscular mycorrhizal fungi. In: Azcón-Aguilar C, Barea JM, Gianinazzi S, Gianinazzi-Pearson V (eds) Mycorrhizas—functional processes and ecological impact. Springer, Berlin, pp 123–135

Pozo MJ, Jung SC, López-Ráez JA, Azcón-Aguilar C (2010) Impact of arbuscular mycorrhizal symbiosis on plant response to biotic stress: the role of plant defence mechanisms. In: Kapulnick Y, Douds DD (eds) Arbuscular mycorrhizas: physiology and function. Springer, Dordrecht, pp 193–207

Read DJ (1998) Plants on the web. Nature 396:22–23

Redecker D, Kodner R, Graham LE (2000) Glomalean fungi from the ordovician. Science 289:1920–1921

Richardson AE, Hocking PJ, Simpson RJ, George TS (2009) Plant mechanisms to optimise access to soil phosphorus. Crop Pasture Sci 60:124–143

Rinaudo V, Bàrberi P, Giovannetti M, van der Heijden M (2010) Mycorrhizal fungi suppress aggressive agricultural weeds. Plant Soil 333:7–20

Rosendahl S (2008) Communities, populations and individuals of arbuscular mycorrhizal fungi. New Phytol 178:253–266

Sanders IR, Croll D (2010) Arbuscular mycorrhiza: the challenge to understand the genetics of the fungal partner. Annu Rev Genet 44:271–292

Schellenbaum L, Berta G, Ravolanirina F, Tisserant B, Gianinazzi S, Fitter AH (1991) Influence of endomycorrhizal infection on root morphology in a micropropagated woody plant species (*Vitis vinifera* L.). Ann Bot 68:135–141

Schliemann W, Ammer C, Strack D (2008) Metabolite profiling of mycorrhizal roots of Medicago truncatula. Phytochemistry 69:112–146

Schüßler A, Walker C (2011) Evolution of the 'plant-symbiotic' fungal phylum, glomeromycota. In: Pöggeler S, Wöstemeyer J (eds) Evolution of fungi and fungal-like organisms. Springer, Berlin, pp 163–185

Schüßler A, Schwarzott D, Walker C (2001) A new fungal phylum, the *glomeromycota*, phylogeny and evolution. Mycol Res 105:1413–1421

Smith SE, Read DJ (2008) Mycorrhizal symbiosis. Academic, New York, NY

Sood SG (2003) Chemotactic response of plant-growth-promoting bacteria towards roots of vesicular-arbuscular mycorrhizal tomato plants. FEMS Microbiol Ecol 45:219–227

St Arnaud M, Elsen A (2005) Interaction of arbuscular mycorrhizal fungi with soil-borne pathogens and non-pathogenic rhizosphere micro-organisms. In: Declerck S, Fortin JA, Strullu D-G (eds) In vitro culture of mycorrhizas. Springer, Berlin, pp 217–231

Thaler JS, Humphrey PT, Whiteman NK (2012) Evolution of jasmonate and salicylate signal crosstalk. Trends Plant Sci 17:260–270

Van der Ent S, Van Wees SCM, Pieterse CMJ (2009) Jasmonate signaling in plant interactions with resistance-inducing beneficial microbes. Phytochemistry 70:1581–1588

Van Hulten M, Pelser M, Van Loon LC, Pieterse CMJ, Ton J (2006) Costs and benefits of priming for defense in Arabidopsis. Proc Natl Acad Sci USA 103:5602–5607

Van Wees SCM, Van der Ent S, Pieterse CMJ (2008) Plant immune responses triggered by beneficial microbes. Curr Opin Plant Biol 11:443–448

Vicari M, Hatcher PE, Ayres PG (2002) Combined effect of foliar and mycorrhizal endophytes on an insect herbivore. Ecology 83:2452–2464

Vierheilig H, Steinkellner S, Khaosaad T, Garcia-Garrido JM (2008) The biocontrol effect of mycorrhization on soilborne fungal pathogens and the autoregulation of the AM symbiosis: one mechanism, two effects? In: Varma A (ed) Mycorrhiza. Springer, Berlin, pp 307–320

Vos C, Claerhout S, Mkandawire R, Panis B, De Waele D, Elsen A (2011) Arbuscular mycorrhizal fungi reduce root-knot nematode penetration through altered root exudation of their host. Plant Soil 354:335–345

Vosátka M, Albrechtová J, Patten R (2008) The international marked development for mycorrhizal technology. In: Varma A (ed) Mycorrhiza: state of the art, genetics and molecular biology, eco-function, biotechnology, eco-physiology, structure and systematics. Springer, Berlin, pp 419–438

Walling LL (2008) Avoiding effective defenses: strategies employed by phloem-feeding insects. Plant Physiol 146:859–866

Walters D, Heil M (2007) Costs and trade-offs associated with induced resistance. Physiol Mol Plant Pathol 71:3–17

Whipps JM (2004) Prospects and limitations for mycorrhizas in biocontrol of root pathogens. Can J Bot 82:1198–1227

Yao MK, Désilets H, Charles MT, Boulanger R, Tweddell RJ (2003) Effect of mycorrhization on the accumulation of rishitin and solavetivone in potato plantlets challenged with *Rhizoctonia solani*. Mycorrhiza 13:333–336

Zamioudis C, Pieterse CMJ (2012) Modulation of host immunity by beneficial microbes. Mol Plant Microbe Interact 25:139–150

Zhu HH, Yao Q (2004) Localized and systemic increase of phenols in tomato roots induced by *Glomus versiforme* inhibits *Ralstonia solanacearum*. J Phytopathol 152:537–542

Part V
Other Endophytic Fungi

Chapter 16
Fungal Endophytes in Plant Roots: Taxonomy, Colonization Patterns, and Functions

Diana Rocío Andrade-Linares and Philipp Franken

Abbreviations

CEs   Clavicipitaceous and endophytes
NCEs   Non-Clavicipitaceous endophytes
DSE(s)   Dark septate endophyte(s)
FE(s)   Fungal endophyte(s)
AM   Arbuscular mycorrhizal

16.1 Introduction

Plants are potential hosts for a broad spectrum of bacteria and fungi that live on their surface as epiphytes or inside plant tissues as endophytes. The term endophyte meaning "within" and "plant" (Greek words *endo* and *phyton*) denotes a broad spectrum of endosymbionts colonizing any plant organ (Schulz and Boyle 2005). Different kinds of symbiotic interactions in natural ecosystems have been roughly categorized into mutualism, commensalism, and parasitism according to the benefits for both partners (Paszkowski 2006), but this categorization has not been well established for the outcome of the relationship between root endophytes and

D.R. Andrade-Linares (✉)
Institut für Biologie, Plant Ecology, Freie Universität Berlin, Altensteinstr. 6, 14195 Berlin, Germany
e-mail: andrade@zedat.fu-berlin.de

P. Franken
Department of Plant nutrition, Leibniz-Institute of Vegetable and Ornamental Crops, Theodor-Echtermeyer-Weg 1,
Altensteinstr. 6, 14195 Großbeeren, Germany
e-mail: Franken@igzev.de

plants (Schulz and Boyle 2005; Junker et al. 2012). This chapter will consider fungal endophytes (FEs) that colonize plant roots without establishing mycorrhizal associations such as the formation of arbuscules. FEs are micromycetes which grow internally in living plant tissues without causing disease symptoms in the host (Petrini 1991; Wilson 1995; Saikkonen et al. 1998; Stone et al. 2000). They are represented by diverse taxonomic groups mainly in the phyla Ascomycota and Basidiomycota (Arnold et al. 2007; Arnold and Lutzoni 2007; Weiss et al. 2004; Selosse et al. 2009; Andrade-Linares et al. 2011a) and show a continuum of interactions with their hosts ranging from positive to neutral or even negative (Johnson et al. 1997; Brundrett 2004; Schulz and Boyle 2005; Rodriguez et al. 2009; Andrade-Linares et al. 2013). FEs have been recognized to be ancient as arbuscular mycorrhizal (AM) fungi (Krings et al. 2007) and could play a similar important role in some ecosystems for plant survival, affecting not only the fitness of single but also the structure of plant communities (Brundrett 2004; Saikkonen et al. 2004; Porras-Alfaro et al. 2008; Rodriguez and Redman 2008; Bultman et al. 2012; Knapp et al. 2012; Torres et al. 2012).

FEs have been categorized into two major groups known as the Clavicipitaceous and non-Clavicipitaceous endophytes (CEs and NCEs, respectively). CEs belong to the *Clavicipitaceae* family (phylum Ascomycota), which colonize systemically shoots and rhizomes of a narrow host range of cool- and warm-season grasses (*Poaceae*) (Clay and Schardl 2002). They are mutualistic symbionts that grow in the intercellular spaces of plant tissues and are transmitted by seeds (Schardl et al. 2004; Kuldau and Bacon 2008). In contrast, the NCEs are phylogenetically diverse and horizontally transmitted, most of them belonging to different orders among the Ascomycota phylum. Due to their facultative saprophytic growth on culture media, they could be isolated from tissues of disinfected leaves, stems, and roots of almost all sampled plants (Petrini 1996; Schulz and Boyle 2005; Arnold and Lutzoni 2007; Rodriguez et al. 2009). Nevertheless, different genera that do not grow well in standard agar media or live as obligate biotrophs have been also detected by molecular techniques evidencing a high biodiversity (Vandenkoornhuyse et al. 2002; Neubert et al. 2006; Duong et al. 2006; Gallery et al. 2007; Porras-Alfaro et al. 2008; Higgins et al. 2010; Knapp et al. 2012; Kohout et al. 2012) According to host range, colonized plant tissue, biodiversity, transmission, fitness benefits, and colonization pattern, NCEs have been categorized into three functional groups (classes 2, 3, and 4, Table 1 in Rodriguez et al. 2009) which show variable fungal lifestyles and therefore different life strategies along the r–K selection continuum.

The main emphasis of this chapter is on root fungal NCEs belonging to class 4. Root-inhabiting NCEs establish symbiotic associations with a wide range of plant species; however, their biodiversity is not well investigated (Rodriguez et al. 2009). Our objective is to summarize the current knowledge about the diversity, colonization characteristics, and potential functions of root endophytic fungi which do not colonize vascular tissues and subsequently the shoots of plants. The dark septate endophytes (DSEs—Ascomycota; Jumpponen and Trappe 1998) as well as the non-mycorrhizal members of the order Sebacinales (Basidiomycota; Weiss et al. 2004) will be also considered in this chapter because they are ubiquitous in healthy roots of different ecosystems indicating a potential benefit to plants.

Their functions have been, however, only studied in some plant-fungal interactions (Barrow and Aaltonen 2001; Addy et al. 2005; Selosse et al. 2009; Schäfer and Kogel 2009; Alberton et al. 2010; Mandyam et al. 2013).

16.2 Taxonomy of Root-Inhabiting NCEs

Class 4 endophytic fungi which colonize only roots belong to different taxonomic groups among Ascomycota, mainly to DSEs, Xylariales, Hypocreales, and non-mycorrhizal members of the order Sebacinales (Basidiomycota). A summary of the principal taxonomic genera and morphological features of some root-inhabiting NCEs is presented in Table 16.1 and Fig. 16.1.

Among the root endophytic Ascomycota, the DSEs present a polyphyletic group of endophytes first described as "mycelium radicus astrovirens" (Merlin 1922). They are characterized by melanized hyphal growth inside roots and on agar (Fig. 16.2) and frequently observed especially in roots of gymnosperms of the family *Pinaceae* (Addy et al. 2005). DSEs are usually found in boreal and alpine habitats but also in arid ecosystems (Porras-Alfaro et al. 2008; Knapp et al. 2012). They produce mitosporic conidia or sterile dark mycelia with microsclerotia in roots which can be also colonized at the same time by ecto- and endomycorrhizal fungi (Jumpponen and Trappe 1998; Mandyam and Jumpponen 2005; Wagg et al. 2008). The accumulation of melanin in hyphae is characteristic for this diverse group and probably reflects a common response to different environmental pressures (Butler and Day 1998; Robinson 2001). The fungal isolates correspond to different anamorphic genera, such as *Phialocephala*, *Scytalidium*, *Oidiodendron*, *Trichocladium*, *Cadophora*, *Exophiala*, *Heteroconium*, *Chloridium*, *Leptodontidium*, and *Cryptosporiopsis*, and are identified by morphological features of their sporulation on agar (Verkley et al. 2003; Addy et al. 2005). This classification is often difficult because many isolates remain as sterile mycelium or sporulate only under specific conditions (Wang and Wilcox 1985). *Phialocephala fortinii* (Wang and Wilcox 1985) is a DSE isolated from *Pinaceae* roots and liverwort rhizoids (Jumpponen et al. 2003). Phylogenetic studies suggest that *P. fortinii* is accompanied by cryptic species that can co-occur in the same root and include a sterile slow growing morphotype (Type 1) which was described as *Acephala applanata* based on RFLP data and culture morphology (Sieber and Grünig 2005).

Phylogenetic analyses have shown the close relationship between *Mollisia* (teleomorphic genus) and the different species of *Phialocephala* and *Cadophora* (Harrington and McNew 2003), a genus which describe *Phialophora*-like hyphomycetes (Domsch et al. 2007). In the phylogeny of DSE of the order Helotiales, combined studies of genetic and morphological approaches are needed to elucidate their relationship and classification. Day et al. (2012) characterized phialide features of species related to *Cadophora*, *Phialocephala*, and *Leptodontidium orchidicola*. Ancestral character reconstructions of conidiogenic cells (conidiophores or phialides) and phylogenetic tree analysis of DSE show that

Table 16.1 Description of root-inhabiting NCEs (class 4) frequently isolated from different host plants

Root endophyte	Order (Phylum)	Isolation	Morphological features—sporulation	Isolation references
Chloridium paucisporum	Helotiales (Ascomycota)	*Pinus resinosa* *Picea rubens* *Betula alleghaniensis*	Colony black with melanized sclerotia. Olive-black conidiophores. Sporulation at 16 °C. Long, thin, solitary phialides. Hyaline conidia cylindrical to ovoid or reniform	Wang and Wilcox (1985) Wilcox and Wang (1987)
Phialophora finlandica Harrington and McNew	Helotiales (Ascomycota)	*P. resinosa* *P. silvestres* *Betula alleghaniensis*	Colony grayish black, floccose. Brown branched conidiophores with collarettes. Ellipsoid single conidia. Sporulation after cold treatment	Wang and Wilcox (1985)
Phialocephala fortinii	Helotiales (Ascomycota)	*Abies alba* *P. resinosa* *Alnus rubra* *Rhododendron* sp. *Salix glauca*	Brownish black, floccose colony. Brown and verrucose aerial hyphae and conidiophores with collarettes. Globose single conidia. Sporulation after cold treatment	Wang and Wilcox (1985)
Leptodontidium orchidicola Sigler and Currah	Helotiales (Ascomycota)	*Platanthera hyperborean* *Colobanthus quitensis* *Artemisia norvegica* *Carex* sp. and *Picea glauca* *Solanum lycopersicum* L.	Grayish brown colony with diffusible dark red pigment. Single minute. Globose to teardrop-shaped solitary conidia directly from dematiaceous hyphae. Conidia unusual. Formation of small black sclerotia on water agar	Grünig et al. (2002) Upson et al. (2009a) Fernando and Currah (1995) Andrade-Linares et al. (2011a)
Acephala applanata	Helotiales (Ascomycota)	*P. sylvestris* *Picea abies*	Dark brown colony of slow growth without aerial mycelium related to *Phialocephala* Kendrick without heads of phialides	Grünig and Sieber (2005)
Mollisia sp. (*M. bolleyi*)	Helotiales (Ascomycota)	*P. abies* *Deschampsia antarctica*	Teleomorph related to *Phialocephala* and *Cadophora* species. Asci in apothecia	Menkis et al. (2005) Upson et al. (2009a)
Rhizoscyphus ericae	Helotiales (Ascomycota)	*Calluna vulgaris* *Erica andevalensis* *Cephaloziella varians* *Vaccinium macrocarpon* *Colobanthus quitensis*	*Meliniomyces* (the anamorph) arthroconidia, non-phialide production	Vralstad et al. (2002) Turnau et al. (2007) Upson et al. (2007) Upson et al. (2009a)

Trichoderma sp.	Hypocreales (Ascomycota)	*S. lycopersicum* L.	Yellowish green fast growth colony. Subglobose amero-gloio-halo-conidia with smooth cell wall	Andrade-Linares et al. (2011a)
		Musa sp.		Xia et al. (2011)
		Saccharum sp.		Romão-Dumaresq et al. (2012)
Bionectria rossmaniae	Hypocreales (Ascomycota)	*S. lycopersicum* L.	Beige-pale pink powdery. Penicillate conidiophores. Conidia one-celled, ellipsoidal	Andrade-Linares et al. (2011a)
Periconia macrospinosa	Microascales	*Andropogon gerardii* (dominant in a tallgrass prairie)	Hyaline and partly pigmented hyphae. Conidiophores erect, brown bearing an apical head of short monoblastic conidiogenous cells with brown, rough-walled globose conidia in acropetal chains	Mandyam et al. (2010)
Doratomyces sp.	Microascales (Ascomycota)	*Arabidopsis thaliana*	Slow growth with hyaline or some pigmented hyphae giving rise to dark synnemata which is a sterile stalk and upper half is covered with flask-shaped anellides giving rise to chains of smooth-walled or verrucose conidia	Junker et al. (2012)
Plectosphaerella cucumerina	Phyllochorales (Ascomycota)	*A. thaliana*	Moist yellowish white. Amero- and didymo-conidia ovoid and hyaline	Junker et al. (2012)
		S. lycopersicum L.		Andrade-Linares et al. (2011a)
Phoma sp.	Pleosporales (Ascomycota)	*A. thaliana*	Pycnidia separate, single ellipsoidal-cylindrical, hyaline conidia. Chlamydospores formation	Junker et al. (2012)
Rhizopycnis vagum	Pleosporales (Ascomycota)	*Pinus halepensis*	Dark septate hyphae. Nonproduction of Pycnidia or conidia. Dematiaceous chlamydospores	Girlanda et al. (2002)
		Rosmarinus officinalis		Xu et al. (2008)
		Dioscorea zingiberensis		
		S. lycopersicum L.		Andrade-Linares et al. (2011a)

(continued)

Table 16.1 (continued)

Root endophyte	Order (Phylum)	Isolation	Morphological features—sporulation	Isolation references
Sebacina vermifera	Sebacinales (Basidiomycota)	Terrestrial orchids *Acianthus reniformis Caladenia* sp. *Eriochilus* sp.	Probasidia subglobose to ovate, long vermiform basidiospores aseptate, or with 1–3 septate. Hyphae without clamp connections with small dolipore in the septa	Warcup (1988)
Piriformospora indica	Sebacinales (Basidiomycota)	*Glomus mosseae* spore from the Indian Thar desert	Beige-yellowish flat colony of fast growth. Hyaline pear-shaped spores. Hyaline hyphae without clamp connections	Verma et al. (1998)
Piriformospora williamsii sp. nov.	Sebacinales (Basidiomycota)	*Glomus fasciculatum* spore	Colony cream-colored to pale yellow growing faster on MEA. Mycelium is mostly flat and submerged into the medium. Spherical multinucleated spores. Hyaline multinucleated hyphae without clamp connections	Williams (1985) Basiewicz et al. (2012)
Microdochium sp.	Xylariales (Ascomycota)	*Andropogon gerardii* (dominant in a tallgrass prairie) *A. thaliana*	Colony pink, red-brown to black. Formation of dark chlamydospore clusters. Conidiogenous cells little differentiated ampulliform. Blastoconidia fusiform hyaline, smooth-walled	Mandyam et al. (2010) Junker et al. (2012)
Xylaria sp.	Xylariales (Ascomycota)	*Dendrobium* sp.	White conidia in asexual state, cylindrical to knobby, branched or lobed black ascocarps with black perithecia embedded in a white stroma	Chen et al. (2013)

16 Fungal Endophytes in Plant Roots: Taxonomy, Colonization Patterns, and Functions 317

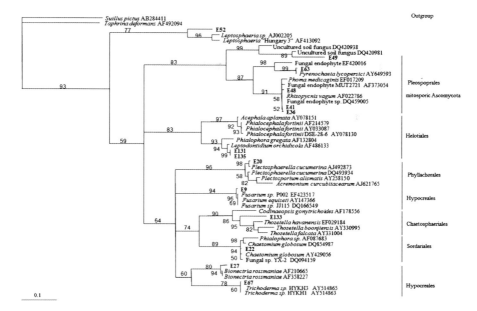

Fig. 16.1 Phylogenetic analysis. Sequences of the internal transcribed spacer (ITS) regions from the endophytes isolated from tomato roots were aligned using ClustalW with ITS sequences from representative members of the Ascomycota and as out-group those from two basidiomycetes (Andrade-Linares et al. 2011a). Based on the alignment, the phylogeny of the sequences was reconstructed with the program PUZZLE and the distance tree was displayed by the program TREEVIEW. Quartet puzzling support values of 1,000 replicates are indicated

fungi belonging to *Phialocephala* with multiple phialides clustered in only one clade (clade A) together with species of *Mollisia*. Phialide development of this group seems to be ancestral, while members of *Cadophora* show a reduction in phialide complexity (clades B and C). Clade B represents species of *Cadophora philandica*, *Chloridium paucisporum*, and three species of *Meliniomyces* (the anamorph of *Rhizoscyphus ericae*) with single or reduced phialides, while *Cadophora* and *Mollisia* species with nonfunctional or aborted phialides cluster together with *L. orchidicola* in clade C. Therefore, the authors state that these morphological features might be related to the dispersion style of these fungi in the nature and propose the placement of *Cadophora hiberna* into *Phialocephala* genus (Clade A) and of *L. orchidicola* into *Cadophora* genus (Day et al. 2012).

In Basidiomycota phylum, the order Sebacinales (Agaricomycotina) shows a diversity of mycorrhizal interactions forming ectomycorrhizas with tree roots (Urban et al. 2003), ericoid mycorrhizas with Ericaceae (Setaro et al. 2006; Selosse et al. 2007), and endomycorrhizas with Orchidaceae and Ranunculaceae (McKendrick et al. 2002; Kottke et al. 2003). Molecular phylogenetic studies divide them into two informally designated subgroups (A and B). Ectomycorrhizal and arbutoid mycorrhizal fungi clustered together in the subgroup A, while ericoid and

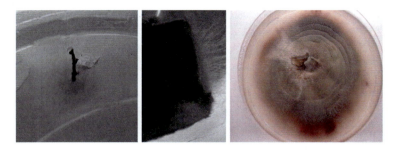

Fig. 16.2 Isolation of root fungal endophytes. At the *left side*, disinfected root fragment on agar showing fungal growth in the lower tip of the fragment and metabolite production (drop) at the opposite side. In the *middle*: hyphal outgrowth from disinfected roots. At the *right side*: fungal colony of one pleospora DSE producing *red pigment* into the agar

cavendishioid mycorrhizal are in subgroup B (Weiss et al. 2004). Endophytic Sebacinales fungi have been detected not only in the subgroup A but also in B (Weiss et al. 2011). *Sebacina vermifera*, *Piriformospora indica*, and a multinucleate *Rhizoctonia*, renamed as *P. williamsii* sp. nov. (Williams 1985; Basiewicz et al. 2012), are mutualistic endophytes recognized in a single different group as sebacinalean endophytic mycobionts (Oberwinkler et al. 2013). They cluster together in the subgroup B (Weiss et al. 2004; Selosse et al. 2009; Qiang et al. 2012a). *S. vermifera* first isolated from roots of terrestrial orchids (Warcup 1988) colonizes roots of different host plants without forming typical mycorrhizal associations. This change inside the host shows the plasticity of Sebacinales to switch their interaction according to its partner or the endophytic state as an ancestral trait (Weiss et al. 2011). For a review on Sebacinales taxonomy, the authors recommend Oberwinkler et al. (2013). *P. indica* might be considered as a model for non-mycorrhizal root endophytes conferring considerable benefits to different plant hosts (Varma et al. 1999; Schäfer and Kogel 2009; Zuccaro et al. 2011; Franken 2012).

16.3 Colonization Patterns of Root-Inhabiting NCEs

The presence of root endophytes beside the AM fungi has been often overlooked probably due to their fine hyaline hyphae. Detection using particular stains such as wheat germ agglutinin-Alexa Fluor 488 (WGA-AF 488) enables the visualization of fungal structures inside roots. WGA-AF488 interacts with fungal chitin which permits the inter- and intracellular localization of hyphae (green) in root tissues (Fig. 16.3; Deshmukh et al. 2006; Andrade-Linares et al. 2011a). Root cell reactions to fungal colonization can be stained with the endocytosis marker FM4-64 for plant membranes activity (red) as well as with Concanavalin

16 Fungal Endophytes in Plant Roots: Taxonomy, Colonization Patterns, and Functions

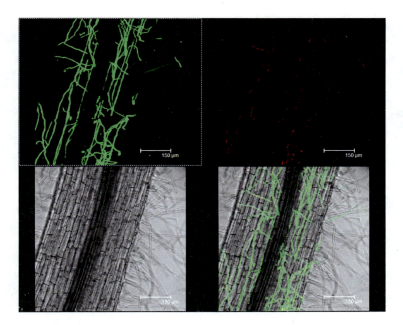

Fig. 16.3 Endophytic colonization of tomato roots 4 weeks after inoculation. Root fragments were stained with WGA-Alexa Fluor 488 and *Congo red* and were evaluated by CLSM. The figure shows fungal growth (*green*) in root and an overlay picture of WGA-AF and bright field microscopy

A (ConA-AF633) to detect papillae formation (red) indicative for the presence of glycoproteins at the fungal penetration site (Zuccaro et al. 2011).

Colonization patterns by root NCEs seem to be different according to the plant host and the cultivation system as it has been best described for the DSE *Phialocephala fortinii* (Peterson et al. 2008). Single hyphae colonize parallel to the main axis of the root, between epidermal and cortical cells, and intracellular colonization is visualized as formation of microsclerotia (O'Dell et al. 1993). Cocultivation in vitro experiments of endophytes and plants on media for plant growth or on water agar allow to follow the root fungal colonization process and to determine the production of symptoms in plant tissues (Fig. 16.4). Up to now it is not clear if fungal endophytes sense root exudates for colonization. At the first contact of the fungus to the root, they produce running hyphae and try to penetrate the root through the intercellular spaces. For instance, *Trichoderma* hyphae have been shown to grow around roots, to form appressorium-like structures, to penetrate through epidermal and cortical cell layers, and to induce deposition of cell wall material and production of phenolic compounds by the surrounding plant cells (Yedidia et al. 1999).

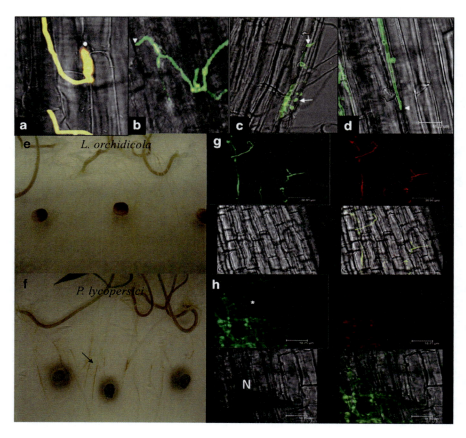

Fig. 16.4 Endophytic colonization of roots and evaluation of symptom production. After the first contact of the fungus with the root (**a–d**), root fragments were stained with WGA-Alexa *Fluor* and *Congo red* and were evaluated by CLSM. The figure shows formation of hyphal single tips (**b, d**; *filled inverted triangle*), appressorium-like structures (**a**; *filled circle*), swollen cells (**c**; *arrow*) attached to the intercellular space as the first fungal contact with the root surface. Tomato seedlings were placed on PNM in the presence of agar plugs of the endophytes (**e, f**). After 2 weeks tomato roots were stained with WGA-Alexa Fluor and evaluated under the microscope. The endophyte *L. orchidicola* did not produce any symptom in root tissue (**e, g**) compared with the pathogen *P. lycopersici* (**f, h**) which produced *brown spots* in roots (**f**; *arrow*). Microscopic observation of the root revealed a high fungal growth with the formation of fungal coils (*asterisk*) and necrotic tissue (N)

The analysis of colonization by root endophytic Ascomycota using confocal laser scanning microscopy (CLSM) showed that they penetrate the roots by different mechanisms (Andrade-Linares et al. 2011a). Adhesion tips as well as appressorium-like structures and swollen cells were visualized on the roots surface of the host tomato. Growing intracellular hyphae showed a constriction at sites where the fungus traverses the cell wall and root colonization was inter- and intracellular from epidermal to cortical cell layers (Fig. 16.4). Complete colonization of the root without infection of vascular tissue was observed 2–4 weeks after inoculation (Fig. 16.3). A pathogenic fungus, isolated as endophyte, showed heavy hyphal growth parallel to the root axis, infected also the vascular cylinder, and necrotized tissues with the formation of intracellular hyphal coils and fluorescence in single cells (Fig. 16.4).

Whether the colonization pattern is controlled by the plant host, remains to be investigated in more detail. The analysis of fungal colonization in wheat roots by two Ascomycota mitosporic fungi clearly showed a shift in colonization patterns from living to dead root cells (Abdellatif et al. 2009). Endophyte fungi in living roots showed low abundance, patchy colonization, and variable values of growth direction indices and produced knot, coil, and vesicle structures. The same endophyte species possessed a highly regular growth in dead roots, and this growth was mostly parallel to the root axis. To analyze the colonization patterns, fluorescent stains or molecular probes can be used for CLSM (see above).

P. indica shows a particular root colonization behavior which has not been observed before for any other root endophyte (Deshmukh et al. 2006; Schäfer et al. 2009; Qiang et al. 2012b). Its interaction with barley and Arabidopsis plants clearly shows two different temporal phases (Lahrmann and Zuccaro 2012). After spore germination and root contact, the fungus grows on the root surface in the first day, penetrates intercellularly the epidermal cell layer, and colonizes living cortical cells and root hairs in the differentiation zone as a biotroph during the first 3 days of interaction where it suppresses callose deposition and MAMP-triggered plant immunity responses (Jacobs et al. 2011). After 3 days, programmed plant cell death is induced and the cell death-associated colonization phase starts. Colonization gradually increases and the life cycle is completed by intracellular production of chlamydospores after 7 days (Deshmukh et al. 2006; Jacobs et al. 2011; Lahrmann and Zuccaro 2012).

16.4 Potential Functions of Root-Inhabiting NCEs

16.4.1 Production of Secondary Metabolites

Endophytes may produce new bioactive substances in vitro and *in planta* due to their constant metabolic interaction with their host plant and the environment which is at the same time important for the outcome of the symbiosis (Schulz et al. 2002; Schulz and Boyle 2005; Suryanarayanan et al. 2009). These secondary metabolites have been isolated and characterized according to their ecological role and their potential application in industry and medicine (Tan and Zou 2001; Strobel and Daisy 2003). They show antibacterial, antifungal, antimalarial, antiviral, and anticancer activities and correspond to diverse structural compounds related to steroids, quinones, alkaloids, phenols, terpenoids, peptides, flavonoids, aliphatics, and antimicrobial compounds (Li et al. 2005; Gunatilaka 2006; Suryanarayanan et al. 2009; You et al. 2009).

FEs with capacity of producing the same metabolites as those produced by their medicinal host plant have shifted the focus of new drug production. These substances could be produced by fungal fermentation which would reduce the production costs (Strobel 2003; Suryanarayanan et al. 2009). For instance, taxon, a diterpene, initially extracted from yew tree (*Taxus brevifolia*) was subsequently turned out to be also produced by the fungal endophytes *Taxomyces andreanae* and *Pestalotiopsis microspora* (Strobel et al. 1996; Li et al. 1998). The evidence of metabolites that are synthesized by both partners has raised the question of horizontal gene transferring and/or a cross talk of metabolic pathways between endophyte and plant (Stierle et al. 1993; Sirikantaramas et al. 2009; Kusari et al. 2012).

Root FEs also synthesize secondary metabolites; however, they have been less studied for their potential impact on plant host and for applied aspects than the leaf or systemic endophytes (Strobel 2003; Kusari et al. 2011). Root fungal endophytes promote not only the production of metabolites by the plastid-located methylerythritol phosphate pathway (MEP) but also the production of alkaloids (Wang et al. 2006; Schäfer et al. 2009; Gao et al. 2011). NCEs belonging to class 2 such as species of *Fusarium* and *Verticillium* have been more extensively analyzed for production of different metabolites against herbivoric insects, nematodes, and fungal pathogens (Bacon and Yates 2006; Sirikantaramas et al. 2009; You et al. 2009). The metabolic activity of NCEs belonging to class 4 is less known (Schulz et al. 1999; Xu et al. 2008; Vinale et al. 2010) and therefore is prompted the screening of new metabolites from them, which are often produced at the first isolation from roots or on agar plates (Fig. 16.2). Some examples to mention are *Larix decidua* root colonization by the DSE *Cryptosporiopsis* sp. and *Phialophora* sp. resulted in increased concentrations of soluble proanthocyanidins (Schulz et al. 1999). The fungus *P. indica* is able to produce indol-3-acetic acid (IAA) and cytokinins in liquid cultures (Sirrenberg et al. 2007; Vadassery et al. 2008) and plant genes related to auxin, gibberellin, and abscisic acid biosynthesis and signaling are upregulated at 3 and/or 7 days after *P. indica* inoculation in barley roots (Schäfer et al. 2009). Finally, the DSE *Phialocephala europaea* is able to produce in vitro extracellular metabolites, two of them identified as sclerin and sclerotinin A which inhibited the growth of the pathogenic oomycete *Phytophthora citricola* (Tellenbach et al. 2013).

16.4.2 Plant Growth

The NCEs from roots are phylogenetically diverse presenting a variable host response and therefore a continuum in the root-fungus interaction (Schulz and Boyle 2005). Positive or negative effects of these FEs seem to be regulated not only by the genotype of the plant and the fungus (Tellenbach et al. 2011; Knapp et al. 2012; Mandyam et al. 2013) but also by non-genetical factors such as the developmental stage and the nutritional status of the fungus and the plant and by environmental conditions during the interaction (Saikkonen et al. 1998). If an interaction is really mutualistic, is difficult to define, without knowing the actual costs and benefits for both partners (Faeth and Fagan 2002).

Plant growth stimulation by fungal root endophytes could be based on the production of particular phytohormones and this view is, e.g., supported by enhanced adventitious root formation in pelargonium and poinsettia cuttings without colonization by *P. indica* (Drüge et al. 2007). The hypothesis that this is due to fungal IAA production was, however, rejected as knockout of the pathway in the endophyte did not result in loss of plant growth-promoting effects (Hilbert et al. 2012). For the closely related *S. vermifera* which also improves plant growth, for instance, in *Nicotiana attenuata* (Barazani et al. 2005) and in *Panicum virgatum* (Ghimire et al. 2009), this promotion seems to be related to interference with ethylene production and this is accompanied by the decrease resistance to herbivore (Barazani et al. 2005). Promotion of plant performance was also observed during generative development. Tomato plants inoculated with DSEs (Fig. 16.5) and *P. indica* showed accelerated flowering and significantly higher fruit biomass at the beginning of the harvests. This enhancement of fruit production, however, disappeared at later harvests (Andrade-Linares et al. 2011b, 2013).

Trichoderma species also stimulate plant growth (Yedidia et al. 2001; Adams et al. 2007; Shoresh et al. 2010). Maize plant colonized by *Trichoderma harzianum* Rifai T22 showed upregulation of proteins involved in stress and defense responses as well as in carbohydrate metabolism and a photosynthetic enhancement which could suggest that this interaction improves plant growth (Shoresh and Harman 2008). The same fungal strain improved tomato seed germination and growth under abiotic and biotic stresses (Mastouri et al. 2010). The role of the auxin in plant growth was demonstrated in Arabidopsis plants cocultivated with *T. virens* and *T. atroviride* (Contreras-Cornejo et al. 2009). Ethylene levels are reduced by 1-aminocyclopropane-1-carboxylate (ACC) deaminase activity on ACC; this enzyme is also produced by *Trichoderma asperellum* T203 which showed root growth-promoting effect in canola (Viterbo et al. 2010).

16.4.3 Plant Nutrition

The contribution of root endophytes to the mineral nutrient uptake and the exchange of nutrients between fungi and plants have not been yet proved with convincing evidence as it was shown for mycorrhizal symbioses. Most experimental data only indirectly suggest the support of nutrient supply by the endophytes. For example, the growth of Chinese cabbage plants inoculated with the DSE *Heteroconium chaetospira* was much better using amino acids than inorganic N (NO_3) as nitrogen source (Usuki and Narisawa 2007). Maize plants colonized by *Trichoderma harzianum* T22 increased yield and growth under low levels of nitrogen fertilizer (Harman 2000), and Upson et al. (2009b) showed an interrelation

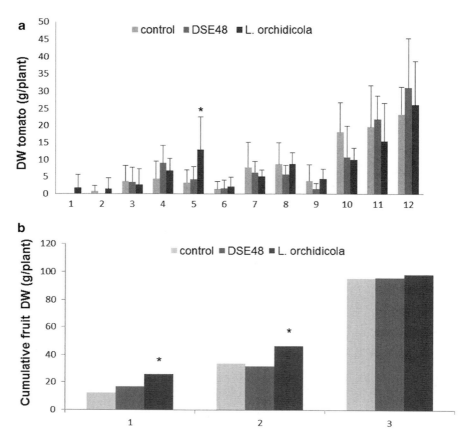

Fig. 16.5 Influence of two DSEs (E48 and *L. orchidicola*) on tomato yield. Twelve harvests were carried out between week 15 and week 22 after endophyte inoculation and dry weight of marketable fruits were estimated for each harvest point (**a**). Due to the asynchrony of fruit production per plant, dates were grouped into harvests beginning (1), middle (2), and end (3) which correspond to 1, weeks 1–2; 2, weeks 3–5; and 3, weeks 6–8 (**b**). Significant differences of endophyte-colonized plants to the respective controls are indicated by *asterisks* above the columns. Statistical comparisons between treatments were performed by one-way ANOVA according to LSD test ($n = 6$; $P = 0.05$)

between plant growth promotion of *Deschampsia antarctica* and the use of organic resources by DSE. Colonized roots of *D. antarctica* possessed enhanced development and increased shoot and root nitrogen and phosphorus contents only when they had been supplied with organic nitrogen (Upson et al. 2009b). Other studies have shown the increase of phosphorus content in plants as *Carex* (Haselwandter and Read 1982) and *Pinus contorta* (Jumpponen et al. 1998) after colonization by DSE. According to the meta-analysis presented by Newsham (2011), under certain conditions, especially with organic N, DSEs enhance plant biomass and N and P contents in shoots. In a more direct approach, the solubilization of tricalcium phosphate (TCP) and rock phosphate (RP) was evaluated in compartment experiments with *Atriplex canescens* (Pursh) plants colonized by *Aspergillus ustus* which showed growth into the phosphate source compartment and produced significant increase of shoot biomass in plants growing in TCP and in less extend in RP (Barrow and Osuna 2002). Recent studies have shown the possible implication of *P. indica* in plant nutrition. Maize plants colonized by the fungus presented increased phosphate in shoots and the role of a fungal high-affinity phosphate transporter (PiPT) was evident (Yadav et al. 2010). Additionally, *Arabidopsis* roots colonized by *P. indica* presented elevated nitrate reductase activity and nitrate uptake (Sherameti et al. 2005) which could indicate improved nitrate supply of the plant by the fungus.

If such interactions are mutualistic, the fungal partner should also have a benefit. If this is a transfer of carbohydrates has up to now only be indicated. In the above-mentioned interaction between Chinese cabbage plants and *H. chaetospira*, carbohydrate sucrose and mannitol were considerably labeled in roots colonized by the fungus, if $^{13}C–CO_2$ was applied (Usuki and Narisawa 2007). Another hint was that *P. indica* colonization induced an enzyme involved in starch degradation (Sherameti et al. 2005) accordingly with lower levels of hexose, starch, and amino acids in these colonized plants which could be transferred from the root to the fungus (Schäfer et al. 2009). Similar to mycorrhizal systems this transfer could be balanced by enhanced CO_2 assimilation in colonized plants (Achatz et al. 2010).

16.4.4 Plant Resistance to Abiotic and Biotic Stressors

Plants growing in natural ecosystem are exposed to adverse conditions such as high temperatures, drought, salinity, and pathogenic attack. Such stress situations could be meliorated by mutualistic symbioses. A well-known case is the habitat-adapted symbiosis of FEs class 2 belonging to the genera *Fusarium, Colletotrichum*, and *Curvularia* which confer plant resistance to salt, heat and drought of tomato, rice, and dunegrass plants (Rodriguez et al. 2008; Redman et al. 2011). These habitat-adapted symbioses have not been yet demonstrated for root FEs class 4. Isolates of the DSE *H. chaetospira* suppressed clubroot by *Plasmodiophora brassicae* in Chinese cabbage plants growing in sterile and non-sterile soil without evidence of mycoparasitism (Narisawa et al. 1998). Further studies showed the ability of this DSE to protect plants against *Verticillium* yellows (Narisawa et al. 2004). The disease index of the pathogenic fungus *Verticillium dahliae* was reduced in tomato plants colonized with the DSE, *L. orchidicola*, protection that seems to depend on plant grow conditions and concentration of the pathogen (Andrade-Linares et al. 2011b). Different species of *Trichoderma* are able to colonize roots improving seed germination, plant growth, water and fertilizer efficiency utilization, and disease resistance, for instance, against *Phytophthora* blight on cucumber and hot pepper (*Capsicum annuum*), *Botrytis* blight on begonia, and *Botryosphaeria* dieback on ericaceous plants (Shoresh et al. 2010; Harman et al. 2004;

Khan et al. 2004; Horst et al. 2005; Hoitink et al. 2006; Bae et al. 2011). Moreover, *Trichoderma* isolates not only promote plant growth and disease resistance but also confer drought tolerance (Shukla et al. 2012) and change the expression of ESTs putatively involved in the production of osmoprotectants and regulatory metabolites (Bae et al. 2009). The water deficit tolerance in tomato seedlings colonized by *T. harzianum* T22 was correlated with higher activity of ascorbate- and glutathione-recycling enzymes and enhanced antioxidant defense mechanism (Mastouri et al. 2012). A similar protective effect was demonstrated in Arabidopsis and cucumber plants colonized by *T. asperelloides*, in which not only genes involved in auxin biosynthesis and response were upregulated but also genes and transcription factors (WRKY) related to plant defense. Additionally, higher amount of antioxidants were also found in *T. asperelloides*-treated plants under salinity conditions (Brotman et al. 2013).

The endophyte *P. indica* confers a broad range of resistance or tolerance to different hosts. For instance, in barley plants the fungus increases tolerance to salt stress as well as resistance to the leaf biotrophic fungus *Blumeria graminis* and the root necrotrophic pathogen *Fusarium culmorum* (Waller et al. 2005) and in tomato plants against *V. dahliae* (Fakhro et al. 2010).

16.5 Conclusions

NCEs colonize the roots of a broad spectrum of hosts, and the outcome of this interaction is regulated by different genetic, physiological, biotic, and abiotic factors. Direct effects on plant due to the fungal colonization of roots are, e.g., improved plant shoot and root growth, enhanced seed germination and adventitious root formation, induced systemic resistance to pathogens, tolerance to abiotic stressors, and enhanced uptake and/or solubilization of mineral nutrients. Probably different mechanisms enable fungi to colonize roots, but plant responses such as expression of defense-related genes, higher photosynthetic capabilities, and enhanced activities of enzymes involved in the antioxidant machinery are similar in different fungus–host combinations. Root NCEs can also be saprophytes, and there must be a metabolic switch between living in the soil and living inside roots as it was demonstrated for *P. indica* (Zuccaro et al. 2011) but need to be analyzed in other already known root endophytes like DSEs and *Trichoderma*. Further studies including transcriptomics, proteomics, and metabolomics are necessary to understand the different facets of root NCE–plant interactions. This must include not only the NCE–plant interaction at early phases of plant growth but also later stages when the plant switches from vegetative to generative development.

In the cross talk between root NCEs and plants, it is also important to consider the possible interaction between root endophytes and other root-associated microorganisms which could, e.g., contribute signaling molecules that trigger otherwise silent biosynthetic pathways in NCEs and plant hosts. The dimension of this cross talking seems indeed to be more complex as recent studies have revealed the common presence of mycoviruses (viruses that infect fungi) in NCEs (Feldman et al. 2012) which could affect fungal growth and interaction with the plant host (Márquez et al. 2007). In addition to this complexity, the role of endohyphal bacteria, already detected in both leaf-inhabiting and root-inhabiting endophytes (Hoffman and Arnold 2010; Sharma et al. 2008), has to be investigated. They can impact and/or be partially responsible for the effects of fungal endophytes on plant performances (Sharma et al. 2008).

References

Abdellatif L, Bouzid S, Kaminskyj S, Vujanovic V (2009) Endophytic hyphal compartmentalization is required for successful symbiotic Ascomycota association with root cells. Mycol Res 113:782–791

Achatz B, von Ruden S, Andrade D, Neumann E, Pons-Kuhnemann J, Kogel KH, Franken P, Waller F (2010) Root colonization by *Piriformospora indica* enhances grain yield in barley under diverse nutrient regimes by accelerating plant development. Plant Soil 333:59–70

Adams P, De-Leij FA, Lynch JM (2007) *Trichoderma harzianum* Rifai 1295-22 mediates growth promotion of Crack willow (*Salix fragilis*) saplings in both clean and metal-contaminated soil. Microb Ecol 54:306–313

Addy HD, Piercey MM, Currah RS (2005) Microfungal endophytes in roots. Can J Bot 83:1–13

Alberton O, Kuyper TW, Summerbell RC (2010) Dark septate root endophytic fungi increase growth of Scots pine seedlings under elevated CO_2 through enhanced nitrogen use efficiency. Plant Soil 328:459–470. doi:10.1007/s11104-009-0125-8

Andrade-Linares DR, Grosch R, Franken P, Karl HR, Kost G, Restrepo S et al (2011a) Colonization of roots of cultivated *Solanum lycopersicum* by dark septate and other ascomycetous endophytes. Mycologia 103:710–721

Andrade-Linares DR, Grosch R, Restrepo S, Krumbein A, Franken P (2011b) Effects of dark septate endophytes on tomato plant performance. Mycorrhiza 21:413–422

Andrade-Linares DR, Anja Müller A, Fakhro A, Schwarz D, Franken P (2013) Impact of *Piriformospora indica* on tomato. In: Varma A (ed) *Piriformospora indica*, soil biology, vol 33. Springer, Berlin, pp 107–117. doi:10.1007/978-3-642-33802-1

Arnold AE, Lutzoni F (2007) Diversity and host range of foliar fungal endophytes: are tropical leaves biodiversity hotspots? Ecology 88(3):541–549

Arnold AE, Henk DA, Eells RA, Lutzoni F, Vilgalys R (2007) Diversity and phylogenetic affinities of foliar fungal endophytes in loblolly pine inferred by culturing and environmental PCR. Mycologia 99:185–206

Bacon CW, Yates IE (2006) Endophytic root colonization by *Fusarium* species: histology, plant interactions, and toxicity. In: Schulz B, Boyle C, Sieber T (eds) Microbial root endophytes. Springer, Heidelberg, pp 133–152

Bae H, Sicher RC, Kim MS, Kim S-H, Strem MD, Melnick RL et al (2009) The beneficial endophyte *Trichoderma hamatum* isolate DIS 219b promotes growth and delays the onset of the drought response in *Theobroma cacao*. J Exp Bot 60(11):3279–3295

Bae H, Roberts DP, Lim HS, Strem MD, Park SC et al (2011) Endophytic *Trichoderma* isolates from tropical environments delay disease onset and induce resistance against *Phytophthora capsici* in hot pepper using multiple mechanisms. Mol Plant Microbe Interact 24(3):336–351

Barazani O, Benderoth M, Groten K, Kuhlemeier C, Baldwin IT (2005) *Piriformospora indica* and *Sebacina vermifera* increase growth performance at the expense of herbivore resistance in *Nicotiana attenuata*. Oecologia 146:234–243

Barrow JR, Aaltonen RE (2001) Evaluation of the internal colonization of *Atriplex canescens* (Prush) Nutt. roots by dark septate fungi and the influence of host physiological activity. Mycorrhiza 11:199–205

Barrow JR, Osuna P (2002) Phosphorus solubilization and uptake by dark septate fungi in fourwing saltbush, *Atriplex canescens* (Pursh) Nutt. J Arid Environ 51(3):449–459

Basiewicz M, Weiss M, Kogel K-H, Langen G et al (2012) Molecular and phenotypic characterization of *Sebacina* vermifera strains associated with orchids, and the description of *Piriformospora williamsii* sp. nov. Fungal Biol 116:204–213

Brotman Y, Landau U, Cuadros-Inostroza A, Takayuki T, Fernie AR et al (2013) *Trichoderma*-plant root colonization: escaping early plant defense responses and activation of the antioxidant machinery for saline stress tolerance. PLoS Pathog 9(3):e1003221. doi:10.1371/journal. ppat.1003221

Brundrett M (2004) Diversity and classification of mycorrhizal associations. Biol Rev 79 (3):473–495

Bultman TL, Aguilera A, Sullivan TJ (2012) Influence of fungal isolates infecting tall fescue on multitrophic interactions. Fungal Ecol 5(3):372–378

Butler MJ, Day AW (1998) Fungal melanins: a review. Can J Microbiol 44:1115–1136

Chen J, Zhang L-C, Xing Y-M, Wang Y-Q, Xing X-K et al (2013) Diversity and taxonomy of endophytic xylariaceous fungi from medicinal plants of *Dendrobium* (*Orchidaceae*). PLoS One 8(3):e58268. doi:10.1371/journal.pone.0058268

Clay K, Schardl C (2002) Evolutionary origins and ecological consequences of endophyte symbiosis with grasses. Am Nat 160S:99–127

Contreras-Cornejo HA, Macias-Rodriguez L, Cortes-Penagos C, Lopez-Bucio J (2009) *Trichoderma virens*, a plant beneficial fungus, enhances biomass production and promotes lateral root growth through an auxin-dependent mechanism in Arabidopsis. Plant Physiol 149:1579–1592

Day MJ, Hall JC, Currah RS (2012) Phialide arrangement and character evolution in the helotialean anamorph genera *Cadophora* and *Phialocephala*. Mycologia 104(2):371–381

Deshmukh S, Hückelhoven R, Schäfer P, Imani J, Sharma M et al (2006) The root endophytic fungus *Piriformospora indica* requires host cell death for proliferation during mutualistic symbiosis with barley. Proc Natl Acad Sci USA 103:18450–18457

Domsch K, Gams W, Anderson T (2007) Compendium of soil fungi, 2nd edn. IHW-Verlag, Eching. ISBN 978-3930167-69-2

Drüge U, Baltruschat H, Franken P (2007) *Piriformospora indica* promotes adventitious root formation in cuttings. Sci Hortic 112:422–426

Duong LM, Jeewon R, Lumyong S, Hyde KD (2006) DGGE coupled with ribosomal DNA gene phylogenies reveal uncharacterized fungal endophytes. Fungal Divers 23:121–138

Faeth SH, Fagan WF (2002) Fungal endophytes: common host plant symbionts but uncommon mutualists. Integr Comp Biol 42:360–368

Fakhro A, Andrade-Linares DR, von Bargen S, Bandte M, Buttner C, Grosch R, Schwarz D, Franken P (2010) Impact of *Piriformospora indica* on tomato growth and on interaction with fungal and viral pathogens. Mycorrhiza 20(3):191–200

Feldman TS, Morsy MR, Roossinck MJ (2012) Are communities of microbial symbionts more diverse than communities of macrobial hosts? Fungal Biol 116:465–477

Fernando AA, Currah RS (1995) *Leptodontidium orchidicola* (Mycelium-Radicis-Atrovirens Complex) aspects of its conidiogenesis and ecology. Mycotaxon 54:287–294

Franken P (2012) The plant strengthening root endophyte *Piriformospora indica*: potential application and the biology behind. Appl Microbiol Biotechnol 96:1455–1464

Gallery RA, Dalling JW, Arnold AE (2007) Diversity, host affinity and distribution of seed-infecting fungi: a case study with Cecropia. Ecology 88:582–588

Gao F, Yong Y, Dai C (2011) Effects of endophytic fungal elicitor on two kinds of terpenoids production and physiological indexes in *Euphorbia pekinensis* suspension cells. J Med Plant Res 5(18):4418–4425

Ghimire SR, Nikki CD, Craven KD (2009) The mycorrhizal fungus, *Sebacina vermifera*, enhances seed germination and biomass production in switchgrass (*Panicum virgatum* L). Bioenergy Res 2:51–58

Girlanda M, Ghignone S, Luppi AM (2002) Diversity of root associated fungi of two Mediterranean plants. New Phytologist 155:481–498

Grünig CR, Sieber TN (2005) Molecular and phenotypic description of the widespread root symbiont *Acephala applanata* gen. et sp. nov., formerly known as dark-septate endophyte type 1. Mycologia 97(3):628–640

Grünig CR, Sieber TN, Rogers SO, Holdenrieder O (2002) Genetic variability among strains of *Phialocephala fortinii* and phylogenetic analysis of the genus *Phialocephala* based on rDNA ITS sequence comparisons. Can J Bot 80:1239–1249

Gunatilaka AAL (2006) Natural products from plant-associated microorganisms: distribution, structural diversity, bioactivity and implication of their occurrence. J Nat Prod 69:509–526

Harman GE (2000) Myths and dogmas of biocontrol—changes in perceptions derived from research on *Trichoderma harzianum* T-22. Plant Dis 84(4):377–393

Harman GE, Howell CR, Viterbo A, Chet I, Lorito M (2004) *Trichoderma* species opportunistic, avirulent plant symbionts. Nat Rev Microbiol 2(1):43–56

Harrington TC, McNew DL (2003) Phylogenetic analysis places the *Phialophora*-like anamorph genus *Cadophora* in the Helotiales. Mycotaxon 87:141–152

Haselwandter K, Read DJ (1982) The significance of a root-fungus association in two *Carex* species of high-alpine plant communities. Oecologia 52:352–354

Higgins KL, Coley PD, Kursar TA, Arnold AE (2010) Culturing and direct PCR suggest prevalent host generalism among diverse fungal endophytes of tropical forest grasses. Mycologia 103:247–260

Hilbert M, Voll LM, Yi Ding Y, Hofmann J, Sharma M, Zuccaro A (2012) Indole derivative production by the root endophyte *Piriformospora indica* is not required for growth promotion but for biotrophic colonization of barley roots. New Phytol 196:520–534. doi:10.1111/j.1469-8137.2012.04275.x

Hoffman M, Arnold AE (2010) Diverse bacteria inhabit living hyphae of phylogenetically diverse fungal endophytes. Appl Environ Microbiol 76:4063–4075

Hoitink HAJ, Madden LV, Dorrance AE (2006) Systemic resistance induced by *Trichoderma* spp.: interactions between the host, the pathogen, the biocontrol agent and soil organic matter quality. Phytopathology 96:186–189

Horst LE, Locke J, Krause CR, McMahon RW, Madden LV, Hoitink HAJ (2005) Suppression of Botrytis blight of begonia by *Trichoderma hamatum* 382 in peat and compost-amended potting mixes. Plant Dis 89:1195–1200

Jacobs S, Zechmann B, Molitor A, Trujillo M, Petutschnig E et al (2011) Broad-spectrum suppression of innate immunity is required for colonization of Arabidopsis roots by the fungus *Piriformospora indica*. Plant Physiol 156:726–740

Johnson NC, Graham JH, Smith FA (1997) Functioning of mycorrhizal associations along the mutualism-parasitism continuum. New Phytol 135:575–585

Jumpponen A, Trappe JM (1998) Dark septate endophytes: a review of facultative biotrophic root-colonizing fungi. New Phytol 140:295–310

Jumpponen A, Mattson KG, Trappe JM (1998) Mycorrhizal functioning of *Phialocephala fortinii* with *Pinus contorta* on glacier forefront soil: interactions with soil nitrogen and organic matter. Mycorrhiza 7:261–265

Jumpponen A, Newsham KK, Neises DJ (2003) Filamentous ascomycetes inhabiting the rhizoid environment of the liverwort *Cephaloziella varians* in Antarctica are assessed by direct PCR and cloning. Mycologia 95:457–466

Junker C, Draeger S, Schulz B (2012) A fine line e endophytes or pathogens in *Arabidopsis thaliana*. Fungal Ecol 5(3):657–662

Khan J, Ooka JJ, Miller SA, Madden LV, Hoitink HAJ (2004) Systemic resistance induced by *Trichoderma hamatum* 382 in cucumber against *Phytophthora* crown rot and leaf blight. Plant Dis 88:280–286

Knapp DG, Pintye A, Kovács GM (2012) The dark side is not fastidious—dark septate endophytic fungi of native and invasive plants of semiarid sandy areas. PLoS One 7(2):e32570. doi:10.1371/journal.pone.0032570

Kohout P, Sýkorova Z, Čtvrtlíková M, Rydlová J, Suda J, Vohník M, Sudová R (2012) Surprising spectra of root-associated fungi in submerged aquatic plants. FEMS Microbiol Ecol 80:216–235

Kottke I, Beiter A, Weiß M, Haug I, Oberwinkler F, Nebel M (2003) Heterobasidiomycetes form symbiotic associations with hepatics: Jungermanniales have sebacinoid mycobionts while *Aneura pinguis* (Metzgeriales) is associated with a *Tulasnella* species. Mycol Res 107:957–968

Krings M, Taylor TN, Hass H, Kerp H, Dotzler N, Hermsen EJ (2007) Fungal endophytes in a 400-million-yr-old land plant: infection pathways, spatial distribution, and host responses. New Phytol 174:648–657

Kuldau G, Bacon C (2008) Clavicipitaceous endophytes: their ability to enhance resistance of grasses to multiple stresses. Biol Control 46(1):57–71

Kusari S, Kosuth J, Cellarova E, Spiteller M (2011) Survival-strategies of endophytic *Fusarium solani* against indigenous camptothecin biosynthesis. Fungal Ecol 4:219–223

Kusari S, Hertweck C, Spiteller M (2012) Chemical ecology of endophytic fungi: origins of secondary metabolites. Chem Biol 19:792–798

Lahrmann U, Zuccaro A (2012) Opprimo ergo sum—evasion and suppression in the root endophytic fungus *Piriformospora indica*. Mol Plant Microbe Interact 25(6):727–737

Li JY, Sidhu R, Ford EJ, Long DM, Hess WM, Strobel GA (1998) The induction of production in the endophytic fungus *Periconia* sp. from *Torreya grandifolia*. J Ind Microbiol Biotechnol 20:259–264

Li HY, Qing C, Zhang YL, Zhao ZW (2005) Screening for endophytic fungi with antitumor and antifungal activities from Chinese medicinal plants. World J Microbiol Biotechnol 21:1515–1519

Mandyam K, Jumpponen A (2005) Seeking the elusive function of the root-colonising dark septate endophytic fungi. Stud Mycol 53:173–189

Mandyam K, Loughin T, Jumpponen A (2010) Isolation and morphological and metabolic characterization of common endophytes in annually burned tallgrass prairie. Mycologia 102(4):813–821

Mandyam KG, Roe J, Jumpponen A (2013) *Arabidopsis thaliana* model system reveals a continuum of responses to root endophyte colonization. Fungal Biol. doi:10.1016/j.funbio.2013.02.001

Márquez LM, Redman RS, Rodriguez RJ, Roossinck MJ (2007) A virus in a fungus in a plant: three-way symbiosis required forthermal tolerance. Science 315:513–515

Mastouri F, Bjorkman T, Harman GE (2010) Seed treatment with *Trichoderma harzianum* alleviates biotic, abiotic, and physiological stresses in germinating seeds and seedlings. Phytopathology 100:1213–1221

Mastouri F, Bjorkman T, Harman GE (2012) *Trichoderma harzianum* enhances antioxidant defense of tomato seedlings and resistance to water deficit. Mol Plant Microbe Interact 25:1264–1271

McKendrick SL, Leake DJ, Taylor DL, Read DJ (2002) Symbiotic germination and development of myco-heterotrophic orchid *Neottia nidus-avis* in nature and its requirement for locally distributed *Sebacina* spp. New Phytol 154:233–247

Menkis A, Vasiliauskas R, Taylor AF, Stenlid J, Finlay R (2005) Fungal communities in mycorrhizal roots of conifer seedlings in forest nurseries under different cultivation systems, assessed by morphotyping, direct sequencing and mycelial isolation. Mycorrhiza 16:33–41

Merlin E (1922) On the mycorrhizas of *Pinus sylvestris* L. and *Picea abies* Karst. A preliminary note. J Ecol 9:254–257

Narisawa K, Tokumasu S, Hashiba T (1998) Suppression of clubroot formation in Chinese cabbage by the root endophytic fungus, *Heteroconium chaetospira*. Plant Pathol 47:210–216

Narisawa K, Usuki F, Hashiba T (2004) Control of *Verticillium* yellows in chinese cabbage by the dark septate endophytic fungus LtVB3. Phytopathology 94:412–418

Neubert K, Mendgen K, Brinkmann H, Wirsel SGR (2006) Only a few fungal species dominate highly diverse mycofloras associated with the common reed. Appl Environ Microbiol 72:1118–1128

Newsham KK (2011) A meta-analysis of plant responses to dark septate root endophytes. New Phytol 10:783–793

O'Dell TE, Massicotte HB, Trappe JM (1993) Root colonization of *Lupinus latifolius* Agardh., and *Pinus contorta* Dougl. by *Phialocephala fortinii* Wang and Wilcox. New Phytol 124:93–100. doi:10.1111/j.1469-8137.1993.tb03800.x

Oberwinkler F, Riess K, Bauer R, Selosse M-A, Weiss M, Garnica S, Zuccaro A (2013) Enigmatic sebacinales. Mycol Prog 12:1–27. doi:10.1007/s11557-012-0880-4

Paszkowski U (2006) Mutualism and parasitism: the yin and yang of plant symbioses. Curr Opin Plant Biol 9(4):364–370

Peterson RL, Wagg C, Pautler M (2008) Associations between microfungal endophytes and roots: do structural features indicate function? Botany 86:445–456

Petrini O (1991) Fungal endophytes of tree leaves. In: Andrews J, Hirano S (eds) Microbial ecology of leaves. Springer, New York, NY, pp 179–197

Petrini O (1996) Ecological and physiological aspects of host-specificity in endophytic fungi. In: Redlin SC, Carris LM (eds) Endophytic fungi in grasses and woody plants. APS Press, St. Paul, MN, pp 87–100

Porras-Alfaro A, Herrera J, Sinsabaugh RL, Odenbach KJ, Lowrey T, Natvig DO (2008) Novel root fungal consortium associated with a dominant desert grass. Appl Environ Microbiol 74:2805–2813. doi: 10.1128/AEM.02769-07

Porras AA, Herrera J, Sinsabaugh RL, Odenbach KJ, Lowrey T, Natvig DO (2008) Novel root fungal consortium associated with a dominant desert grass. Appl Environ Microbiol 74:2805–2813

Qiang X, Weiss M, Kogel KH, Schäfer P (2012a) *Piriformospora indica* a mutualistic basidiomycete with an exceptionally large plant host range. Mol Plant Pathol 13:508–518

Qiang XY, Zechmann B, Reitz MU, Kogel KH, Schafer P (2012b) The mutualistic fungus Piriformospora indica colonizes Arabidopsis roots by inducing an endoplasmic reticulum stress-triggered caspase-dependent cell death. Plant Cell 24:794–809

Redman RS, Kim YO, Woodward CJDA, Greer C, Espino L et al (2011) Increased fitness of rice plants to abiotic stress via habitat adapted symbiosis: a strategy for mitigating impacts of climate change. PLoS One 6(7):e14823. doi:10.1371/journal.pone.0014823

Robinson CH (2001) Cold adaptation in arctic and antarctic fungi. New Phytol 151:341–353

Rodriguez RJ, Redman RS (2008) More than 400 million years of evolution and some plants still can't make it on their own: plant stress tolerance via fungal symbiosis. J Exp Bot 59:1109–1114

Rodriguez RJ, Henson J, Van Volkenburgh E, Hoy M, Wright L et al (2008) Stress tolerance in plants via habitat-adapted symbiosis. ISME J 2(4):404–416

Rodriguez RJ, White JF, Arnold AE, Redman RS (2009) Fungal endophytes: diversity and functional roles. New Phytol 182:314–330

Romão-Dumaresq AS, de Araújo WL, Talbot NJ, Thornton CR (2012) RNA interference of endochitinases in the sugarcane endophyte *Trichoderma virens* 223 reduces its fitness as a biocontrol agent of pineapple disease. PLoS One 7(10):e47888. doi:10.1371journal.pone.0047888

Saikkonen K, Faeth SH, Helander M, Sullivan TJ (1998) Fungal endophytes: a continuum of interactions with host plants. Annu Rev Ecol Syst 29:319–343

Saikkonen K, Wali P, Helander M, Faeth SH (2004) Evolution of endophyte-plant symbioses. Trends Plant Sci 9(6):275–280

Schäfer P, Kogel KH (2009) The sebacinoid fungus *Piriformospora indica*: an orchid mycorrhiza which may increase host plant reproduction and fitness. In: Deising H (ed) Plant relationships, vol 5, 2nd edn, The mycota. Springer, Berlin, pp 99–112

Schäfer P, Pfiffi S, Voll LM, Zajic D, Chandler PM, Waller F et al (2009) Manipulation of plant innate immunity and gibberellin as factor of compatibility in the mutualistic association of barley roots with *Piriformospora indica*. Plant J 59:46–474

Schardl CL, Leuchtmann A, Spiering MJ (2004) Symbioses of grasses with seedborne fungal endophytes. Annu Rev Plant Biol 55:315–340

Schulz B, Boyle C (2005) The endophytic continuum. Mycol Res 109:661–686

Schulz B, Rommert AK, Dammann U, Aust HJ, Strack D (1999) The endophyte-host interaction: a balanced antagonism? Mycol Res 103:1275–1283

Schulz B, Boyle C, Draeger S, Römmert AK, Krohn K (2002) Endophytic fungi: a source of novel biologically active secondary metabolites. Mycol Res 106:996–1004

Selosse MA, Setaro S, Glatard F, Richard F, Urcelay C, Weiß M (2007) Sebacinales are common mycorrhizal associates of Ericaceae. New Phytol 174:864–878

Selosse MA, Dubois MP, Alvarez N (2009) Do Sebacinales commonly associate with plant roots as endophytes? Mycol Res 113:1062–1069

Setaro S, Weiss M, Oberwinkler F, Kottke I (2006) Sebacinales form ectendomycorrhizas with *Cavendishia nobilis*, a member of the Andean clade of Ericaceae, in the mountain rain forest of southern Ecuador. New Phytol 169(2):355–365

Sharma M, Schmid M, Rothballer M, Hause G, Zuccaro A et al (2008) Detection and identification of bacteria intimately associated with fungi of the order Sebacinales. Cell Microbiol 10:2235–2246

Sherameti I, Shahollari B, Venus Y, Altschmied L, Varma A, Oelmüller R (2005) The endophytic fungus *Piriformospora indica* stimulates the expression of nitrate reductase and the starch-degrading enzyme glucan-water dikinase in tobacco and Arabidopsis roots through a homeodomain transcription factor that binds to a conserved motif in their promoters. J Biol Chem 280:26241–26247

Shoresh M, Harman GE (2008) The molecular basis of shoot responses of maize seedlings to *Trichoderma harzianum* T22 inoculation of the root: a proteomic approach. Plant Physiol 147:2147–2163

Shoresh M, Harman GE, Mastouri F (2010) Induced systemic resistance and plant responses to fungal biocontrol agents. Annu Rev Phytopathol 48:1–23

Shukla N, Awasthi RP, Rawat L, Kumar J (2012) Biochemical and physiological responses of rice (*Oryza sativa* L.) as influenced by *Trichoderma harzianum* under drought stress. Plant Physiol Biochem 54:78–88

Sieber TN, Grünig CR (2005) Biodiversity of fungal root-endophyte communities and populations in particular of the dark septate endophyte *Phialocephala fortinii*. In: Schulz B, Boyle C, Sieber TN (eds) Microbial root endophytes. Springer, Berlin, 2006

Sirikantaramas S, Yamazaki M, Saito K (2009) A survival strategy: the coevolution of the camptothecin biosynthetic pathway and self-resistance mechanism. Phytochemistry 70:1894–1898

Sirrenberg A, Goebel C, Grond S, Czempinski N, Ratzinger A et al (2007) *Piriformospora indica* affects plant growth by auxin production. Physiol Plant 131:581–589

Stierle A, Strobel GA, Stierle D (1993) Taxol and taxane production by *Taxomyces andreanae*, an endophytic fungus of Pacific yew. Science 260:214–216

Stone JK, Bacon CW, White JF (2000) An overview of endophytic microbes: endophytism defined. In: Bacon CW, White JF (eds) Microbial endophytes. Dekker, New York, NY, pp 3–30

Strobel GA (2003) Endophytes as sources of bioactive products. Microbes Infect 5:535–544

Strobel G, Daisy B (2003) Bioprospecting for microbial endophytes and their products. Microbiol Mol Biol Rev 67:491–502

Strobel G, Yang X, Sears J, Kramer R, Sidhu RS, Hess WH (1996) Taxol from *Pestalotiopsis microspora*, an endophytic fungus of *Taxus wallichiana*. Microbiology 142:435–440

Suryanarayanan TS, Thirunavukkarasu N, Govindarajulu MB, Sasse F et al (2009) Fungal endophytes and bioprospecting. Fungal Biol Rev 23:9–19

Tan RX, Zou WX (2001) Endophytes: a rich source of functional metabolites. Nat Prod Rep 18:448–459

Tellenbach C, Grünig CR, Sieber TN (2011) Negative effects on survival and performance of Norway spruce seedlings colonized by dark septate root endophytes are primarily isolate-dependent. Environ Microbiol 13:2508–2517

Tellenbach C, Sumarah MW, Grüning CR, Miller JD (2013) Inhibition of *Phytophthora* species by secondary metabolites produced by the dark septate endophyte *Phialocephala europaea*. Fungal Ecol 6:12–18

Torres MS, White JF, Zhang X, Hinton DM, Bacon CW (2012) Endophyte-mediated adjustments in host morphology and physiology and effects on host fitness traits in grasses. Fungal Ecol 5(3):322–330

Turnau K, Henriques FS, Anielska T, Renker C, Buscot F (2007) Metal uptake and detoxification mechanisms in *Erica andevalensis* growing in a pyrite mine tailing. Environ Exp Bot 61:117–123

Upson R, Read DJ, Newsham KK (2007) Widespread association between the ericoid mycorrhizal fungus *Rhizoscyphus ericae* and a leafy liverwort in the sub- and maritime Antarctic. New Phytol 176:460–471

Upson R, Newsham K, Bridge PD, Pearce DA, Read DJ (2009a) Taxonomic affinities of dark septate root endophytes of *Colobanthus quitensis* and *Deschampsia antarctica*, the two native Antarctic vascular plant species. Fungal Ecol 2:184–196

Upson R, Read DJ, Newsham KK (2009b) Nitrogen form influences the response of *Deschampsia antarctica* to dark septate root endophytes. Mycorrhiza 20:1–11

Urban A, Weib M, Bauer R (2003) Ectomycorrhizas involving sebacinoid mycobionts. Mycol Res 107(1):3–14

Usuki F, Narisawa K (2007) A mutualistic symbiosis between a dark septate endophytic fungus, *Heteroconium chaetospira*, and a nonmycorrhizal plant, Chinese cabbage. Mycologia 99:175–184

Vadassery J, Ritter C, Venus Y, Camehl I, Varma A et al (2008) The role of auxins and cytokinins in the mutualistic interaction between Arabidopsis and *Piriformospora indica*. Mol Plant Microbe Interact 21(10):1371–1383, 24(3): 336–351

Vandenkoornhuyse P, Baldauf SL, Leyval C, Straczec J, Young JPW (2002) Extensive fungal diversity in plant roots. Science 295:2051

Varma A, Verma S, Sudha Sahay N, Butehorn B, Franken P (1999) *Piriformospora indica*, a cultivable plant-growth-promoting root endophyte. Appl Environ Microbiol 65:2741–2744

Verkley GJM, Zijlstra JD, Summerbell RC, Berendse F (2003) Phylogeny and taxonomy of root-inhabiting *Cryptosporiopsis* species, and *C. rhizophila* sp. nov., a fungus inhabiting roots of several Ericaceae. Mycol Res 107(6):689–698

Verma S, Varma A, Rexer K-H, Hassel A, Kost G, Sarbhoy A et al (1998) *Piriformospora indica*, gen. et sp. nov., a new root-colonizing fungus. Mycologia 90:896–903

Vinale F, Ghisalberti EL, Flematti G, Marra R, Lorito M, Sivasithamparam K (2010) Secondary metabolites produced by a root-inhabiting sterile fungus antagonistic towards pathogenic fungi. Lett Appl Microbiol 50(4):380–385. doi:10.1111/j.1472-765X.2010.02803.x

Viterbo A, Landau U, Kim S, Chernin L, Chet I (2010) Characterization of ACC deaminase from the biocontrol and plant growth-promoting agent *Trichoderma asperellum* T203. FEMS Microbiol Lett 305:42–48

Vralstad T, Myhre E, Schumacher T (2002) Molecular diversity and phylogenetic affinities of symbiotic root-associated ascomycetes of the Helotiales in burnt and metal polluted habitats. New Phytol 155(1):131–148

Wagg C, Pautler M, Massicotte HB, Peterson RL (2008) The co-occurrence of ectomycorrhizal, arbuscular mycorrhizal, and dark septate fungi in seedlings of four members of the *Pinaceae*. Mycorrhiza 18:103–110

Waller F, Achatz B, Baltruschat H, Fodor J, Becker K et al (2005) The endophytic fungus *Piriformospora indica* reprograms barley to salt-stress tolerance, disease resistance, and higher yield. Proc Natl Acad Sci USA 102:13386–13391

Wang CJK, Wilcox HE (1985) New species of ectendomycorrhizal and pseudomycorrhizal fungi: *Phialophora finlandia*, *Chloridium paucisporum* and *Phialocephala fortinii*. Mycologia 77(6):951–958

Wang JW, Zheng LP, Tan RX (2006) The Preparation of an elicitor from a fungal endophyte to enhance artemisinin production in hairyroot cultures of *Artemisia annua* L. Chin J Biotechnol 22:829–834

Warcup JH (1988) Mycorrhizal associations of isolates of *Sebacina vermifera*. New Phytol 110:227–231

Weiss M, Selosse M-A, Rexer K-H, Urban A, Oberwinkler F (2004) Sebacinales: a hitherto overlooked cosm of heterobasidiomycetes with a broad mycorrhizal potential. Mycol Res 108:1003–1010

Weiss M, Sykorova Z, Garnica S, Riess K, Martos F et al (2011) Sebacinales everywhere: previously overlooked ubiquitous fungal endophytes. PLoS One 6:e16793

Wilcox HE, Wang CJK (1987) Mycorrhizal and pathological associations of dematiaceous fungi in roots of 7-month-old tree seedlings. Can J Forest Res 17:884–899

Williams PG (1985) Orchidaceous rhizoctonias in pot cultures of vesicular- arbuscular mycorrhizal fungi. Can J Bot 63:1329–1333

Wilson D (1995) Endophyte—the evolution of a term, and clarification of its use and definition. Oikos 73:274–276

Xia X, Lie T, Qian X, Zheng Z, Huang Y, Shen Y (2011) Species diversity, distribution, and genetic structure of endophytic and epiphytic *Trichoderma* associated with banana roots. Microb Ecol 61(3):619–625. doi:10.1007/s00248-010-9770-y

Xu L, Zhou L, Zhao J, Li J, Li X, Wang J (2008) Fungal endophytes from *Dioscorea zingiberensis* rhizomes and their antibacterial activity. Lett Appl Microbiol 46:68–72

Yadav V, Kumar M, Deep DK, Kumar H, Sharma R et al (2010) A phosphate transporter from the root endophytic fungus *Piriformospora indica* plays a role in phosphate transport to the host plant. J Biol Chem 285:26532–26544

Yedidia I, Benhamou N, Chet I (1999) Induction of defense responses in cucumber plants (*Cucumis sativus* L.) by the biocontrol agent *Trichoderma harzianum*. Appl Environ Microbiol 65(3):1061–1070

Yedidia I, Shrivasta AK, Kapulnik Y, Chet I (2001) Effect of *Trichoderma harzianum* on microelement concentration and increased growth of cucumber plants. Plant Soil 235:235–242

You F, Han T, Wu J, Huang B, Qin L (2009) Antifungal secondary metabolites from endophytic *Verticillium* sp. Biochem Syst Ecol 37(3):162–165

Zuccaro A, Lahrmann U, Güldener U, Langen G, Pfiffi S, Biedenkopf D et al (2011) Endophytic life strategies decoded by genome and transcriptome analyses of the mutualistic root symbiont *Piriformospora indica*. PLoS Pathog 7:e1002290

Chapter 17
Endophytic Yeasts: Biology and Applications

Sharon Lafferty Doty

17.1 Introduction

Compared to the large body of literature on endophytic filamentous fungi and bacteria, relatively little research has been done on the biodiversity of endophytic yeast (Table 17.1). The surface of plant foliage is often colonized by epiphytic yeasts. However, few yeast species seem to colonize the plant interior. In the majority of published reports, the plant tissues were surface-sterilized and then plated on malt dextrose agar or potato dextrose agar. Whether the lack of isolated yeast strains is due to an actual habitat limitation or is the result of the isolation techniques or culture media that may favor filamentous fungi and bacteria remains to be determined.

Some of the most commonly isolated endophytic yeast genera include *Cryptococcus*, *Debaryomyces*, *Sporobolomyces*, and *Rhodotorula*. Since yeasts are generally considered to be nonmotile, their entry into the plant interior without flagella or hyphae would seem challenging. Some may enter through the stomata or wounds on the plant surfaces. Some species excrete cutinase that can attack the plant cuticle. Others have lipolytic and pectinolytic activities that could enable the yeasts to gain entry into the plant. The specific mechanisms by which endophytic yeast colonize plants have yet to be determined.

S.L. Doty (✉)
School of Environmental and Forest Sciences, College of the Environment, University of Washington, Seattle, WA 98195-2100, USA
e-mail: sldoty@uw.edu

Table 17.1 Summary of endophytic yeast isolations

Host plant (common name)	Yeast genus	Tissue	References
Allium (onion)		Bulbs	Isaeva et al. (2010)
Alstonia scholaris-India	19 unidentified	Leaves	Mahapatra and Banerjee (2010)
Anthriscus sylvestris (chervil)		Storage roots	Isaeva et al. (2010)
Arctium tomentosum (burdock)		Storage roots	Isaeva et al. (2010)
Armoracia rusticana (horseradish)		Storage roots	Isaeva et al. (2010)
Aronia melanocarpa (chokeberry)		Fruits	Isaeva et al. (2010)
Beta vulgaris (beet)		Storage roots	Isaeva et al. (2010)
Cichorium intybus (chicory)		Rhizomes	Isaeva et al. (2010)
Citrus sinensis (sweet orange)	*Rhodotorula* *Cryptococcus* *Pichia*	All	Gai et al. (2009)
Crataegus oxyacantha (hawthorn)		Fruits	Isaeva et al. (2010)
Cucurbita pepo (pumpkin)		Fruits	Isaeva et al. (2010)
Daucus sativus (carrot)		Storage roots	Isaeva et al. (2010)
Daucus sativus (carrot)	*Pichia*		Rodriguez et al. (2011)
Euonymus verrucosa (euonymus)		Fruits	Isaeva et al. (2010)
Malus domestica (apple)-Brazil	*Rhodotorula* *Candida* *Pichia*	Leaves	Camatti-Sartori et al. (2005)
Maize	*Williopsis*	Roots	Nassar et al. (2005)
Orchidaceae (orchids)-Brazil	*Rhodotorula* *Candida* *Besingtonia*		Vaz et al. (2009)
Picea mariana (black spruce)	Black yeasts	Needles	Johnson and Whitney (1992)
Pinus sylvestris (Scots pine)	*Hormonema* *Rhodotorula*	Buds	Pirttila et al. (2003)
Pinus tabulaeformis-China	*Rhodotorula*	Twigs	Zhao et al. (2002)
Populus trichocarpa (black cottonwood)	*Rhodotorula*	Branches	Xin et al. (2009)
P. trichocarpa × deltoides (poplar hybrid)	*Rhodotorula*	Branches	Xin et al. (2009)
Quercus rober (English oak)	*Cryptococcus*	Acorns	Isaeva et al. (2009)
Quercus (oak and elm)	*Filobasidium*	Canopy	Unterseher et al. (2007)
Salix	*Ogataea*	Leaf galls	Glushakova et al. (2010)
Sequoia sempervirens (coast redwood)	*Debaryomyces* *Trichosporon* Red yeast	Shoots	Middelhoven (2003)
Solanum cernuum (panaceia)	*Candida* *Cryptococcus* *Kwoniella* *Meyerozyma*	Leaves, stem	Vieira et al. (2012)

17.2 Endophytic Yeasts of Forests

Most of the published studies of endophytic yeasts involve trees and plants of forests. One of the earliest studies focused on the endophytic fungal population in black spruce (*Picea mariana*) buds and needles (Johnson and Whitney 1992). Out of 400 buds examined, only one endophytic fungal isolate was detected; however, 914 were isolated from needles, suggesting that colonization occurs after bud flush. The majority of the fungal isolates belonged to a small number of taxa, and only 11 % of the total were yeasts. The number of endophytes increased with age with only 4 % of needle samples containing endophytes in the current year of growth to 90 % of the samples in the third year. The timing may have ecological significance since spruce bud worm, a major pest of black spruce, primarily feeds on the buds and the new growth when the endophytic fungi are essentially absent.

Pine twigs of *Pinus tabulaeformis* in Beijing had few endophytic yeast isolates (Zhao et al. 2002). Out of a thousand plates of twig fragments from 30 different trees, only 3 yeast colonies grew on the malt extract plates, and no filamentous fungi were evident. These were initially identified as *Rhodotorula minuta* but nucleotide differences in some of the ribosomal gene regions indicated that it was a novel species. The proposed novel species, *Rhodotorula pinicola*, was closely related to a yeast species isolated from Antarctica.

The coast redwood, like the pine, had few associated fungi in a study of the shoots of *Sequoia sempervirens* (Middelhoven 2003). Three yeast species were isolated from young shoots: *Debaryomyces hansenii*, *Trichosporon pullulans*, and an unidentified red Basidiomycete. Only *D. hansenii* was isolated from older shoots. The author suggested that the overall low biodiversity of redwood forests correlated with this low level of associated yeast biodiversity.

Scots pine, the main source of timber in Northern Europe, was screened for endophytes as a cause of browning in micropropagation (Pirttila et al. 2003). Using in situ hybridization techniques, *Rhodotorula minuta* was identified in 40 % of the meristematic bud tissues, in the cells of scale primordial, in epithelial cells of resin ducts, and in the cells of developing stems. A strikingly different result was seen from culturing experiments, however, where only a few cultures contained this yeast. Therefore, studies that depend on growth of the yeast from the plant samples could be greatly underestimating the prevalence of endophytic yeast. The high percentage of growing tissues with yeast may suggest that the yeast has a role in plant growth.

Using a canopy crane, researchers isolated endophytic fungi from leaves of a temperate forest dominated by oak and elm (Unterseher et al. 2007). The authors noted a variety of habitat preferences of the fungal endophytes including host species, light regimes, and season. Of the 626 isolates representing 48 taxa of fungi, however, only one was a yeast species, *Filobasidium uniguttulatum*.

The endophytic yeast population of English oak was surveyed from the acorns (Isaeva et al. 2009). Prior to germination, the only yeast species present was *Candida railenensis*. The population increased at germination to as high as

10^7 CFU/g. During germination, the cotyledons were colonized by common epiphytic and litter yeasts: *Cryptococcus albidus* (14 %), *Rhodotorula glutinis* (5 %), and *Cystofilobasidium capitatum* (7 %), while the majority remained *C. railenensis*.

Three pink-pigmented yeasts were isolated from *Populus* (cottonwood) trees (Xin et al. 2009). *Rhodotorula graminis* was isolated from *Populus trichocarpa* growing in its native riparian environment while two different strains of *Rhodotorula mucilaginosa* were isolated from hybrid poplar in greenhouses in Oregon and Washington states. All three strains produced high levels of the phytohormone, indoleacetic acid, and grew well on common plant sugars including sucrose, glucose, and xylose but differed in other carbon source utilization. In a survey of *P. deltoides* trees using molecular methods to identify endophytes regardless of culturability, a wide variety of bacterial and fungal species were identified but few yeasts (Gottel et al. 2011). *Cryptococcus aerius* was listed and the *Pucciniomycotina urediniomycete*.

Endophytic yeasts that utilize methanol were isolated from leaf galls of willow plants (Glushakova et al. 2010). On the leaf surface were *Cryptococcus*, *R. glutinous*, and *Cystofilobasidium* yeast species but the interior was dominated by a novel species, *Ogataea cecidiorum*. The strain, related most closely to *Pichia* species, fermented glucose after a delay of a few days but was unable to ferment the plant sugars, sucrose, galactose, and xylose.

In a screen of Brazilian orchids for endophytic fungi, over a dozen strains were isolated (Vaz et al. 2009). Both Ascomycota (*Candida* sp.) and Basidiomycota (*R. mucilaginosa*) yeasts were represented in the orchids of the tropical ecosystem.

A survey of endophytic fungi with antimicrobial activity in the Indian medicinal tree, *Alstonia scholaris*, was conducted (Mahapatra and Banerjee 2010). Of the 1,152 fungal isolates from over a thousand plant segments, 19 were yeast strains that were not further identified. In the Brazilian medicinal plant commonly referred to as panaceia (*Solanum cernuum*), 21 yeasts were isolated out of a collection of 246 endophytic fungi (Vieira et al. 2012). The number of yeast isolated varied with the seasons with 18 isolated in winter and only 3 isolated in summer.

The presence of endophytic yeasts was demonstrated to be a widespread phenomenon in a study that focused on the plant storage tissues as the most likely location in plants to harbor yeasts (Isaeva et al. 2010). Twenty-nine plant species of forests and meadows were screened. The most common species were *Cryptococcus albidus* (22 isolates), *Hanseniaspora guilliermondii* (34), *Metschnikowia pulcherrima* (11), *Rhodotorula* spp. (4), plus 19 other species at lower frequencies. In the subsurface plant storage organs, *Rhodotorula* was the most prevalent genus.

17.3 Endophytic Yeasts of Agricultural Plants

Endophytic yeasts have been identified in a range of crop plants including apples, maize, carrot, and sweet orange. In a study of Brazilian apples, a comparison was done to assess the overall biodiversity of endophytic fungi in apple trees grown under three different cultivation systems: conventional, integrated, and organic (Camatti-Sartori et al. 2005). A variety of yeast species were identified including *Sporobolomyces roseus*, *S. pararoseus*, *R. mucilaginosa*, *D. hansenii*, *Candida* spp., *Cryptococcus* spp., and *Pichia* spp. *Rhodotorula* was exclusively in the leaves, while *Candida* and *Pichia* were primarily in the fruits. The trees cultivated using organic methods had a higher endophytic diversity than trees under both conventional and integrated management. For example, 38 yeast strains were isolated from the latter while 108 were isolated from the organically grown trees. It was proposed that the use of fungicides may have been at least partly responsible for the reduced diversity in the nonorganic cultivated trees.

In a screen for endophytic yeasts of maize roots, 24 strains were isolated (Nassar et al. 2005). Some of the strains produced the phytohormones IAA or IPYA. The highest phytohormone producer was identified as *Williopsis saturnus*. A nonproducing strain was identified as *Rhodotorula glutinis*.

Endophytic yeasts were isolated from sweet orange plants (*Citrus sinensis*) in Brazil (Gai et al. 2009). A total of 24 strains were isolated and then placed into three groups based on morphological characteristics. Group A included *Candida*, *Pichia*, and *Aureobasidium* spp. Group B was comprised of *Cryptococcus* spp. *R. mucilaginosa* comprising Group C was detected in all of the samples, while *Pichia guilliermondii* was mostly in plants infected with the pathogen *Xylella fastidiosa*. Uninfected plants had the highest concentration of endophytic yeasts with *Cryptococcus flavescens* being the dominant species. When the three groups were added to the roots of axenic seedlings, the concentration of endophytes ranged from 10^6 to 10^9 CFU/g in 45 days. Roots, stems, and leaves were colonized but the yeasts were mostly observed in the stomata and none were on the plant surface. To investigate the correlation of *Pichia* with disease, cell-free extracts of *Pichia* cultures and *Cryptococcus* cultures were added to in vitro cultures of the pathogen. The *Cryptococcus* extract had no effect but the *Pichia* extract enhanced the growth of *Xylella*.

17.4 Applications of Endophytic Yeasts

The majority of the research on endophytic yeasts has focused on ecological studies; however, endophytic yeasts have many potential uses in the production of valuable biochemicals, antimicrobials, and biocatalysts and in plant growth enhancement. Since they would be adapted to the types of sugars, both hexoses and pentoses, present in plants, and would be resistant to phytochemicals that may be toxic to other microbes, endophytic yeasts make strong candidates for use in biofuel production. Brewer's yeast, *Saccharomyces cerevisiae*, can ferment only six carbon sugars, making lignocellulosic fermentation inefficient since a high percentage of hemicellulose is comprised of five carbon sugars. Furthermore, it is sensitive to phytochemicals released during the pretreatment phase of biofuel production. The *Rhodotorula* yeasts isolated from poplar trees fermented xylose and glucose to xylitol and ethanol, respectively (Bura et al. 2012). Xylitol is a valuable alternative sweetener that has fewer calories than sucrose and can be used by diabetics without raising blood-sugar levels. A genetic analysis of the poplar yeast strains revealed that the xylose metabolism genes had unique and complex structures with multiple introns compared to the xylose genes of other yeasts (Xu et al. 2011). A comparison of the xylose gene expression levels of two poplar yeasts showed that the better xylose fermenter had higher gene expression, suggesting an avenue for further improving strains. The poplar yeasts were also shown to be resistant to some of the common fermentation inhibitors (Vajzovic et al. 2012). The poplar yeast performed best, producing the highest levels of ethanol and xylitol, when fermenting hydrolysate from poplar and Douglas-fir (Bura et al. 2012), indicating that it is well adapted to the sugars and inhibitors found in trees.

Some endophytic yeast strains have antimicrobial properties that may be useful in pharmaceutical applications. The work by Vaz et al. (2009) on endophytic fungi of Brazilian orchids showed that 3 of the 13 yeast isolates had antimicrobial activity. Crude extracts of these three yeasts, *R. mucilaginosa*, *Besingtonia* sp., and *Candida parapsilosis*, presented minimum inhibitory concentration values of >1 mg/ml against three pathogenic targets. Although the extracts had less potent activity than extracts of some of the other endophytic fungi, the research points to the possibility of endophytic yeasts as a largely untapped resource. Considering that yeasts are more easily cultured than filamentous fungi, they may provide a good avenue for the production of bioactive compounds. Endophytes of another Brazilian plant, the medicinal plant panaceia, also showed promise as a source of antimicrobials (Vieira et al. 2012). In the work of Viera and colleagues, five bioactive endophytic yeast strains were identified: *Candida carpophila*, two *Cryptococcus* sp., *Kwoniella mangroviensis*, and *Meyerozyma guilliermondii*. In both studies, the target microbes were human pathogens. Analysis of the potential abilities of endophytic yeast to protect the plant host from pathogens remains to be explored.

Endophytic yeasts also have potential for industrial-scale production of therapeutics (Corchero et al. 2013). Since they are adapted to a variety of carbon sources in plants, they could utilize low-cost biomass, and being recently isolated from the environment, they may be more hardy and robust than traditional industrial yeast strains. Being eukaryotic gives the endophytic yeast advantages over bacteria in bioproduction and bioprocessing of recombinant proteins.

Endophytic yeasts have the potential to be used in biosynthesis. As industry moves toward biochemical production using natural resources rather than nonrenewable fossil fuels, there is a need to develop new methods using microorganisms adapted to phytochemicals. Endophytes utilize compounds produced by their plant hosts, making bioactive compounds that directly affect the host or are used in defense against pathogens or herbivores. More novel structures are produced by endophytes than by soil isolates (Wang and Dai 2011). Biocatalyst production was tested by Rodriguez et al. (2011). Among the endophytic yeast strains that were identified, a *Pichia* sp. strain from carrot had the best biocatalyst activity, with higher bioreduction of the α-alkyl-β-ketoesters than the endophytic bacteria.

A few endophytic yeast strains produced phytohormones and promoted increased plant growth. Eight of the *W. saturnus* endophytic yeast strains of maize roots produced the auxins IAA and IPYA (Nassar et al. 2005). Three endophytic yeasts of poplar, *Rhodotorula graminis*, and two strains of *R. mucilaginosa* also produced IAA (Xin et al. 2009). Some of the maize endophyte strains and all of the poplar strains required L-tryptophan as a precursor for production of IAA, diverting the host nitrogen storage into increased root mass. The maize endophytes promoted plant growth, significantly increasing dry weights and length of roots and shoots. Not only can endophytic yeast increase the growth of the host plant species from which it was isolated, some have broader host ranges. The endophytic yeast strain WP1 from poplar strongly promoted the growth of bell pepper with a 60 % increase in root and shoot mass, earlier flowering and fruit set, and significantly increased fruit yields (Khan et al. 2012). In a study to test the impacts of the poplar endophytes on sweet corn, WP1 inoculation increased maize root mass by 84 % and shoot mass by 41 % and increased the maize leaf area (Knoth et al. 2012). In a current field trial, inoculation of poplar with the endophytic yeast also increased plant height compared to uninoculated controls (Knoth et al. unpublished). These studies clearly demonstrate the potential of using endophytic yeast to promote plant growth.

17.5 Conclusions

Relatively few research articles discuss the presence of yeasts within plants even when overall endophytic diversity is being explored. The work by Pirttila et al. (2003) pointed to the problem of underestimation of endophytic yeasts within plant samples. Isolation methods that are less biased toward filamentous fungi or bacteria may be necessary to get an accurate assessment of the endophytic yeast populations.

Endophytic yeasts are clearly a valuable and underutilized resource. Endophytic yeasts have potential uses in industry for biochemical production. Endophytic bacteria and filamentous fungi are being actively explored for their uses in biosynthesis, biotransformation, and biodegradation (Wang and Dai 2011), while endophytic yeasts have received far less attention. Given the ease by which yeast can be cultured compared to filamentous fungi, endophytic yeast offers significant potential for a variety of biosynthesis applications. Their use in biocontrol of plant pathogens has not been well explored. Considering the significant impacts that endophytic yeast can have on plant growth, there is an opportunity for using these natural yeasts to improve agriculture and biomass production for bioenergy in an environmentally sustainable way.

References

Bura R, Vajzovic A, Doty SL (2012) Novel endophytic yeast Rhodotorula mucilaginosa strain PTD3 I: production of xylitol and ethanol. J Ind Microbiol Biotechnol 39:1003–1011

Camatti-Sartori V, Silva-Ribeiro RTD, Valdebenito-Sanhueza RM, Pagnocca FC, Echeverrigaray S, Azevedo JL (2005) Endophytic yeasts and filamentous fungi associated with southern Brazilian apple *(Malus domestica)* orchards subjected to conventional, integrated or organic cultivation. J Basic Microbiol 45:397–402

Corchero JL, Gasser B, Resina D, Smith W, Parrilla E, Vazquez F, Abasolo I, Giuliani M, Jantti J, Ferrer P, Saloheimo M, Mattanovich D, Schwartz S Jr, Tutino L, Villaverde A (2013) Unconventional microbial systems for the cost-efficient production of high-quality protein therapeutics. Biotechnol Adv 31(2):140–153

Gai CS, Lacava PT, Maccheroni W Jr, Glienke C, Araujo WL, Miller TA, Azevedo JL (2009) Diversity of endophytic yeasts from sweet orange and their localization by scanning electron microscopy. J Basic Microbiol 49:441–451

Glushakova AM, Maximova IA, Kachalkin AV, Yurkov AM (2010) Ogataea cecidiorum sp. nov., a methanol-assimilating yeast isolated from galls on willow leaves. Antonie Van Leeuwenhoek 98:93–101

Gottel NR, Castro HF, Kerley M, Yang Z, Pelletier DA, Podar M, Karpinets T, Uberbacher E, Tuskan GA, Vilgalys R, Doktycz MJ, Schadt CW (2011) Distinct microbial communities within the endosphere and rhizosphere of *Populus deltoides* roots across contrasting soil types. Appl Environ Microbiol 77:5934–5944

Isaeva OV, Glushakova AM, Yurkov AM, Chernov IY (2009) The yeast Candida railenensis in the fruits of English Oak (*Quercus robur* L.). Microbiology 78:355–359

Isaeva OV, Glushakova AM, Garbuz A, Kachalkin AV, Chernov IY (2010) Endophytic yeast fungi in plant storage tissues. Biol Bull 37:26–34

Johnson JA, Whitney NJ (1992) Isolation of fungal endophytes of black spruce (*Picea mariana*) dormant buds and needles from New-Brunswick Canada. Can J Bot 70:1754–1757

Khan Z, Guelich G, Phan H, Redman RS, Doty SL (2012) Bacterial and yeast endophytes from poplar and willow promote growth in crop plants and grasses. Agronomy 2012:Art890280. doi:10.5402/2012/890280

Knoth J, Kim S-H, Ettl G, Doty SL (2012) Effects of cross host species inoculation of nitrogen-fixing endophytes on growth and leaf physiology of maize. GCB Bioenergy 5(4):408–418

Mahapatra S, Banerjee D (2010) Diversity and screening for antimicrobial activity of endophytic fungi of *Alstonia scholaris*. Acta Microbiol Immunol Hung 57:215–223

Middelhoven WJ (2003) The yeast flora of the coast redwood, *Sequoia sempervirens*. Folia Microbiol 48:361–362

Nassar AH, El-Tarabily KA, Sivasithamparam K (2005) Promotion of plant growth by an auxin-producing isolate of the yeast Williopsis saturnus endophytic in maize (*Zea mays* L.) roots. Biol Fertil Soils 42:97–108

Pirttila AM, Pospiech H, Laukkanen H, Myllyla R, Hohtola A (2003) Two endophytic fungi in different tissues of Scots pine buds (*Pinus sylvestris* L.). Microb Ecol 45:53–62

Rodriguez P, Reyes B, Barton M, Coronel C, Menendez P, Gonzalez D, Rodriguez S (2011) Stereoselective biotransformation of -alkyl–keto esters by endophytic bacteria and yeast. J Mol Catal B Enzym 71:90–94

Untersher M, Reiher A, Finstermeier K, Otto P, Morawetz W (2007) Species richness and distribution patterns of leaf-inhabiting endophytic fungi in a temperate forest canopy. Mycol Prog 6:201–212

Vajzovic A, Bura R, Doty SL (2012) Novel endophytic yeast Rhodotorula mucilaginosa strain PTD3 II: production of xylitol and ethanol in the presence of inhibitors. J Ind Microbiol Biotechnol 39:1453–1463

Vaz ABM, Mota RC, Bomfim MRQ, Vieira ML, Zani CL, Rosa CA, Rosa LH (2009) Antimicrobial activity of endophytic fungi associated with Orchidaceae in Brazil. Can J Microbiol 55:1381–1391

Vieira ML, Hughes AF, Gil VB, Vaz AB, Alves TM, Zani CL, Rosa CA, Rosa LH (2012) Diversity and antimicrobial activities of the fungal endophyte community associated with the traditional Brazilian medicinal plant Solanum cernuum Vell (Solanaceae). Can J Microbiol 58:54–66

Wang Y, Dai CC (2011) Endophytes: a potential resource for biosynthesis, biotransformation, and biodegradation. Ann Microbiol 61:207–215

Xin G, Glawe D, Doty SL (2009) Characterization of three endophytic, indole-3-acetic acid-producing yeasts occurring in Populus trees. Mycol Res 113:973–980

Xu P, Bura R, Doty SL (2011) Genetic analysis of D-xylose metabolism by endophytic yeast of *Populus*. Genetics Mol Biol 34:471–478

Zhao J-H, Bai F-Y, Guo L-D, Jia J-H (2002) Rhodotorula pinicola sp. nov., a basidiomycetous yeast species isolated from xylem of pine twigs. FEMS Yeast Res 2:159–163

Index

A
Abscisic acid (ABA), 152, 156, 218, 222, 273, 279
Actinobacteria, 196
Actinorhizal plants, 89
Alnus, 104
Aluminum, 106
1-Aminocyclopropane-1-carboxylate (ACC), 167, 262
AM-plant-herbivore interaction, 297
Antibiosis, 173
Appressorium, 292
Aquaporins, 276–278
Arabidopsis, 172
Arbuscular mycorrhizal (AM) fungi, 216, 217, 241, 253, 271, 273, 291
  hyphae, 219
  symbiosis, 92, 221, 290
Arbuscules, 253, 271
Ascomycota, 313
Ascospores, 178
Autoregulation, 14, 222–223
Auxin, 92
Azoarcus, 204
Azospirillum, 153, 156, 255

B
Bacillus, 152, 167, 168, 171
Bacteroid, 17
Basidiomycota, 312
Biofertilizer, 226–227
Biologic nitrogen fixation (BNF), 244
Biopesticide, 181
Biovars, 73
Boron, 106
Botrytis, 301
Bradyrhizobium, 4
Bradyrhizobium, 28, 81, 157, 200
Burkholderia, 153

C
Cadophora, 313
Calmodulin-dependent protein kinase (CCaMK), 12
Candida, 337
Carbon
  costs, 235
  loss, 236
  resources, 243
Carbonic anhydrase, 59
Carotenoid cleavage dioxygenases (CCD), 218
Casuarina, 106
CCD8 enzyme, 226
Celery, 258
Chalcone synthase, 91
Chitinase, 169
Clavicipitaceae, 312
CLE peptides, 223
Clustering coefficient, 37
CO_2 enrichment, 110
Comparative analysis, 45–47
Computational modeling, 31
Computational simulations, 37–41
Cryptococcus, 338
Cryptococcus flavescens, 339
Cultivated endophytic bacteria, 197–199

D

Dark CO_2-fixation, 62–63
Datisca, 124
Debaryomyces, 335
Debaryomyces hansenii, 339
Deschampsia, 324
Desulfonema limicola, 206
Desulfovibrio gigas, 206
2,4-Diacetylphloroglucinol (2,4-DAPG), 177
Diversity of endophytic populations, 197–200
DNA-DNA homology, 140
DNA-DNA hybridization, 124

E

Ectomycorrhizal fungi, 115–116
Elaeagnus, 130
Endophytic nitrogen fixing bacteria, 206
Endophytic yeast, 335
Endosymbiosis, 92
Ensifer, 70
Ethylene, 300
Extraradical mycelium (ERM), 292

F

Firmicutes, 196
Fixed C, 237
Flavonoids, 91
Flux balance analysis (FBA), 40
Frankia, 103, 109, 123, 129
Frankia diversity, 124
Frankiales, 123
Frankia phylogeny, 138–139
Functional classification, 43
Fusarium, 165, 294, 322

G

Gibberellins, 154
Gigaspora, 263
Glomeromycota, 291
Glomus, 254
    G. intraradices, 261, 262
    G. mosseae, 259
Gluconacetobacter, 204
Gram-negative, 182
Gram-positive, 170
Gross C assimilation, 236

H

Herbaspirillum, 206
Hopanoid lipids, 128

Hydrogen cyanide (HCN), 178
Hyphae, 253
Hyphal network, 255

I

Indole-3-acetic acid (IAA), 154, 341
Induced systemic resistance (ISR), 169
Infection thread, 93
Invertase, 57–58
Ion balance, 274–276

J

Jasmonic acid (JA), 222, 245
    JA-dependent, 300

K

α-Ketobutyrate, 167
Koch's postulates, 129

L

Legumes, 3, 53, 71
Lotus, 247

M

Magnaporthe, 166
Malate, 58–62
MAP kinases, 301
Mathematical representation, 34–37
Medicago, 4, 71, 246, 262
Medicago sulfate transporters (MtSULTRs), 262
Mesorhizobium, 4, 28
Metabolic networks, 31–34
Metabolic reconstruction, 42–43
Metagenomic analysis, 196
Methanotrophic enrichments, 202
Methanotrophs, 201
Methylosinus trichosporium, 202
Microbe-associated molecular patterns (MAMPs), 289
Microbial endophytes, 196
Micronutrients, 262–263
Micropropagation, 337
Myc factors, 217, 220
Mycorrhiza-induced resistance (MIR), 294
Mycorrhizal fungi, 233, 258
Mycorrhizal legumes, 241
Myrica, 126
Myricaceae, 104

Index

N
N-acetylglucosamine, 220
NaCl, 275
Nickel, 105
nifH genes, 197
nifH microarray, 206
Nitric oxide, 107
Nitrogen, 256–259
    content, 239
    fixation, 6, 27, 105–109
    mineralization, 257, 258
    uptake, 258
NodD, 28
Nod factor (NF), 6, 91
    signaling, 9
Nodule, 15, 53

O
Oomycetes, 165, 299
Orobanchaceae, 224
Osmoregulation, 273–274

P
Panphenome, 77
Parasponia, 97, 124
Peribacteroid membrane, 29
Phaseolus, 4
Phaseolus vulgaris, 261
Phenotypic variability, 77–80
Phialocephala, 313
Phloem-sucking insects, 297
PHO2, 261
Phosphatase, 235, 257
Phosphoenolpyruvate (PEP) carboxylase, 58
Phosphorus, 105, 259–260
Photosynthesis, 244
Pht1, 260
Phylogeographic, 74
Phythopthora, 322
Picea, 337
Pichia, 339, 341
Piriformospora, 318
Plant growth promoting (PGP), 195
Plant growth-promoting rhizobacteria (PGPR), 151, 255
pmoA gene, 202
pmoA microarray, 201
Populus, 338
Prenodule, 95
Proline, 274

Proteobacteria, 197
Pseudomonas, 167, 176
Pseudomonas fluorescens, 178
Puccinia, 166
Pucciniomycotina urediniomycete, 338
Pythium, 294

Q
Quorum sensing (QS), 170

R
Reactive oxygen species (ROS), 278
Rhizobia, 239
Rhizobium, 4, 27, 72
Rhizobium etli, 30
Rhizoctonia, 294, 301, 318
Rhizophagus irregularis, 277
Rhizosphere, 3, 215
Rhodotorula, 335
Rhodotorula minuta, 337
Rice, 204
Root exudates, 91
Root-soil interfaces, 289
Root system architecture (RSA), 224

S
Saccharomyces cerevisiae, 339
Salicylic acid (SA), 170, 300
Salinity, 106–107
Sclerotinia, 166
Sebacinales, 312
Sinorhizobium, 4, 28, 69
Sinorhizobium-Medicago, 73
Sinorhizobium meliloti, 30
Soil moisture, 112–113
Soil pH, 277
Solanum, 168, 338
Solanum tuberosum, 261
Solution space, 39
Specific P absorption rate (SPAR), 242
Sporulation phenotype, 126–128
16S rRNA, 139
Striga, 225, 298
Strigolactones, 217–219, 224
Subtilase, 95
Sucrose, 54–58
Sugarcane, 205
Symbiosis (SYM) pathway, 220
Symbiosis receptor kinase *(SYMRK)*, 92

T

Taxomyces, 322
Thlaspi, 255
TILLING, 57
Tissue construction cost, 238–239
Transcriptomic responses, 77
Tricarboxylic acid (TCA) cycle, 245
Trichoderma, 319, 323

V

Verticillium, 294, 322, 325
Volatile organic compounds (VOCs), 169

W

Water potential, 272
Williopsis saturnus, 339

Printed by Publishers' Graphics LLC